Springer Tracts in Nature-Inspired Computing

The book series is aimed at providing an exchange platform for researchers to summarize the latest research and developments related to nature-inspired computing in the most general sense. It includes analysis of nature-inspired algorithms and techniques, inspiration from natural and biological systems, computational mechanisms and models that imitate them in various fields, and the applications to solve real-world problems in different disciplines. The book series addresses the most recent innovations and developments in nature-inspired computation, algorithms, models and methods, implementation, tools, architectures, frameworks, structures, applications associated with bio-inspired methodologies and other relevant areas.

The book series covers the topics and fields of Nature-Inspired Computing, Bio-inspired Methods, Swarm Intelligence, Computational Intelligence, Evolutionary Computation, Nature-Inspired Algorithms, Neural Computing, Data Mining, Artificial Intelligence, Machine Learning, Theoretical Foundations and Analysis, and Multi-Agent Systems. In addition, case studies, implementation of methods and algorithms as well as applications in a diverse range of areas such as Bioinformatics, Big Data, Computer Science, Signal and Image Processing, Computer Vision, Biomedical and Health Science, Business Planning, Vehicle Routing and others are also an important part of this book series.

The series publishes monographs, edited volumes and selected proceedings.

More information about this series at http://www.springer.com/series/16134

Preface

The applications of wireless sensor network (WSN) increase rapidly due to three basic goals such as availability, confidentially, and integrity. WSN is a collection of a large number of sensor nodes. Each sensor node contains a battery having limited energy capacity. The purpose of this node is to receive, process, and send the data and information. Based on this purpose, there are several challenges in WSN such as deployment, design constraints, energy constraint, limited bandwidth, node costs, and security. The factors that influence these challenges are coverage, dependability, range, reliability, scalability, security, speed, etc. It causes several types of uncertainties and imprecise information. Hence, there is a need for some innovative intelligent nature-inspired techniques in the area of WSN. So, the above-mentioned issues are estimated efficiently. Therefore, the proposed book effectively helps the academicians, researchers, computer professionals, industry people, and valued users.

Objective of the Book

This book contains some nature-inspired algorithms for optimizing several issues of WSN. The main aim of this book is to solve or innovate different problems of WSN in terms of several applications. It is edited for academicians, researchers, computer professionals, industry people, and valued users.

Organization of the Book

The book contains 14 chapters that are organized in three parts as follows. Before starting the parts, Chap. 1 describes the overview of WSN with its several applications, challenges, and algorithms. **Part I** contains five chapters that outline some bio-inspired algorithms in WSN. **Part II** contains three chapters that highlight some swarm optimization techniques in WSN. **Part III** contains five chapters that illustrate some multi-objective optimizations in WSN.

Part I: Bio-inspired Optimization (Chaps. 2–6)

This part describes some bio-inspired optimization algorithms such as genetic algorithm (GA), artificial neural network (ANN), and artificial immune system (AIS) in the context solving fault analysis and diagnosis, and traffic management in WSN. Short descriptions of these chapters are given as follows.

Chapter 2

This chapter contains a fitness function which is used to balance the load of the cluster head during data transmission. It helps to analyze with some existing related algorithms and compare their performance in terms of different metrics like energy efficiency, number of alive nodes, and packet delivery ratio.

Chapter 3

This chapter contains an efficient technique, i.e., direct diffusion for finding an optimum path by recovering dead nodes. It helps to enhance the network lifetime by reducing data packet loss as well as energy consumption.

Chapter 4

This chapter outlines a GA-based traffic management technique. It helps to enhance the network lifetime efficiently by maximizing the green signal of the network. Finally, it is compared with some existing techniques in terms of some features. The final comparison shows that the proposed method outperformed the existing methods.

Chapter 5

This chapter highlights an intelligent technique for fault diagnosis in WSN using deep learning technique. It helps to reduce several faults such as low battery, calibration, and sensor aging in terms of internal and external influences environmental conditions.

Chapter 6

This chapter outlines a fault diagnosis system in WSN. The issue of fault diagnosis in WSN can be comparable in many aspects with an AIS. Different approaches of AIS are discussed in this chapter, which can be applied in WSN for fault diagnosis. An overall view of the biological immune system is also explained in detail in this chapter.

Part II: Swarm Optimization (Chaps. 7–9)

This part indicates some swarm optimization techniques such as African buffalo optimization (ABO), particle swarm optimization (PSO), and modified swarm intelligence technique for solving the issues of routing, network parameters optimization, and energy estimation in WSN. Short descriptions of these chapters are given as follows.

Chapter 7

This chapter outlines a new routing method using AGO optimization technique in WSN. ABO is a nature-inspired combinatorial optimization technique based on the behavior of African buffaloes. Here, ABO acts as the main controller of the WSN and manages all the sensor nodes in correspondence with the base station. It also helps to transfer data packets from the source node to the sink node efficiently.

Chapter 8

This chapter indicates a distributed source localization algorithm in WSN. This algorithm helps to estimate the strategies of the directions of the nodes. Modified particle swarm optimization is proposed to optimize the multimodal maximum likelihood function in a distributed scenario of the network.

Chapter 9

This chapter illustrates an effective optimization method that is inspired by some phenomenon found in nature. One of them is quasi-oppositional harmony search algorithm. Although it is under developing stage, it is still a powerful optimization technique. It has the potential ability to solve various engineering optimization problems.

Part III: Multi-objective Optimization (Chaps. 10–14)

This part describes some multi-objective optimizations techniques using GA, PSO, ANN, teaching-learning-based-optimization (TLBO), and combination of few stated algorithms. The above-mentioned parts provide efficient and optimal solutions to different issues of WSN by using above-mentioned nature-inspired algorithms. Short descriptions of these chapters are given as follows.

Chapter 10

This chapter outlines a comprehensive survey of the recent intelligent-based hierarchical routing protocols in WSN that are developed based on PSO, AIS, GA, ant colony optimization (ACO), and fuzzy logic. These protocols are reviewed in detail according to different metrics such as node deployment, control manner, network architecture, clustering attributes, protocol operation, path establishment, communication paradigm, energy model, protocol objectives and applications.

Chapter 11

This chapter aims to study and analyze the various existing works in terms of survey for sensor node deployment, load balancing, and energy utilization in WSN using some bio-inspired algorithms like genetic algorithm and PSO algorithm to tackle some of the crucial problems of WSN. The chapter also provides a list of some models that are used in sensor node distribution and data aggregation.

Chapter 12

In this chapter, some multi-objective bio-inspired algorithms are discussed for WSN. This algorithm perceived a lot of attention from researchers in the current years; the domain-specific understanding still needs to be improved for its establishment. So, in this chapter, bio-inspired algorithms are discussed concisely with their importance in the field of WSN.

Chapter 13

This chapter highlights an intelligent method for cluster-head selection in WSN using TLBO. This optimization consists of two basic elements such as teacher and student-based natural relation between both entities. The TLBO helps to optimize several conflicting objectives of the network efficiently in terms of learning method.

Chapter 14

This chapter outlines a detailed study and challenges for reliable, low latency communication in WSN for pervasive healthcare applications using nature-inspired algorithm. It helps to resolve several issues such as dropped network connections, power loss, and failure in the transmission of critical alarms in several healthcare facilities.

Kolkata, India Debashis De
Kolkata, India Amartya Mukherjee
Berhampur, India Santosh Kumar Das
Kolkata, India Nilanjan Dey

List of Reviewers

Abhishek Kumar, Subharti Institute of Technology & Engineering (SITE), Swami Vivekananda Subharti University, Meerut, U.P.
Ambedkar Kanapala, IIT(ISM), Dhanbad, Jharkhand-826004
Arijit Karati, National Sun Yat-sen University Kaohsiung-80424, Taiwan
Arvind Kumar, AIACTR, Geeta Colony, Delhi
Asish Kumar Roy, NIST Berhampur, Odisha
Bhabani Sankar Gouda, NIST Berhampur, Odisha
Chandan Kumar Shiva, SR Engineering College, Warangal, Telangana
Debashis Das, Techno India University, Kolkata
Gaytri Kumari Gupta, Jamshedpur Women's College, Jamshedpur
Harendra Kumar, Government Polytechnic Bilaspur, GEC Campus, Koni Bilaspur
Harsh Nath Jha, Webel IT Park, Phase 2, Palashdiha, Durgapur, W.B.
Imran Rasheed, IIT(ISM), Dhanbad, Jharkhand-826004
Joydev Ghosh, National Research Tomsk Polytechnic University (TPU), Russia
Kanhu Charan Gouda, NIST Berhampur, Odisha
Kanupriya Kashyap, NIT Kurukshetra
Madhuri Malakar, NIT Rourkela
Mahendra Prasad, IIT(ISM), Dhanbad
Nabajyoti Mazumdar, Central Institute of Technology, Kokrajhar, Assam
Neha Verma, Subharti Institute of Technology & Engineering (SITE), Swami Vivekananda Subharti University, Meerut, U.P.
Rakesh Ranjan Swain, NIST Berhampur, Odisha
Ranjit Kumar, IIT(ISM), Dhanbad, Jharkhand-826004
Rimjhim, IIT Patna
Ruchika Padhi, NIST Berhampur, Odisha
Samiran Bera, IIT(ISM), Dhanbad
Samiran Gupta, Asansol Engineering College, Asansol, W.B.
Shweta, IBM Pvt Ltd., 137, Galway Drive, Apt 302 Mooresville, North Caro-lina, USA
Siba Prasada Tripathy, NIST Berhampur, Odisha
Soumen Nayak, SOA University, Bhubaneswar-751030

Sourav Samanta, University Institute of Technology, The University of Burdwan, W.B.
Sunil Kumar Gautam, Institute of Advanced Research, Gandhinagar, Gujarat
Susmita Mahato, NIST Berhampur, Odisha
Trilochan Panigrahi, NIT Goa
Vishwas Mishra, Subharti Institute of Technology & Engineering (SITE), Swami Vivekananda Subharti University, Meerut, U.P.

Contents

About the Editors

Debashis De earned his M.Tech. from the University of Calcutta in 2002 and his Ph.D. (Engineering) from Jadavpur University in 2005. He is the Professor and Director in the Department of Computer Science and Engineering of the West Bengal University of Technology, India, and Adjunct Research Fellow at the University of Western Australia, Australia. He is a senior member of the IEEE, a life member of CSI, and a member of the International Union of Radio Science. He worked as R&D engineer for Telektronics and programmer at Cognizant Technology Solutions. He was awarded the prestigious Boyscast Fellowship by the Department of Science and Technology, Government of India, to work at the Heriot-Watt University, Scotland, UK. He received the Endeavour Fellowship Award during 2008–2009 by DEST Australia to work at the University of Western Australia. He received the Young Scientist Award both in 2005 at New Delhi and in 2011 at Istanbul, Turkey, from the International Union of Radio Science, Head Quarter, Belgium. His research interests include wireless sensor network, mobile cloud computing, green mobile networks, and nanodevice designing for mobile applications. He has published in more than 200 peer-reviewed international journals in IEEE, IET, Elsevier, Springer, World Scientific, Wiley, IETE, Taylor Francis and ASP, seventy international conference papers, and four researches monographs in Springer, CRC, NOVA, and ten textbooks published by Pearson education.

Amartya Mukherjee is an Assistant Professor at Institute of Engineering & Management, Salt Lake, Kolkata, India. He holds M.Tech. in computer science and engineering from the National Institute of Technology, Durgapur, India. His primary research interest includes embedded application development, robotics, unmanned aircraft systems, Internet of things, intelligent sensor networks, and ad-hoc networks. He has various publications in the fields of robotics, embedded systems, and IoT in IEEE, Springer, World Scientific, CRC Press, IGI Global. His book "Embedded Systems and Robotics with Open Source Tools" is one of the bestselling books in CRC Press (Taylor & Francis Group).

Santosh Kumar Das received his Ph.D. degree in computer science and engineering from Indian Institute of Technology (ISM), Dhanbad, India, in 2018 and completed his M.Tech. degree in computer science and engineering from Maulana Abul Kalam Azad University of Technology (erstwhile WBUT), West Bengal, India, in 2013. He is currently working as an Assistant Professor at School of Computer Science and Engineering, National Institute of Science and Technology (Autonomous), Institute Park, Pallur Hills, Berhampur, Odisha, India, 761008. He is having more than eight years of teaching experience. He has authored/edited one book in Springer, and published more than 27 research articles. His research interests mainly focus on ad-hoc and sensor network, artificial intelligence, soft computing, and mathematical modelling.

Nilanjan Dey is an Assistant Professor in the Department of Information Technology at Techno India College of Technology, Kolkata. He has completed his Ph.D. in 2015 from Jadavpur University. He is a Visiting Fellow of Wearables Computing Laboratory, Department of Biomedical Engineering University of Reading, UK, the Visiting Professor of College of Information and Engineering, Wenzhou Medical University, P.R. China, and Duy Tan University, Vietnam. He has held honorary position of Visiting Scientist at Global Biomedical Technologies Inc., CA, USA (2012–2015). He is the Editor-in-Chief of International Journal of Ambient Computing and Intelligence, IGI Global, the Series Co-Editor of Springer Tracts in Nature-Inspired Computing, Springer, Advances in Ubiquitous Sensing Applications for Healthcare (AUSAH), Elsevier, and the Series Editor of Intelligent Signal Processing and Data Analysis, CRC Press. He has authored/edited more than 40 books with Elsevier, Wiley, CRC, and Springer, and published more than 350 research articles. His main research interests include medical imaging, machine learning, bio-inspired computing, and data mining. He is a life member of Institute of Engineers (India). He is the Indian ambassador of International Federation for Information Processing (IFIP) – InterYIT (International Young ICT Professionals group).

Chapter 1
Wireless Sensor Network: Applications, Challenges, and Algorithms

Debashis De, Amartya Mukherjee, Santosh Kumar Das and Nilanjan Dey

1 Introduction

In the last few decades, the applications of wireless sensor networks (WSNs) increase rapidly [1–6]. It consists of a set of sensor nodes and base station (BS) which distributed in an environment to achieve a specific goal. BS is a main coordinator system having high-energy and high-processing capacities [7, 8]. The sensor node consists of several elements such as internal and external antenna, microcontroller, battery. All the connected sensors are autonomous, and they directly or indirectly connected with the BS. The sensor nodes are used to detect some information from either physical condition or environmental condition. This information consists of light, heat, pressure, etc. Figure 1 depicts the WSN. In this figure, different sensor nodes are connected with one radio range to another radio range. Each sensor node resides within a radio range. The figure contains different colored sensor nodes. This different colored indicates the variation of radio range with network resources. The collection of sensor nodes is connected with sink node, and sink node is further connected with

D. De
Maulana Abul Kalam Azad University of Technology, Haringhata, Nadia,
West Bengal 721249, India
e-mail: dr.debashis.de@gmail.com

A. Mukherjee
Institute of Engineering & Management, Salt Lake, Kolkata 700091, India
e-mail: mamartyacse1@gmail.com

S. K. Das (✉)
National Institute of Science and Technology (Autonomous), Institute Park, Pallur Hills,
Berhampur, Odisha 761008, India
e-mail: sunsantosh2014@gmail.com

N. Dey
Techno India College of Technology, Kolkata 700156, India
e-mail: neelanjan.dey@gmail.com

D. De et al. (eds.), *Nature Inspired Computing for Wireless Sensor Networks*,
Springer Tracts in Nature-Inspired Computing,
https://doi.org/10.1007/978-981-15-2125-6_1

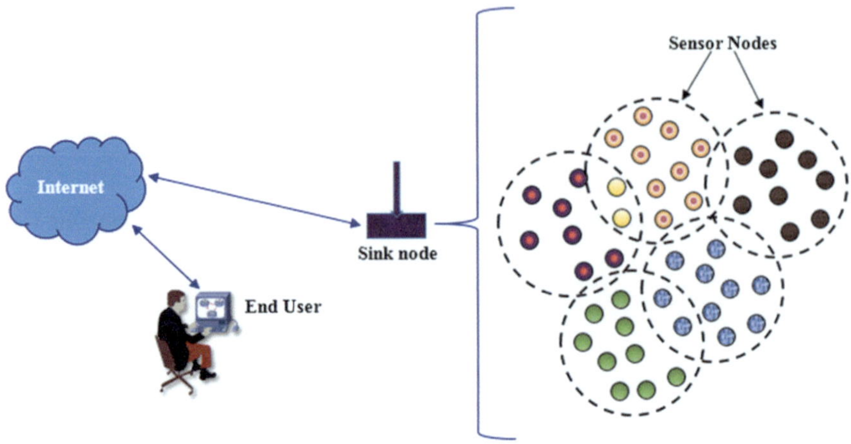

Fig. 1 Wireless sensor network

end user. This end user is connected with the Internet connection. The sink node is also known as BS which is responsible to collect data and information from all the sensor nodes and to transmit summarized information to the end user. The end user may be a person or software or hardware that designs for a specific purpose. Actually, end user directly or indirectly indicates "end goal." This goal indicates exporting the useful information to the customer that fulfills the actual design purpose of the WSN. There are several features of WSN as follows:

(a) The topology of the network change very frequently.
(b) The purpose of the sensor nodes to broadcast communication.
(c) The number of sensor nodes in a network can be several orders.
(d) Sensor nodes are densely deployed.
(e) Sensor nodes are prone to failures.
(f) Each sensor node consists of the limited capacity of the battery.
(g) The nodes having the limited computational capacity as well as memory capacity.
(h) Sensor nodes may not have global identification due to large overhead.

Based on above-mentioned features, there are several types of WSNs that are described as follows:

(i) **Mobile WSN**: This network is also known as mobile WSN (MWSN). It consists of a set of sensor nodes that move from one location to another location to detect the phenomena of the physical environment. After sensing the target element, it communicates to the desired station. In static WSN, data is distributed using fixed routing, but where data is distributed and communicates with dynamic routing. There are several limitations to this WSN.

(ii) **Multimedia WSN**: This network is used to monitor and track the several activities and events of the multimedia devices of software. It deployed pre-planned manner in the environment so that information coverage is guaranteed. This type of network consists of some intelligent equipments like microphone and camera that help to communicate one user to another user with the help of compression, correlation, process, and information retrieval.

(iii) **Terrestrial WSN**: This network is deployed in two ways: (i) pre-plan and (ii) ad hoc ways. In first case, prior requirements and information are required for analysis and diagnosis the network components. In ad hoc mode, the system is efficient and dynamic. In this system, nodes are placed with the help of place and scatter randomly.

(iv) **Underground WSN**: In this network, the sensor nodes are placed in different locations of the underground like mine or cave. The basic purpose of this deployment is to monitor different underground conflicting conditions. Although all the sensor nodes are deployed into the undergrounds, one node, i.e., sink node, is placed in the upper ground to receive sensor data and information. After sensing and processing these data and information, it deployed to the BS.

(v) **Underwater WSN**: This network consists of some sensor nodes and few vehicle nodes. In this network, sensor nodes are autonomous which move here and there into water. The purpose of the vehicle is to collect data and information from the sensor nodes and sends to the BS which is located outside of water in ground.

2 Applications of Wireless Sensor Networks

The basic purpose of the sensor nodes in WSN is to monitor environmental or physical conditions such as sound, pressure, temperature, and after monitoring, and it collects the data and sends it to the BS. Basically, WSN is designed for military applications where it is rapidly used for battlefield surveillance. Due to several advantages, nowadays WSN is used for several purposes. One of the modern rapidly growing facilities of WSN is Internet of things (IoT)-based wireless body area network (WBAN) in healthcare system [9, 10]. It paved the gap of the traditional system like to a visit hospital. IoT allows facilities like communicating, sensing, processing with biomedical, and physical parameters [11, 12]. Cloud computing also provides some advantages to the WBAN because it has large processing and storage infrastructure. It helps to process the data and information as offline as well as online by body sensor streams [13–16]. Apart from military applications and IoT, there are several applications of WSN as follows [17]:

i. Animal movement tracking
ii. Civil structure monitoring system
iii. Commercial application

iv. Consumer application
v. Education system
vi. Entertainment application
vii. Environment monitoring system
viii. Industrial and consumer applications
ix. Industrial process monitoring and control
x. Precision agriculture system
xi. Safety and security system
xii. Security and surveillance system
xiii. Smart building application
xiv. Smart grids and energy control system
xv. Transportation and logistics
xvi. Urban terrain tracking system.

3 Challenges in Wireless Sensor Networks

Although there are several applications and its uses of WSN, it has several challenges as follows:

(a) **Complicated configuration**: Its configuration is more complicated compared to the wired network. So, if any trouble or issue is detected, then it is more difficult to find it and provides its solution.
(b) **Costly**: It is more costly than the wired network. So, it is not easily available everywhere based on the requirements.
(c) **Coverage problems**: In WSN, different sensor nodes are deployed randomly based on pre-planned manner. The radio range of some of the nodes is the same, and some of the other nodes are different. So, due to the variety of sensor networks, coverage problem is raised.
(d) **Distraction**: In modern situations, this network keeps distracting by several wireless devices like Bluetooth.
(e) **Energy efficiency**: The sensor nodes consist of the limited capacity of batteries that are insufficient during any mission or operation because it needed a recharge after some time interval. So, its capacity is low than wired network.
(f) **Low communication speed**: The communication speed of WSN is low compared to the wired network. So, it takes more time to survive or collect information.
(g) **Low latency**: Basically, WSN is designed for immediate operation or action. So, its framework detects and notifies immediately. Due to this reason, its latency is low.
(h) **Scalability**: In WSN, different operations are performed based on situation and user requirements. Depends on the change of requiring extra hardware system, scalability is also changed.

(i) **Security**: The WSN is not secure compared to other networks or wired network. It is easily troubled by its environment. So, hackers easily hack the required information.
(j) **Transmission media**: In multi-hop communication, some of the sensor nodes are connected with a medium. Hence, several types of issues arise such as high error rate, fading, overhead.

The above-mentioned challenges cause several types of limitations in terms of uncertainties in context of privacy and security. Security means safeguarding of the data or data set or different related information, and privacy indicates safeguarding of the user's information or identity [18]. To solve different types of problems such as outlier detection or anomaly detection, fault diagnosis, intrusion detection, mobility prediction, several works have been proposed [19, 20]. Most of the outlier detection and prediction algorithms are outperformed with the help of machine learning algorithms [21, 22] because machine learning algorithms help to enhance the critical situation in order to optimize its way [23–25]. Some of the basic steps that follow by machine learning algorithms are (i) feature selection and output labeling, (ii) sample collection, (iii) offline training, and (iv) online classification. Hence, nature-inspired computing is an efficient and intelligent technique to reduce and control the above-mentioned issues.

4 Nature-Inspired Computing

Nature-inspired algorithm is an effective way to solve complex real-life problem that totally depends on the natural phenomena [26–28]. It has several types such as bio-inspired, swarm intelligence, chemical reaction. Each algorithm is different in terms of parameters and constraints, but the goal is the same to produce a favorable solution. In [29], the authors described some basic information of optimization techniques that relates to nature-inspired computing. This information is illustrated in this section as Figs. 2, 3, 4, and 5 and Pseudocode 1. Figure 2 shows the types of optimization where three types of optimizations are illustrated such as (i) minimization, (ii) maximization, and (iii) on-target. Minimization indicates to minimize the objective or goal. Maximization indicates to maximize the objective or goal. And on-target indicates either minimize or maximize the objective or goal based on specific optimset. Here, optimset indicates goal value.

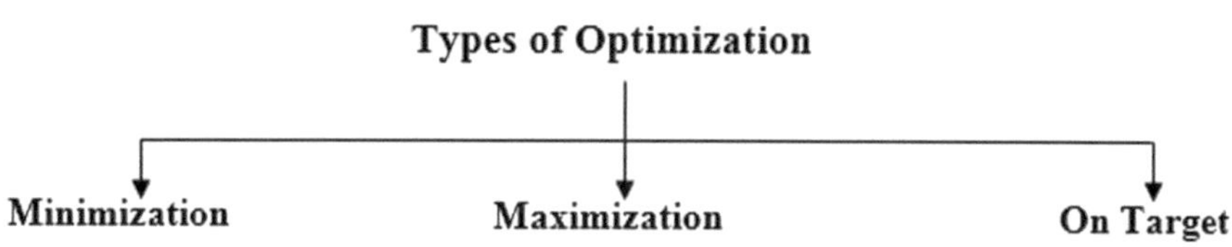

Fig. 2 Types of optimization

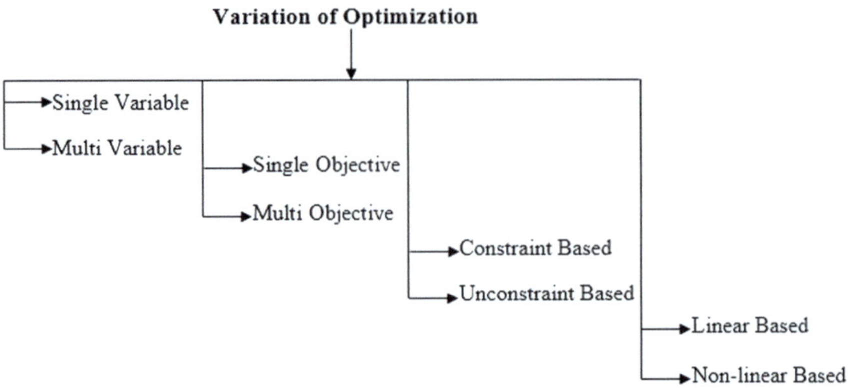

Fig. 3 Variation of optimization

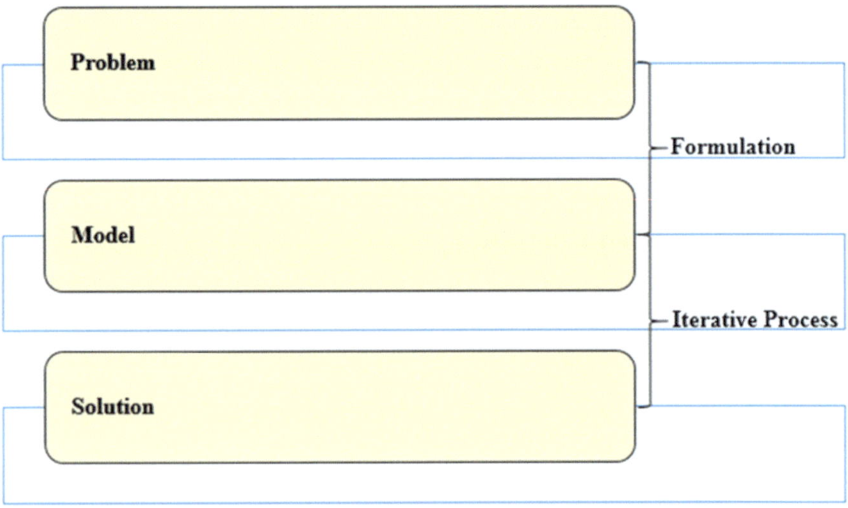

Fig. 4 Basic steps of optimization

The variation of optimization is shown in Fig. 3, where eight types of variations are shown such as (i) single variable, (ii) multi-variables, (iii) single objective, (iv) multi-objectives, (v) constraint-based, (vi) unconstraint-based, (vii) linear-based, and (viii) nonlinear-based. Single variable optimization indicates where the decision variable is only one. Multi-variable optimizations indicate more than one decision variable. Single objective optimization indicates where only one objective function plays the main role. Multi-objective optimization indicates more than one objective are considered and the nature of the one objective to another is conflicting. Constraint-based optimization is the optimization technique where multiple constraints are consisting to satisfy the goal. Unconstraint-based optimization indicates

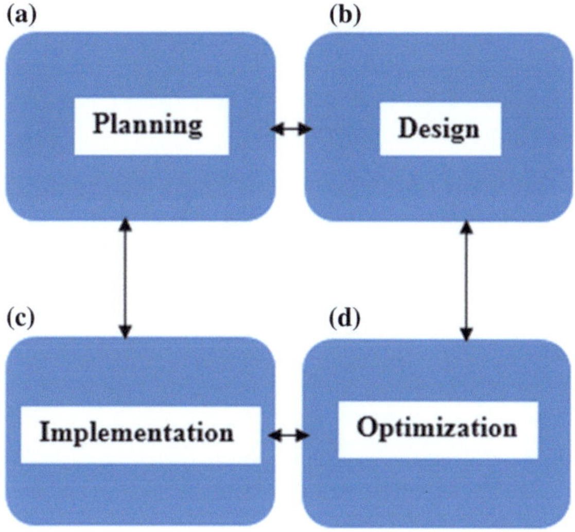

Fig. 5 Iterative system of optimization

where no constraints are available for satisfying the goal. Linear-based optimization indicates where the nature of the objective and constraint is linear. Nonlinear-based optimization indicates where the nature of the objective and constraints is nonlinear.

The above-mentioned types of optimizations and variations help to complete the process into three stages: (i) problem, (ii) model, and (iii) solution which are shown in Fig. 4.

This is an iterative process where "problem" and "model" are connected with "formulation," and "model" and "solution" are connected by "iterative process." This is a cycle process that produces the optimal solution and its circulation which is shown in Fig. 5. This system contains four stages: (a) planning, (b) design, (c) implementation, and (d) optimization. Planning is used to plan the stated problem. The design indicates to organize the problem by objective and constraints with related variables. Optimization indicates to produce an optimal solution based on the previous steps. The whole stages are connected with a cycle process.

The pseudocode of nature-inspired computing is shown in Pseudocode 1. Hence, after concluding Figs. 2, 3, 4, 5 and Pseudocode 1, nature-inspired computing is inspired by nature and then evaluates its concepts and constructs an efficient algorithm. Finally, solve it to produce the optimal result which is shown in Fig. 6 as a base structure of the nature-inspired computing.

Pseudocode 1 Optimization in nature-inspired computing

Step 1: Start
Step 2: Identify the requirement

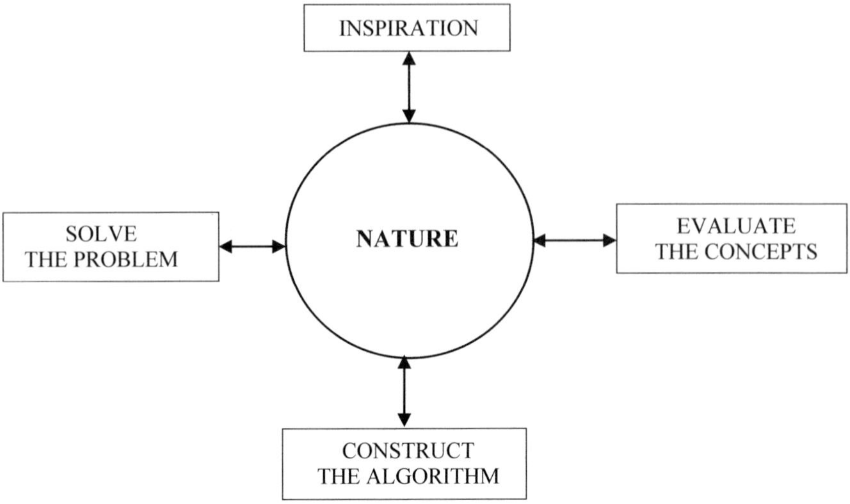

Fig. 6 Base structure of the nature-inspired computing

Step 3: Outline the problem
Step 4: Search required information
Step 5: Identify the objective function
Step 6: Identify the constraints
Step 7: Identify the evaluation criteria
Step 8: Apply optimization technique
Step 9: Generate alternative solutions
Step 10: Analyze the results
Step 11: Make decision
Step 12: Specify the outcome design
Step 13: Deployment
Step 14: Stop.

5 Nature-Inspired-Based Algorithms

Nature-inspired technique in WSN emphasizes the implementation of the biological systems that can form the basic platform to get a globally optimized decision of the sensor nodes. Moreover, the nature-inspired technique is used in network design, path planning, and security mechanism implementation for dynamic nodes. It also helps for network packet routing as unicast, multicast, and broadcast. It has several advantages as follows:

a. Ability to explore local solution
b. Ability to handle the objective cost
c. Ease of implementation
d. Flexibility, simplicity, and robustness
e. High flexibility for adjustment
f. Large application for complex functions
g. Population of solution.

In the last decade, several nature-inspired algorithms are rapidly increased in the field of WSN such as ant colony optimization (ACO), particle swarm optimization (PSO), genetic algorithm (GA), swarm intelligence (SI), artificial neural network (ANN), African buffalo optimization (ABO), and artificial immune system (AIS). Some of the works have been described in this section.

5.1 ACO-Based Algorithms

In the last decade, several works have been proposed based on ACO technique [30–33]. Some of them as follows: Sun et al. [34] proposed a technique for the optimization of the node's deployment for WSN using ACO. In this technique, the authors apply two techniques like culture algorithm with ant colony algorithm. The proposed technique is used to solve coverage and connectivity with both issues and enhances the network lifetime of the WSN. Sharma and Grover [35] proposed a model in WSN for energy-efficient routing protocol. The key technique of this model is modified ACO which is an extension of earlier ACO. The proposed technique handles the utilization of the node batteries of the WSN. This efficient utilization helps to control non-uniform signal transformation during operation. Kaur and Mahajan [36] proposed a metaheuristic-based efficient routing protocol for WSN. The proposed method is the fusion of two metaheuristic techniques as ACO and PSO. This efficient routing protocol is based on the energy capacity of the nodes. The proposed method helps in data aggregation of the network by reducing the residual energy capacity of the nodes. Liao et al. [37] proposed a method for sensor nodes deployment in WSN using ACO. The proposed technique helps to increase the efficient monitoring of network by expanding coverage areas. This method helps to increase the search space of the network and enhances the network lifetime of the WSN. Ho et al. [38] proposed a diffusion technique for WSN using ACO. The main aim of this technique is to design an efficient routing table by optimizing two constraints like processing time and energy consumption of the data transmission. Finally, it also provides route backup provision for avoiding waste energy consumption and data processing time. Sun et al. [39] proposed a routing method for multi-objective optimization in WSN. The main technique in this proposed algorithm is used ACO for two basic purposes: (i) reducing the energy consumption of the nodes, and (ii) enhancing the security of the data transmission. It provides multi-objective optimization for residual energy of the nodes and the trust value of the paths. By combining both, it helps to enhance the network lifetime as well as network metrics.

5.2 PSO-Based Algorithms

In the last decade, several works have been proposed based on PSO technique [40, 41]. Some of them as follows: Parvin and Vasanthanayaki [42] proposed a PSO-based energy-efficient model for WSN. The main aim of this proposed model is to monitor the habitat and vehicle of the environment. This algorithm is based on a distributed optimization technique. Moreover, this method also proposed a clustering for the categorization of the nodes. Phoemphon et al. [43] proposed a localization method for WSN. The proposed method is a hybrid method using a fusion of fuzzy logic and PSO. The proposed method is used to detect the global position of the sensor nodes. It helps to reduce cost as well as battery energy of the nodes. The fuzzy logic is used to reduce the uncertainty of the network where PSO is used to inspire the neighbor nodes. Sun et al. [44] proposed a multi-objective technique for WSN. This technique is used for the binary PSO for multi-objective optimization. It helps to optimize several objectives such as load balancing, energy consumption, and task execution time. The several constraints are designed for each objective function. All objective functions help to enhance the network performance. Cao et al. [45] proposed a deployment optimization technique for WSN using PSO. This method is used parallelism in the context of distributed system. The proposed method is based on NP-hard problems. So, it uses PSO as a cooperative co-evolutionary system. It helps to reduce computational time and enhances message passing as well as data transmission. Yan et al. [46] proposed a positioning system for optical WSN using PSO. The main parameter of this technique is residual energy. Residual energy indicates the difference between initial energy and energy used during sending/receiving data packets. In this proposal, decision-maker first considers the location of the nodes and then optimizes it. Finally, it helps to reduce the energy consumption of the nodes.

5.3 GA-Based Algorithms

In the last decade, several works have been proposed based on GA technique [47–50]. In this subsection, some of them are as follows: Hanh et al. [51] proposed an intelligent and efficient routing protocol for area coverage in WSN. The basic key element of this routing protocol is GA which is a metaheuristic algorithm. This algorithm helps to design a fitness function or objective function by the approximation method for maximizing coverage of the network. Finally, this algorithm helps to produce the best performance in terms of solution quality. Somauroo and Bassoo [52] proposed GA based efficient variants algorithm for 3D WSN. The proposed method uses chain-based technique as PEGASIS technique. This technique is used to enhance the network lifetime of the WSN. Moreover, this approach is also used to design the clustering of the network that helps to reduce a number of sensor nodes in a chain. This process helps to reduce delay in data transmission. Wang et al. [53] proposed an energy-efficient routing and clustering method for WSN. The proposed method is based on GA technique. In this proposed routing protocol, optimal solution is obtained by comparing the previous round with the current round. In each iteration round, it is improved in terms of fitness function and their constraints. Finally, it helps to improve overall network performance. Al-Shalabi et al. [54] proposed an efficient multi-hop routing method in WSN using GA. This is basically optimal multi-path finding system. It is based on clustering method in which the basic aim is to find the optimal path between sources that are cluster head and destination node known as BS. Kumar et al. [55] proposed a GA-based distributed zone approach in WSN. The proposed method is based on a green communication system that includes two major things: first is communication optimization and second is energy optimization. The main aim of this proposal is to find the optimal path between source nodes to sink node. It also helps to enhance the convergence speed of the solution. Finally, it outperforms the existing methods. Barekatain et al. [56] proposed a hybrid method for energy-aware routing in WSN. This hybrid method is the fusion of GA and k-means. The combined method is used to design the clustering of the network. This clustering method is used to reduce energy consumption of the network and finds an optimal cluster head. At last, it produces better results and outperform the existing methods.

5.4 SI-Based Algorithms

In this subsection, some swarm-based techniques are illustrated. Saleem et al. [57] proposed a detailed proposal for the survey in WSN and for swarm intelligence technique. This proposal is totally based on intelligent routing protocols, and it also highlights the future prediction and the direction in the context of swarm intelligence routing methods. This survey discusses the general principle and taxonomy of reverse engineering, desirable properties and decentralizes control. Zahedi et al. [58] proposed a clustering-based hybrid method for WSN. This method is a fusion of fuzzy logic and swarm intelligence. The role of swarm intelligence is to inspire the different clustering of the network. And the role of fuzzy logic is to reduce uncertainty of the network by controlling the conflicting parameters of the network. Bruneo et al. [59] proposed a modeling technique for WSN using swarm intelligence method. The main aim of this proposal is to optimize route of the network between source node to sink node by using different conflicting parameters. The key element of this network is Markovian agent that helps to trap the issues and enhances the message passing system. Ari et al. [60] proposed an efficient clustering technique for WSN using swarm intelligence. In this technique, the honeybee nature-inspired method is used as swarm intelligence for managing clustering of the network. The proposed method is based on multi-objective optimization technique where each objective employs with linear programming and manages the trade-off between energy efficiency and cost function of the network. Sreelaja and Pai [61] proposed a sinkhole attack detection technique for WSN using swarm intelligence. In this method, ACO is used as a swarm intelligence technique. The proposed method is based on voting method which is used to generate a notification for identifying attacks within the network. Finally, it outperforms the network metrics in terms of validation and simulation. Li and Shen [62] proposed a multi-robots method for WSN using swarm behavior technique. The proposed method is based on autonomous, homogenous, self-organize, and decentralize methods. The proposed method efficiently manages and controls the emergent behaviors of the network. Finally, it helps to design efficient interaction channel of the network that helps to cooperate and manage different strategies of the swarms.

5.5 ANN-Based Algorithms

In the last decade, several works have been proposed based on ANN technique [49, 63, 64]. Some of them as follows: Gholami et al. [65] proposed ANN-based localization method for WSN. The proposed method first analyzes the ambient conditions of the network and investigates blockage parts of the network. This process helps to estimate different types of uncertainties of the network efficiently. Finally, it validates in a practical environment and this validation shows that outperform network metrics. Alarifi and Tolba [66] proposed an adaptive neural network-based optimization technique for WSN. This is based on reinforcement-based learning technique. So, it uses the adaptive Q-learning method. It helps to design clustering in the network that helps to reduce the energy consumption of the nodes and data collection from sensor nodes to the BS. Eldhose and Jisha [67] proposed a node aggregation method for WSN. This is based on the clustering method where each active cluster is decided by ANN. This neural network method helps to process the data packet periodically and sends it to the BS efficiently. Chang et al. [68] proposed a frailty detection system for WSN using ANN. It is totally home-based system. The main purpose of this proposal is to automatically manage the personal information of a home. It consists of several information such as ePad, eChair, eScale. Finally, it helps to predict frailty and test overall system performance. Serpen and Gao [69] proposed a perceptron method for WSN using ANN. This method is based on embedded technique. This method is based on parallel as well as distributed technique which works as neurocomputing method. This prediction method helps to enhance the network metrics as well as network lifetime. Li and Zhao [70] proposed a method WSN based on ANN. This method is used for wavefront correction system. The basic aim of this proposal is to estimate the error of distortion by using online measurement correction technique. Finally, this method is validated numerically as well as simulated to outperform the results.

5.6 ABO-Based Algorithms

In this subsection, some ABO-based methods are highlighted. Jebaraj and Keshavan [71] proposed a hybrid method for WSN. This hybrid network is based on a fusion of GA and ABO. It uses two basic parameters such as energy and delay for processing the beacon message of the WSN. So, it helps to manage the collision of the network. Finally, it helps to create and manage the feasible and optimal route of the network. Padmapriya and Maheswari [72] proposed a channel optimization technique for WSN. The proposed method is based on hybrid technique with the fusion of support vector machine (SVM), ABO, and GA. The SVM technique is used for the classification; ABO technique is used for the network optimization; and GA technique is used for the performance evaluation.

5.7 AIS-Based Algorithms

In this subsection, some artificial immune-based approaches are highlighted. Alaparthy et al. [73] proposed a proposal for details study of immune model of WSN. The proposed method is based on artificial neural network that uses immune system. It uses intelligent method for intrusion detection and managing system. This method basically includes clonal selection, danger theory, positive selection, negative selection systems. Li et al. [74] proposed a method for fault tolerance in WSN using immune mechanism. This method is based on multi-path routing system. It is more reliable and efficient proposal. The proposed method is based on guiding principle of artificial ants. These artificial ants use pheromone information of the network for enhancing the network performance. Li et al. [75] proposed a fault-tolerant method for coverage optimization in WSN. In this method, two basic things are checked: first is transmission reliability and second is transmission stability. It establishes reliable packet transmission methods through artificial immune system. It helps to efficiently deploying the nodes of the network and covers maximum coverage of the network. Abo-Zahhad et al. [76] proposed a centralize method for maximum coverage as well as energy conservation in mobile WSN. The basic key elements of this proposal are (i) artificial immune system and (ii) Voronoi diagram, the fusion of both known as immune–Voronoi system. The proposed method is completed in two phases: first is network lifetime and second is coverage issue in the network.

Moreover, sometimes fuzzy logic is used to reduce uncertainty of the network by the help of multi-value logic. This logic is inherent with the above-mentioned nature-inspired algorithms to make the process more efficient and robustness. Here, some of the algorithms are described bases on fuzzy logic in the context of linear and nonlinear optimizations. In [29, 77–84], the authors described several works based on ad hoc network. These works are based on energy-efficient routing. Some of them are used fuzzy logic or some of them are used its extension, or some of them are used fuzzy logic with optimization. Apart from ad hoc network, fuzzy logic is also used in

WSN. Some of the proposals are described as follows. In [85, 86], the authors were used soft computing technique for enhancing the network performance. Fuzzy logic is used in these proposals as a soft computing technique. In these proposals, fuzzy logic is used to design input and output parameters of the route using triangular membership functions. Finally, it helps to evaluate all feasible routes as well as optimal route of the network. Amri et al. [87] proposed a fuzzy-based method for node localization system in WSN. The main aim of this method to precisely finds the location of the sensor nodes in WSN. It also helps in the precision enhancement of the network. This enhancement of the network is helped in energy-saving and network lifetime enhancement of the WSN. Mazinani et al. [88] proposed a fuzzy-based clustering method in WSN. It is a multi-cluster-based routing technique which is used as threshold value for determining the cluster head. The proposed method helps to avoid the control message and possibility of collision in WSN. Finally, it helps to find cluster head and enhances network lifetime as well as messages transmission. Das et al. [85] proposed a soft computing-based routing protocol for WSN. In this proposal, the authors basically used fuzzy logic technique to reduce the uncertainty of the network using fuzzy inference system. There are two basic parameters which are considered here: first is energy and second is distance to evaluate the output parameter reward. It helps to reduce uncertainty related to the route selection of the WSN.

6 Conclusion

The use of WSN is increased rapidly in day-by-day applications. Beside of this, several issues also are increasing in different areas which are difficult to solve in an efficient way. This chapter highlights different applications and challenges of WSN. It also guides to the researchers about different metaheuristic methods or nature-inspired algorithms. It illustrates several existing works of the last decade including their few advantages and disadvantages. This chapter also helps to analyze the basic concepts of the nature-inspired algorithm with its inherent elements. Moreover, this chapter also guides how these algorithms are used for solving a particular problem in a specific area.

References

1. Dener M (2017) WiSeN: a new sensor node for smart applications with wireless sensor networks. Comput Electr Eng 64:380–394
2. Kochhar A, Kumar N (2019) Wireless sensor networks for greenhouses: an end-to-end review. Comput Electron Agric 163:104877
3. Boukerche A, Sun P (2018) Connectivity and coverage based protocols for wireless sensor networks. Ad Hoc Netw 80:54–69

4. Dey N, Ashour AS, Shi F, Fong SJ, Sherratt RS (2017) Developing residential wireless sensor networks for ECG healthcare monitoring. IEEE Trans Consum Electron 63(4):442–449
5. Binh HTT, Hanh NT, Dey N (2018) Improved cuckoo search and chaotic flower pollination optimization algorithm for maximizing area coverage in wireless sensor networks. Neural Comput Appl 30(7):2305–2317
6. Roy S, Karjee J, Rawat US, Dey N (2016) Symmetric key encryption technique: a cellular automata based approach in wireless sensor networks. Procedia Comput Sci 78:408–414
7. Barnawi AY, Mohsen GA, Shahra EQ (2019) Performance analysis of RPL protocol for data gathering applications in wireless sensor networks. Procedia Comput Sci 151:185–193
8. Fong S, Li J, Song W, Tian Y, Wong RK, Dey N (2018) Predicting unusual energy consumption events from smart home sensor network by data stream mining with misclassified recall. J Ambient Intell Humaniz Comput 9(4):1197–1221
9. Skiadopoulos K, Tsipis A, Giannakis K, Koufoudakis G, Christopoulou E, Oikonomou K, Stavrakakis I (2019) Synchronization of data measurements in wireless sensor networks for IoT applications. Ad Hoc Netw 89:47–57
10. Elhayatmy G, Dey N, Ashour AS (2018) Internet of Things based wireless body area network in healthcare. In: Internet of things and big data analytics toward next-generation intelligence. Springer, Cham, pp 3–20
11. Karati A, Biswas GP (2019) Provably secure and authenticated data sharing protocol for IoT-based crowdsensing network. Trans Emerg Telecommun Technol 30(4):e3315, 1–22
12. Karati A, Islam SH, Karuppiah M (2018) Provably secure and lightweight certificateless signature scheme for IIoT environments. IEEE Trans Ind Inform 14(8):3701–3711
13. Panda SK, Jana PK (2019) An energy-efficient task scheduling algorithm for heterogeneous cloud computing systems. Clust Comput 22(2):509–527
14. Panda SK, Jana PK (2018) Normalization-based task scheduling algorithms for heterogeneous multi-cloud environment. Inf Syst Front 20(2):373–399
15. Panda SK, Pande SK, Das S (2018) Task partitioning scheduling algorithms for heterogeneous multi-cloud environment. Arab J Sci Eng 43(2):913–933
16. Karati A, Amin R, Islam SH, Choo KKR (2018) Provably secure and lightweight identity-based authenticated data sharing protocol for cyber-physical cloud environment. IEEE Trans Cloud Comput 1–14
17. Mukherjee A, Dey N, Kausar N, Ashour AS, Taiar R, Hassanien AE (2019) A disaster management specific mobility model for flying ad-hoc network. In: Emergency and disaster management: concepts, methodologies, tools, and applications. IGI Global, pp 279–311
18. Karati A, Islam SH, Biswas GP (2018) A pairing-free and provably secure certificateless signature scheme. Inf Sci 450:378–391
19. Jain PK, Pamula R (2019) Two-step anomaly detection approach using clustering algorithm. International conference on advanced computing networking and informatics. Springer, Singapore, pp 513–520
20. Mishra G, Agarwal S, Jain PK, Pamula R (2019) Outlier detection using subset formation of clustering based method. International conference on advanced computing networking and informatics. Springer, Singapore, pp 521–528
21. Kumari P, Jain PK, Pamula R (2018) An efficient use of ensemble methods to predict students academic performance. In: 2018 4th international conference on recent advances in information technology (RAIT). IEEE, pp 1–6
22. Punam K, Pamula R, Jain PK (2018) A two-level statistical model for big mart sales prediction. In: 2018 international conference on computing, power and communication technologies (GUCON). IEEE, pp 617–620
23. Das SP, Padhy S (2018) A novel hybrid model using teaching–learning-based optimization and a support vector machine for commodity futures index forecasting. Int J Mach Learn Cybern 9(1):97–111
24. Das SP, Padhy S (2017) Unsupervised extreme learning machine and support vector regression hybrid model for predicting energy commodity futures index. Memetic Comput 9(4):333–346

25. Das SP, Padhy S (2017) A new hybrid parametric and machine learning model with homogeneity hint for European-style index option pricing. Neural Comput Appl 28(12):4061–4077
26. Dey N (ed) (2017) Advancements in applied metaheuristic computing. IGI Global
27. Dey N, Ashour AS, Bhattacharyya S (2019) Applied nature-inspired computing: algorithms and case studies
28. Dey N, Ashour AS (2016) Antenna design and direction of arrival estimation in meta-heuristic paradigm: a review. Int J Serv Sci Manag Eng Technol (IJSSMET) 7(3):1–18
29. Das SK, Tripathi S (2019) Energy efficient routing formation algorithm for hybrid ad-hoc network: a geometric programming approach. Peer-to-Peer Netw Appl 12(1):102–128
30. Kaliannan J, Baskaran A, Dey N, Ashour AS (2016) Ant colony optimization algorithm based PID controller for LFC of single area power system with non-linearity and boiler dynamics. World J Model Simul 12(1):3–14
31. Kaliannan J, Baskaran A, Dey N (2015) Automatic generation control of thermal-thermal-hydro power systems with PID controller using ant colony optimization. Int J Serv Sci Manag Eng Technol (IJSSMET) 6(2):18–34
32. Jagatheesan K, Anand B, Dey N, Ashour AS (2018) Effect of SMES unit in AGC of an interconnected multi-area thermal power system with ACO-tuned PID controller. In: Advancements in applied metaheuristic computing. IGI Global, pp 164–184
33. Jagatheesan K, Anand B, Dey KN, Ashour AS, Satapathy SC (2018) Performance evaluation of objective functions in automatic generation control of thermal power system using ant colony optimization technique-designed proportional–integral–derivative controller. Electr Eng 100(2):895–911
34. Sun X, Zhang Y, Ren X, Chen K (2015) Optimization deployment of wireless sensor networks based on culture–ant colony algorithm. Appl Math Comput 250:58–70
35. Sharma V, Grover A (2016) A modified ant colony optimization algorithm (mACO) for energy efficient wireless sensor networks. Opt-Int J Light Electron Opt 127(4):2169–2172
36. Kaur S, Mahajan R (2018) Hybrid meta-heuristic optimization based energy efficient protocol for wireless sensor networks. Egypt Inform J 19(3):145–150
37. Liao WH, Kao Y, Wu RT (2011) Ant colony optimization based sensor deployment protocol for wireless sensor networks. Expert Syst Appl 38(6):6599–6605
38. Ho JH, Shih HC, Liao BY, Chu SC (2012) A ladder diffusion algorithm using ant colony optimization for wireless sensor networks. Inf Sci 192:204–212
39. Sun Z, Wei M, Zhang Z, Qu G (2019) Secure routing protocol based on multi-objective ant-colony-optimization for wireless sensor networks. Appl Soft Comput 77:366–375
40. Chatterjee S, Hore S, Dey N, Chakraborty S, Ashour AS (2017) Dengue fever classification using gene expression data: a PSO based artificial neural network approach. In: Proceedings of the 5th international conference on frontiers in intelligent computing: theory and applications. Springer, Singapore, pp 331–341
41. Jagatheesan K, Anand B, Samanta S, Dey N, Ashour AS, Balas VE (2017) Particle swarm optimisation-based parameters optimisation of PID controller for load frequency control of multi-area reheat thermal power systems. Int J Adv Intell Parad 9(5–6):464–489
42. Parvin JR, Vasanthanayaki C (2019) Particle swarm optimization-based energy efficient target tracking in wireless sensor network. Measurement 106882
43. Phoemphon S, So-In C, Niyato DT (2018) A hybrid model using fuzzy logic and an extreme learning machine with vector particle swarm optimization for wireless sensor network localization. Appl Soft Comput 65:101–120
44. Sun Z, Liu Y, Tao L (2018) Attack localization task allocation in wireless sensor networks based on multi-objective binary particle swarm optimization. J Netw Comput Appl 112:29–40
45. Cao B, Zhao J, Lv Z, Liu X, Kang X, Yang S (2018) Deployment optimization for 3D industrial wireless sensor networks based on particle swarm optimizers with distributed parallelism. J Netw Comput Appl 103:225–238
46. Yan Z, Goswami P, Mukherjee A, Yang L, Routray S, Palai G (2019) Low-energy PSO-based node positioning in optical wireless sensor networks. Optik 181:378–382

47. Karaa WBA, Ashour AS, Sassi DB, Roy P, Kausar N, Dey N (2016) Medline text mining: an enhancement genetic algorithm based approach for document clustering. In: Applications of intelligent optimization in biology and medicine. Springer, Cham, pp 267–287
48. Dey N, Ashour A, Beagum S, Pistola D, Gospodinov M, Gospodinova E, Tavares J (2015) Parameter optimization for local polynomial approximation based intersection confidence interval filter using genetic algorithm: an application for brain MRI image de-noising. J Imaging 1(1):60–84
49. Chatterjee S, Sarkar S, Hore S, Dey N, Ashour AS, Shi F, Le DN (2017) Structural failure classification for reinforced concrete buildings using trained neural network based multi-objective genetic algorithm. Struct Eng Mech 63(4):429–438
50. Chatterjee S, Sarkar S, Dey N, Ashour AS, Sen S (2018) Hybrid non-dominated sorting genetic algorithm: II-neural network approach. In: Advancements in applied metaheuristic computing. IGI Global, pp 264–286
51. Hanh NT, Binh HTT, Hoai NX, Palaniswami MS (2019) An efficient genetic algorithm for maximizing area coverage in wireless sensor networks. Inf Sci 488:58–75
52. Somauroo A, Bassoo V (2019) Energy-efficient genetic algorithm variants of PEGASIS for 3D wireless sensor networks. Appl Comput Inform
53. Wang T, Zhang G, Yang X, Vajdi A (2018) Genetic algorithm for energy-efficient clustering and routing in wireless sensor networks. J Syst Softw 146:196–214
54. Al-Shalabi M, Anbar M, Wan TC, Alqattan Z (2019) Energy efficient multi-hop path in wireless sensor networks using an enhanced genetic algorithm. Inf Sci
55. Kumar S, Kumar V, Kaiwartya O, Dohare U, Kumar N, Lloret J (2019) Towards green communication in wireless sensor network: GA enabled distributed zone approach. Ad Hoc Netw 101903
56. Barekatain B, Dehghani S, Pourzaferani M (2015) An energy-aware routing protocol for wireless sensor networks based on new combination of genetic algorithm & k-means. Procedia Comput Sci 72:552–560
57. Saleem M, Di Caro GA, Farooq M (2011) Swarm intelligence based routing protocol for wireless sensor networks: survey and future directions. Inf Sci 181(20):4597–4624
58. Zahedi ZM, Akbari R, Shokouhifar M, Safaei F, Jalali A (2016) Swarm intelligence based fuzzy routing protocol for clustered wireless sensor networks. Expert Syst Appl 55:313–328
59. Bruneo D, Scarpa M, Bobbio A, Cerotti D, Gribaudo M (2012) Markovian agent modeling swarm intelligence algorithms in wireless sensor networks. Perform Eval 69(3–4):135–149
60. Ari AAA, Yenke BO, Labraoui N, Damakoa I, Gueroui A (2016) A power efficient cluster-based routing algorithm for wireless sensor networks: honeybees swarm intelligence based approach. J Netw Comput Appl 69:77–97
61. Sreelaja NK, Pai GV (2014) Swarm intelligence based approach for sinkhole attack detection in wireless sensor networks. Appl Soft Comput 19:68–79
62. Li W, Shen W (2011) Swarm behavior control of mobile multi-robots with wireless sensor networks. J Netw Comput Appl 34(4):1398–1407
63. Chatterjee S, Sarkar S, Dey N, Ashour AS, Sen S, Hassanien AE (2017) Application of cuckoo search in water quality prediction using artificial neural network. Int J Comput Intell Stud 6(2–3):229–244
64. Hore S, Chatterjee S, Sarkar S, Dey N, Ashour AS, Balas-Timar D, Balas VE (2016) Neural-based prediction of structural failure of multistoried RC buildings. Struct Eng Mech 58(3):459–473
65. Gholami M, Cai N, Brennan RW (2013) An artificial neural network approach to the problem of wireless sensors network localization. Robot Comput-Integr Manuf 29(1):96–109
66. Alarifi A, Tolba A (2019) Optimizing the network energy of cloud assisted internet of things by using the adaptive neural learning approach in wireless sensor networks. Comput Ind 106:133–141
67. Eldhose EK, Jisha G (2016) Active cluster node aggregation scheme in wireless sensor network using neural network. Procedia Technol 24:1603–1608

68. Chang YC, Lin CC, Lin PH, Chen CC, Lee RG, Huang JS, Tsai TH (2013) eFurniture for home-based frailty detection using artificial neural networks and wireless sensors. Med Eng Phys 35(2):263–268
69. Serpen G, Gao Z (2014) Complexity analysis of multilayer perceptron neural network embedded into a wireless sensor network. Procedia Comput Sci 36:192–197
70. Li Z, Zhao X (2017) BP artificial neural network based wave front correction for sensor-less free space optics communication. Opt Commun 385:219–228
71. Jebaraj NS, Keshavan HR (2018) Hybrid genetic algorithm and african buffalo optimization (HGAABO) based scheduling in ZigBee network. Int J Appl Eng Res 13(5):2197–2206
72. Padmapriya R, Maheswari D (2017) Channel allocation optimization using african buffalo optimization-super vector machine for networks. Asian J Inf Technol 16(10):783–788
73. Alaparthy VT, Amouri A, Morgera SD (2018) A study on the adaptability of immune models for wireless sensor network security. Procedia Comput Sci 145:13–19
74. Li H, Chen Q, Ran Y, Niu X, Chen L, Qin H (2017) BIM2RT: BWAS-immune mechanism based multipath reliable transmission with fault tolerance in wireless sensor networks. Swarm Evol Comput
75. Li H, Wang S, Gong M, Chen Q, Chen L (2017) IM2DCA: immune mechanism based multipath decoupling connectivity algorithm with fault tolerance under coverage optimization in wireless sensor networks. Appl Soft Comput 58:540–552
76. Abo-Zahhad M, Sabor N, Sasaki S, Ahmed SM (2016) A centralized immune-Voronoi deployment algorithm for coverage maximization and energy conservation in mobile wireless sensor networks. Inf Fusion 30:36–51
77. Das SK, Tripathi S (2018) Intelligent energy-aware efficient routing for MANET. Wireless Netw 24(4):1139–1159
78. Yadav AK, Das SK, Tripathi S (2017) EFMMRP: design of efficient fuzzy based multi-constraint multicast routing protocol for wireless ad-hoc network. Comput Netw 118:15–23
79. Das SK, Tripathi S (2018) Adaptive and intelligent energy efficient routing for transparent heterogeneous ad-hoc network by fusion of game theory and linear programming. Appl Intell 48(7):1825–1845
80. Das SK, Tripathi S (2017) Energy efficient routing formation technique for hybrid ad hoc network using fusion of artificial intelligence techniques. Int J Commun Syst 30(16):e3340, 1–16
81. Das SK, Yadav AK, Tripathi S (2017) IE2M: Design of intellectual energy efficient multicast routing protocol for ad-hoc network. Peer-to-Peer Netw Appl 10(3):670–687
82. Das SK, Samanta S, Dey N, Kumar R, Design frameworks for wireless networks. Lecture Notes in Networks and systems. Springer, pp 1–439. ISBN: 978-981-13-9573-4
83. Das SK, Tripathi S (2020) A nonlinear strategy management approach in software-defined ad hoc network. In: Design frameworks for wireless networks. Springer, Singapore, pp 321–346
84. Samantra A, Panda A, Das SK, Debnath S (2020) Fuzzy petri nets-based intelligent routing protocol for ad hoc network. In: Design frameworks for wireless networks. Springer, Singapore, pp 417–433
85. Das SK, Kumar A, Das B, Burnwal AP (2013) Ethics of reducing power consumption in wireless sensor networks using soft computing techniques. Int J Adv Comput Res 3(1):301
86. Das SK, Das B, Burnawal AP (2014) Intelligent energy competency routing scheme for wireless sensor network. Int J Res Comput Appl Robot 2(3):79–84
87. Amri S, Khelifi F, Bradai A, Rachedi A, Kaddachi ML, Atri M (2017) A new fuzzy logic based node localization mechanism for wireless sensor networks. Future Gener Comput Syst
88. Mazinani A, Mazinani SM, Mirzaie M (2019) FMCR-CT: an energy-efficient fuzzy multi cluster-based routing with a constant threshold in wireless sensor network. Alex Eng J 58(1):127–141

Part I
Bio-inspired Optimization

Chapter 2
A GA-Based Fault-Aware Routing Algorithm for Wireless Sensor Networks

Nabajyoti Mazumdar and Hari Om

1 Introduction

Considering the frequency of occurrence of natural disasters like volcano, landslide, icescape, etc., the environment monitoring applications have gained substantial significance in recent years. The main environmental monitoring infrastructure of IoT system is wireless sensor networks (WSN) [1], where sensor nodes are positioned in the monitoring region; each is used in the monitoring environment to measure the metrological and hydrological data within its sensing range such as heat, light, and humidity. A node has limited sensing range; therefore, the collaborative functioning of an enormous amount of nodes enables WSN to monitor a large target area and communicate the ambient condition of the monitoring environment to the remote station (RS) for disaster management [2–5]. Figure 1 shows a network model, where numbers of nodes are positioned in a target area and they send their data to the remote station, the RS is connected to the Internet to communicate the environmental information of the monitoring region to disaster management centers. The actions of the disaster management unit are entirely reliant on the accuracy of information conveyed by the WSN. The main challenge of IoT system in environmental monitoring is to set up a reliable infrastructure at hazard environment considering different constraints of a WSN. The foremost constraint of WSN is the limited battery-powered sensor nodes which cannot be recharged or replaced once deployed [6–9]. Thus, the sensor nodes energy should be used efficiently to prolong their lifetime. A sensor

N. Mazumdar (✉)
Department of Information Technology, Central Institute of Technology, Kokrajhar, BTAD, Kokrajhar, Assam 783370, India
e-mail: nabajyoti@cse.ism.ac.in

H. Om
Department of Computer Science and Engineering, IIT (ISM) Dhanbad, Dhanbad, Jharkhand 826004, India
e-mail: hariom4india@gmail.com

D. De et al. (eds.), *Nature Inspired Computing for Wireless Sensor Networks*, Springer Tracts in Nature-Inspired Computing,
https://doi.org/10.1007/978-981-15-2125-6_2

Fig. 1 Wireless sensor network model

node that drained its energy may possibly result in sensing hole in the target area. Therefore, if number of sensor nodes used up their energy early, then it will result in a huge portion of the target area exposed which degrades the reliability of the monitoring IoT system. A sensor node's main source of energy consumption is its radio transceiver. Therefore, to prolong sensors' lifetime, energy-efficient communication protocols are highly desired in WSN. Many energy-saving methods have been suggested for WSN in latest years [10, 11], among these clustering is found to be best suited for WSN. Figure 2 shows the network architecture of a clustered WSN. Here, the cluster head (CH) of a cluster collects data packets from the cluster members (CMs), and forward to the remote station (RS) via other CHs or directly based on its connectivity to the RS. Clustering in WSN has several advantages like

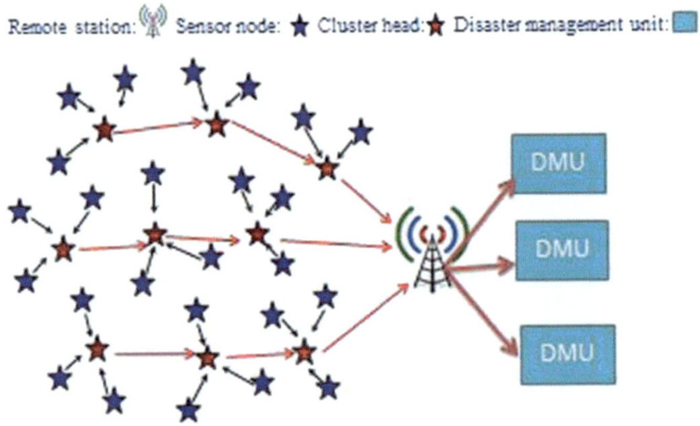

Fig. 2 Clustered WSN model

- It reduces the energy consumption of the WSN.
- Communication bandwidth is conserved.
- Enhances the scalability of the network.

However, in clustered WSN architecture, CHs are burden with aided responsibilities like data aggregation, data routing, etc., resulting in early depletion of their energy. To avoid such imbalanced energy depletion, many researchers [12–17] adopted the CH rotation mechanism. Furthermore, in [18–21], some researchers used special nodes called advanced nodes to carry out the responsibilities of CHs. These advanced nodes are equipped with relatively higher powered battery to sustain the workload of CHs. But, these advanced nodes are also battery run consequently their energy too needs to be proficiently consumed to prolong their lifetime. However, as WSN is generally applied under harsh environment so unpredictable events like a communication link failure, energy depletion of sensor nodes, or environmental hazards may occur which could result in some sensor node failure. The WSN performance degrades rapidly with the failure of sensors; furthermore, the effect is more severe if the dead node is a CH. A CH failure not only interrupts communication with its member sensor nodes, but also with other CHs as they are engaged in transmitting cluster information to the RS via other CHs. WSN's fault tolerance for CH failure is, therefore, an issue as important as energy efficiency.

Again, as the environmental monitoring infrastructure of IoT is responsible for early warning to the disaster management unit, so the routing procedures considered for WSN must be fault tolerant to recover from CHs failure and route sensed information to the RS. The routing protocols intended for WSN should, therefore, not only be energy effective, but should also be fault aware. Now, for N number of CHs in the network and an average k number of candidates next-hop routing CH for each CH, there are k^N possible routing path in the network. This value is very large for a large scale WSN; this infers the computational cost of routing tree construction in WSN using a brute force method. So it is extremely desirable to discover an optimization strategy such as genetic algorithm (GA) to find an almost ideal routing alternative [22].

This paper proposes a meta-heuristic method based on GA to design a fault-aware routing algorithm called FAR which considers different routing challenges which may occur due to failure of some CHs. We also present an efficient chromosome generation algorithm such that each chromosome represents a routing solution. A fitness function is derived to estimate the quality chromosomes. Finally, the performance FAR is evaluated with various existing related algorithms to highlight its effectiveness. The rest of the chapter is planned as follows. Section 2 provides literature survey. In Sect. 3, we provide the system model assumed. Section 4 describes the proposed protocol (FAR). The performance of FAR is evaluated in Sect. 5. Lastly, Sect. 6 concludes by highlighting some future direction.

2 Related Work

In recent years, several studies on energy-saving techniques for WSN are done predominantly related to clustering and routing [10–19, 23–27]. The LEACH [13] is the most popular clustering protocol proposed for WSN which forms clusters uniformly using a randomize method. Later, many modified versions of LEACH have been proposed afterward. PEGASIS [17] and TEEN [16] are improved versions of LEACH where CHs are selected considering different parameters like energy, distance from neighbors, node degree, etc. In HEED [14], CHs are nominated considering the remaining energy of sensor nodes. All these algorithms are designed for homogeneous WSN. In homogeneous WSN, considering the workload on CHs it is obvious that CHs will deplete their energy much earlier compared to non-CH nodes which results early death of some nodes. In order to overcome such problem, researchers [18–21] have focused on heterogeneous WSN, where some higher energy provisioned advanced nodes are deployed. The main challenge of clustering in heterogeneous WSN is to maximize the CHs lifetime. Now, considering the computational cost of finding optimal solutions for load-balanced clustering using brute force methods, evolutionary algorithms have been used recently [20, 26, 28]. LEACH-C [26] is a centralized form of LEACH, where RS is used to form clusters. In [28], a GA-based clustering protocol is proposed where the objective is to select CH such that intracluster distance and CHs to RS distance is minimized.

Like clustering, an intensive research has been done on a cluster-based routing. In EHE-LEACH [18], CHs use direct communication to RS if the distance is below a predefined threshold value otherwise considers multi-hop communication to RS using other CHs. In MHRM [24], a CH selects its next-hop CH to route its data, such that the hop count to RS is minimal. Again, ERA [25] is another routing algorithm which uses a cost function considering energy of CH, distance, etc., to select next-hop CH. In [29], a GA-based approach is discussed for sensor-to-CH binding, where different possible connectivity between sensor nodes and CHs were considered such that each sensor will be bonded to a CH directly or via other intermediate nodes. GADA-LEACH [30] is a GA-based distance-aware routing protocol for WSNs where the fitness function for CHs selection comprises energy, distance of CH from RS, and CH distance from its neighbors. The selected CHs then communicate with RS via a relay node.

Recently, swarm intelligence-based optimization methods have gained popularity for solving the clustering and routing problem in WSNs. Swarm Intelligence has many advantages like ease of implementation, quick convergence, high-quality solutions, etc. In [31], a PSO-based clustering and routing protocol are proposed where the clustering method elects CHs such that it maximizes energy efficiency, network coverage, and the routing protocol connects each CH to RS to produce a complete routing solution.

None of the clustering and routing protocols discussed so far have considered the fault-tolerance aspect in their work. In [32], authors discussed a fault-tolerant method where during clustering each cluster has two CHs one primary and the other secondary CH. Here, if primary CH fails, secondary CH will perform the CH responsibility. But, such assumption may not be feasible under heterogeneous WSN as all nodes are positioned arbitrarily. In [21], authors have proposed a fault-tolerant routing protocol where CHs can reconfigure the routing path dynamically in case of failure of any CH during data routing. DEFTR [33] presents a fault-tolerant routing protocol where a CH chooses an alternative secondary routing path if the energy status of the primary path falls below a predefined threshold. Moreover, to enhance fault-tolerance and energy efficiency, a CH re-election routine is executed among the cluster members of a cluster if current CH's energy falls below a certain threshold. In ICE [34], during intercluster communication closest neighbor nodes of two clusters are used as an intermediate node to forward data. Thus, if these intermediate nodes die, the other nodes take over. But, the involvement of such intermediate nodes adds to more energy depletion as well as enhances the message complexity. In [35], a PSO-based algorithm is proposed for WSNs where the routing path is formed such that it maximizes the CHs lifetime. Further, during routing, if the next-hop CH fails, then the CH selects another CH toward the RS having maximum lifetime.

3 System Model and Terminologies

We consider a heterogeneous sensor network model for environment monitoring IoT system. It comprises of resource-limited ordinary sensor nodes and resource-rich advanced nodes. Based on the capability of these nodes, the network is separated into two layers, where the lower layer is formed by ordinary sensor nodes while the upper layer is made of advanced nodes. The task of lower layer ordinary sensor node is environment sensing and transmitting the sensed data to the advanced nodes, whereas the advanced nodes are responsible for aggregating the received packets, then communicating it to the RS. The number of advanced nodes N_{AN} is much lesser than the number of ordinary sensor nodes N_{S}.

The radio transceiver is the main source of energy consumption in WSN. The energy model assumed in this work is analogous to the one considered in [13]. Here, the energy spent while transmitting a packet of 'r' bits between node 'i' and node 'j' over a range '$d_{i,j}$' is

$$E_{\mathrm{tx}}(i,j) = \begin{cases} (E_{\mathrm{elec}} + \varepsilon_{\mathrm{fs}} \times d_{i,j}^2) \times r, & d_{i,j} < d_0 \\ (E_{\mathrm{elec}} + \varepsilon_{\mathrm{mp}} \times d_{i,j}^4) \times r, & d_{i,j} \geq d_0 \end{cases} \quad (1)$$

where E_{elec} is transmitting energy coefficient, $\varepsilon_{\mathrm{fs}}$ and $\varepsilon_{\mathrm{mp}}$ represent the amplification coefficient in free space and multi-fading model and $d_0 = \sqrt{\frac{\varepsilon_{\mathrm{fs}}}{\varepsilon_{\mathrm{mp}}}}$. Likewise, energy consumed in receiving a packet of 'r' bits by a node 'i' is given by

$$E_{\mathrm{r}}(S_i) = E_{\mathrm{elec}} \times r \quad (2)$$

A fault model is also adopted to evaluate the possibility of a node failure. A sensor node failure may happen due to several causes like battery energy depletion, hardware component failure, etc. The failure of sensor nodes, especially CHs cause network partition while dynamically changing in network topology. This paper considers Weibull distribution [36, 37] to evaluate the reliability of a sensor node as it provides various failure patterns of devices like sensors over time. The Weibull distribution function which shows the failure probability of a device at time t is expressed as follows:

$$f(t) = \frac{\beta}{\eta}\left(\frac{t}{\eta}\right)^{\beta-1} e^{-\left(\frac{t}{\eta}\right)^{\beta}} \tag{3}$$

where β and η are shape and scaling parameter. Here, $\beta < 1$ means failure rate decreases over time, $\beta = 1$ means the constant failure rate, and $\beta > 1$ means failure rate increases over time. In WSN, the failure chance of nodes increases over time, so we consider $\beta > 1$.

Terminologies

Following are the terminologies and definitions used.

- $S = \{S_1, S_2, \ldots, S_n\}$ is a set of sensor nodes.
- $\text{CH} = \{\text{CH}_1, \text{CH}_2, \ldots, \text{CH}_k\}$ is a set of cluster heads.
- $\text{Dist}(i, j)$ denotes the distance between nodes i and j.
- $\text{Dist}_{\text{RS}}(i)$ *is t*he distance of node i from the RS.
- $\text{Energy}_{\text{res}}(i)$ is node i's residual energy.
- The communication range of a node i is denoted by R_{com}.
- The cluster members of CH I is denoted by CM(i).
- The set of candidate CHs those are within the sensing radius of sensor node S_i is denoted by $\text{Cand\_CH}(S_i) = \{G_j: \text{Dist}(S_i, G_j) \leq R_{\text{com}}\}$.

Based on the connectivity among CHs, we form the following definitions:

Definition 1 Downstream CH set (DS_CH) of a CH i is the CHs in its transmission range toward the remote station. A CH j is an element of DS_CH(i) if *f*, $(\text{Dist}(i, j) \leq R_{\text{com}}) \wedge (\text{Dist}_{\text{RS}}(j) < \text{Dist}_{\text{RS}}(i))$. Note that if $|\text{DS\_CH}(i)| \neq 0$ then CH i will select a CH from its DS_CH(i) set to forward its packet toward remote station. It is illustrated in Fig. 3, where CH 10 has three CHs namely 1, 2, and 5 within 10's communication range and closer to remote station compared to 10. So, CHs 1, 2, and 5 are members of DS_CH(10) and CH 10 will select one of them to forward its data.

Definition 2 Upstream CH set (US_CH) of a CH i is the CHs in its transmission range, but further from the remote station and have at least one downstream CH other than i. A CH j is an element of US_CH(i) if *f*, $(\text{Dist}(i, j) \leq \text{R}_{\text{com}}) \wedge (\text{Dist}_{\text{RS}}(j) \geq \text{Dist}_{\text{RS}}(i)) \wedge |\text{DN\_CH}(j)| > 1$. Note that, if a CH i do not have any downstream CH to forward its data packet, then the CH i can use its upstream CHs

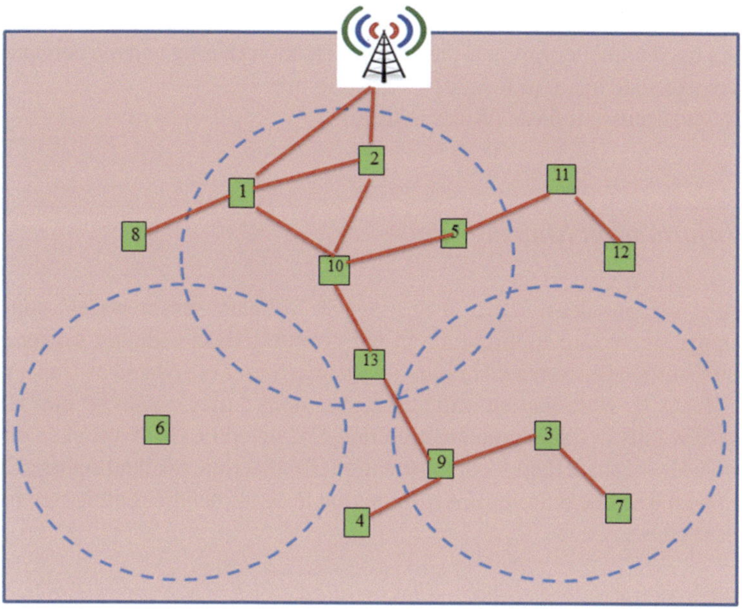

Fig. 3 Connectivity among CHs

to relay data to the remote station. Consider Fig. 3, here, CH 3 does not have any downstream CH but it has CHs 9 and 7 in backward direction. Now, since 7 does not have any downstream CH so only 9 will be considered as upstream CH of 3.

Definition 3 A CH is called Unconnected CH (UnCon_CH) if it does not have a downstream or upstream CH in its transmission range. Mathematically, a CH i is a UnCon_CH if f, $|\text{DS\_CH}(i)| == 0 \wedge |\text{US\_CH}(i)| == 0$. In Fig. 3, CH 6 is an Unconnected CH as it has neither a downstream or upstream CH within its range.

4 Proposed Algorithm

The working of the proposed algorithm can be divided into three phases, namely, information sharing, network setup, and steady phase. In information sharing phase, each node shares its information like its id, its position, residual energy, etc., with the RS. It is used during the network setup phase to form the routing path. In this phase, the network is distributed in clusters and a routing tree is constructed using CHs of each cluster to forward their data packets to the RS. The network setup phase is followed by a steady phase in which CHs assigns a time slot for each of its CMs to send their sensed data to the respective CH using time division multiple access (TDMA). Then, CHs aggregate these received packets and forward it to the

RS following the routing tree constructed in network setup phase. The steady phase runs for a fixed number of rounds then the information sharing and the network setup phase are executed again to form a new routing tree.

Now, we discuss all these phases in detail.

4.1 Information Sharing Phase

Initially, a CH broadcast HELLO message in ordinary sensor nodes' communication range to declare its presence in the network. If an ordinary sensor node i accepts this message from a CH j then its $\text{Cand}_{\text{CH}}(i)$ set is updated as $\text{Cand}_{\text{CH}}(i) = \{\text{Cand}_{\text{CH}}(i) \cup j\}$. Note that an ordinary sensor node i may obtain several HELLO message if it falls in the communication range of several CHs. Now, each ordinary node and CH will share their local information like position, residual energy, Cand_{CH} set, etc., with RS. The RS uses this information to construct the routing tree for CHs in the next phase.

4.2 Network Setup Phase

In this phase, initially, the network is divided into a set of clusters such that each sensor node is connected to a CH. This clustering can be achieved using any standard clustering algorithm proposed for WSN.

Then, a CH-CH routing path is constructed such that each CH can route its cluster data to the RS. The main challenge of designing a routing algorithm is to route data of unconnected CHs and CHs having no downstream CH within their communication range. Initially, as all CHs are alive, so there will be very less chance of any unconnected CH but w.r.t. time as CHs deplete their energy some CHs will die early which may result in some unconnected CHs.

In our proposed routing algorithm, we attempt to design a proximate optimum solution for load balancing of CHs using genetic algorithm (GA) so that CHs lifetime is maximized and each CH can route its data to RS. In the proposed routing algorithm, we consider both connected and unconnected CHs to establish a path to route their data to the RS. Here, the preliminary population of chromosomes is generated arbitrarily such that each chromosome of the population gives a valid routing path from CHs to RS. Then, fitness evaluation and population updating are repeated till the maximum number of iterations is reached or fitness value reached a saturation point. All these steps are illustrated as follows:

Initial population generation:

The preliminary population is an arbitrarily generated chromosome set such that all chromosomes provide a valid routing solution. At this point, each chromosome's length is equivalent to the network's number of CHs. The generated chromosomes are such that a value j at the ith location means that CH i is bonded with CH j (if $i \in Con_{set}$) or with sensor j (if $i \in UnCon_{set}$). Now, depending on the connectivity of CH i with other CHs an Option(i) set created where the Option(i) set contains the set of possible nodes which can be used to route data packets of CH i. The Option(i) set of CH i is formed considering the following cases:

Case 1 (CH i has downstream CH): This is the simplest case as CH i has downstream CHs, so it will select one of them as its relay node. Here, Option(i) set of i will be all elements of DS_CH(i).
Case 2 (CH without any downstream CH): In this case the CH i has no next-hop CH toward RS but may have some upstream CH within its communication range. Now, CH i will consider the upstream CHs having downstream CH as a routing agent and add such CH to its Option(i) set.
Case 3 (CH i has neither a downstream nor upstream CH): There may be a scenario where a CH i has no downstream or upstream CH to route its data packets to RS. Such scenario may occur due to the failure of CHs within its range. In this situation, cluster members (CMs) of i which has other connected CH within its communication range can be used to forward CH i's data packets, so such CMs are added to Option(i) set.

Thus, in all above cases, CHs can be linked with RS by selecting the appropriate member from the option set of each CH.

Now, we present the pseudocode to generate a chromosome considering the above-mentioned cases in Algorithm 1.

Algorithm 1

---

1. For each CH i
    1.1. Compute x[i]=(rand%100)/100
    1.2. If($Dist_{RS}(i)<R_{com}$)
        $Option(i)=\{RS\}$
        End If
    1.2. Else If $(DS\_CH(i) \neq 0)$ then
        $Option(i) = DS\_CH(i)$
        End If
    1.3. Else if $(DS\_CH(i) == 0 \;\&\&\; US\_CH(i) \neq 0)$ then
        For each $j \in US\_CH(i)$
            If $(DS\_CH(j) \neq 0)$ then
            $Option(i) = Option(i) \cup j$
            End If
        End For

1.4. Else if $(DS\_CH(i) == 0 \; \&\& \; US\_CH(i) == 0)$ then
For each $j \in CM(i)$
If ($Cand_{CH}(j) > 1$)
For each CH $k \in Cand_{CH}(j)$
If (DS_CH(k)≠0 || US_CH(k)≠0)
Flag[j]=1
End if
End For
End if
If (Flag[j] ==1)
$Option(i) = Option(i) \cup j$
End if
End For

2. For each CH i
$\eta = Ceil(x[i] \times |Option(i)|)$
$k = Index(Option(i), \eta)$
Set k at the ith position of the chromosome
End For

Now, we illustrate Algorithm 1 with an example.

Example 1 Figure 4 shows a WSN having 15 CHs and 5 ordinary sensor nodes, where rectangles represent CHs and circles represent ordinary sensor nodes. The edges between the vertices mean they are within each other's communication range. Here, all CHs except C_7 have connectivity with other CHs, so in order to explain the routing path of C_7 we have considered it cluster members. Now, we illustrate chromosome representation considering the previously mentioned cases. Consider CH C_1, it has C_9, C_{12}, and C_{14} as downstream CHs, i.e., DS_CH(C_1) = {C_9, C_{12}, C_{14}}, so Option(C_1) = DS_CH(C_1), i.e., {C_9, C_{12}, C_{14}}. As shown in Table 1, the randomly generated number $x[C_1]$ is 0.38 and Ceil($x[C_1] \times$ |Option(C_1)|) = Ceil(0.38 × 3) = 2 means C_1 will forward its data packet through second member of Option(C_1) set, i.e., C_{12}. Again, consider CH C_{13} which has no downstream CH toward RS within its communication range, but has C_2, C_9, and C_{11} as upstream CH in the opposite direction. Here, C_2 and C_9 have downstream CH but C_{11} does not have any downstream CH other than C_{13}. So, only C_2 and C_9 are considered as a possible routing node and is added to Option(C_{13}) = {C_2, C_9}. From Table 1, it can be observed that $x[C_{13}] = 0.95$ and Ceil($x[C_{13}] \times$ |Option(C_{13})|) = Ceil(0.95×2) = 2. So, C_{13} is linked to second member of Option[C_{13}] set, i.e., C_9. Again for CH C_7, the random number $x[C_7] = 0.44$, but C_7 does not have any downstream or upstream CH within its communication range, i.e., C_7 is an unconnected CH. So, C_7 will consider its CMs having other connected CHs within their communication range to forward data packets. From Fig. 4, we can observe that sensor S_1 has C_3, sensor S_2 has C_9 and C_3, and sensor S_3 has C_{11} within their communication range. But since S_4 and S_5 do not have any other CH except C_7 within their range, so S_4 and S_5 will not be considered as a candidate routing agent. But, S_1, S_2, and S_3

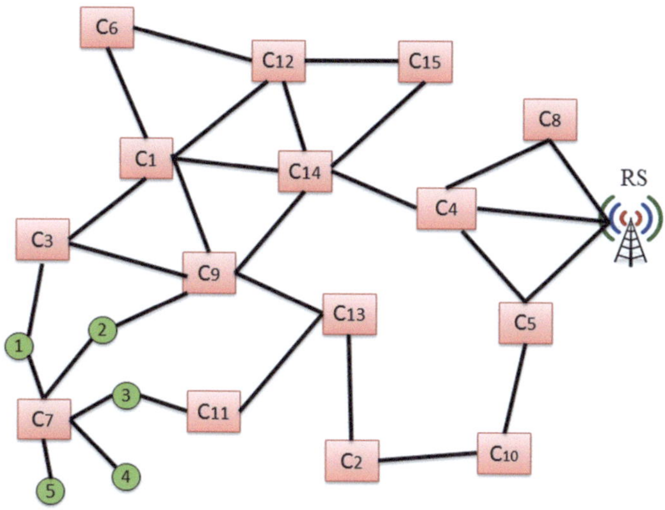

Fig. 4 Connectivity model

have other connected CH within their range so they are considered as a candidate routing agent and added to Option(C_7) set, i.e., Option(C_7) = $\{S_1, S_2, S_3\}$. Now, Ceil($x[C_7] \times |\text{Option}(C_7)|$) = Ceil($0.44 \times 3$) = 1. So, C_7 will use second element of Option(C_7), i.e., S_2 as routing agent, where routing agent S_2 will forward the data packet to either C_3 or C_9 whichever is closer to it. In this way, all CHs are linked to another CH/sensor node to construct a routing tree rooted at the RS as shown in Table 1. Figure 5 displays the final chromosome from Table 1.

Fitness evaluation:

A chromosome's fitness value implies its quality based on the goals. Thus, a fitness function evaluates each population chromosome. The objective of the proposed algorithm is to form a routing tree to connect each CH to RS such that the routing loads among CHs are evenly balanced. The load on a CH is mostly added by the communication load, which is the energy consumed while receiving and forwarding data of its CMs and other CHs. So, the load on a CH i contributed by its CMs is,

$$L_C(i) = (E_{\text{tx}} + E_{\text{r}}) \times r_c^i$$

and load contributed due to forwarding other CHs data is

$$L_f(i) = \sum\left((E_{\text{tx}} + E_{\text{r}}) \times r_c^j\right)$$

where E_{tx} and E_{r} are the energy spent in sending and receiving one-bit data which can be calculated using Eqs. 2 and 3. Again, r_c^i is the cluster data size of CH i and r_c^j is the data size of a forwarding CH j which selected i as a next-hop relay node.

Table 1 Next-hop CH mapping

CH(C_i)	$X[i]$	Option(C_i)	|Option(C_i)|	Ceil($X[i] \times$ |Option(C_i)|)	Bonded node
C_1	0.38	$\{C_9, C_{12}, C_{14}\}$	3	2	C_{12}
C_2	0.67	$\{C_{10}, C_{13}\}$	2	2	C_{13}
C_3	0.17	$\{C_1, C_9\}$	2	1	C_1
C_4	0.55	$\{RS\}$	1	1	RS
C_5	0.84	$\{RS\}$	1	1	RS
C_6	0.48	$\{C_1, C_2\}$	2	1	C_1
C_7	0.44	$\{S_1, S_2, S_3\}$	3	2	S_2
C_8	0.91	$\{RS\}$	1	1	RS
C_9	0.74	$\{C_{13}, C_{14}\}$	2	2	C_{14}
C_{10}	0.42	$\{C_5\}$	1	1	C_5
C_{11}	0.29	$\{C_{13}\}$	1	1	C_{13}
C_{12}	0.62	$\{C_{14}, C_{15}\}$	2	2	C_{15}
C_{13}	0.95	$\{C_2, C_9\}$	2	2	C_9
C_{14}	0.63	$\{C_4, C_{15}\}$	2	2	C_{15}
C_{15}	0.79	$\{C_{12}, C_{14}\}$	2	2	C_{14}

CH i ↓	C_1	C_2	C_3	C_4	C_5	C_6	C_7	C_8	C_9	C_{10}	C_{11}	C_{12}	C_{13}	C_{14}	C_{15}
Next hop CH of i	C_{12}	C_{13}	C_1	RS	RS	C_1	S_2	RS	C_{14}	C_5	C_{13}	C_{15}	C_9	C_{15}	C_{14}

Fig. 5 Sample chromosome

So, the total load on CH i is

$$L_{\text{tot}}(i) = L_C(i) + L_f(i) \tag{4}$$

Now using Eq. 4, CHs load is computed and compared to validate whether the loads are stable or not by computing the standard deviation of the loads. It is calculated as follows:

$$\text{std} = \sqrt{\frac{1}{N_{\text{CH}}} \sum_{i=1}^{N_{\text{CH}}} \left(\text{Avg}_L - L_{\text{tot}}(i)\right)^2} \tag{5}$$

where $\text{Avg}_L = \frac{1}{N_{\text{CH}}} \sum_{i=1}^{N_{\text{CH}}} L_{\text{tot}}(i)$ and N_{CH} are the total number of CHs present. So, Eq. 5 is the function of fitness, where a better chromosome is indicated by lower std value. It is to be noted that lower std value indicates smaller deviation of loads among the CHs, thus, the fitness function assures uniform load distribution among the CHs.

Population updating:

Here, some chromosomes from the existing population are selected using Roulette-wheel selection method as parents to generate new chromosomes. Note that chromosomes with higher fitness will have higher chances for selection in Roulette-wheel method. The selected chromosomes undergo crossover and mutation process to produce new child chromosomes as shown below.

Crossover: During crossover, a point is randomly selected and the information of two selected parent chromosomes is swapped after that point. Figure 6 shows the crossover operation.

Note that the offspring created by the crossover process is also a valid routing solution. Because during the crossover, we swap the parent chromosomes, such that the value of each gene is interchanged only. As the parents chromosomes are valid so the produced offspring must also be valid.

Mutation: If a gene in a chromosome is selected for mutation at that time we simply substitute another valid node id to that gene as shown in Fig. 7. The valid node ids are present in the Option set of the CH conforming to that gene. The probability

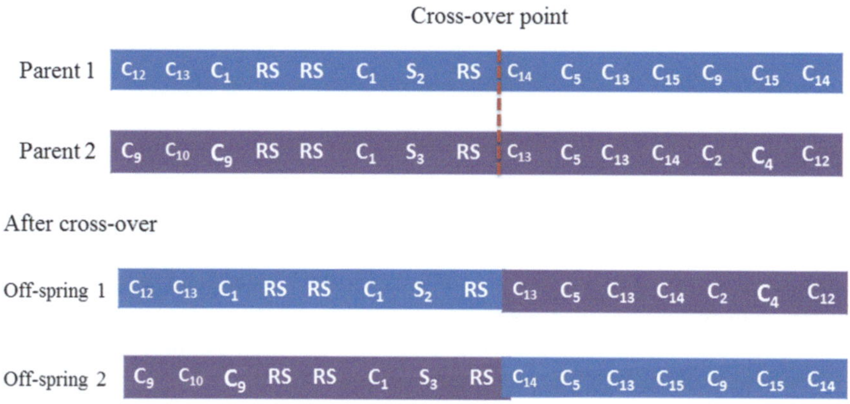

Fig. 6 Chromosome representation before crossover and after crossover

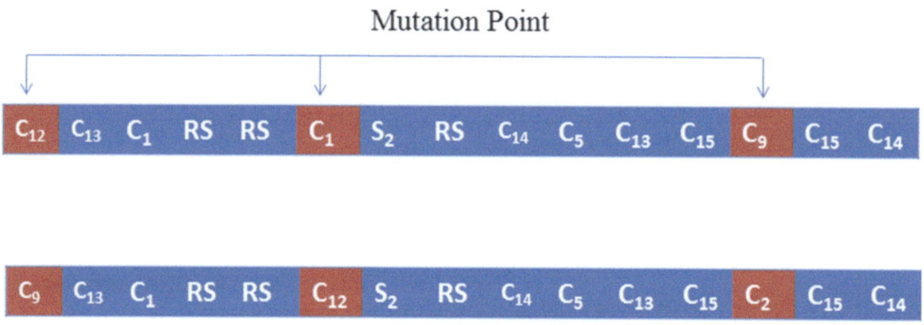

Fig. 7 Chromosome representation before mutation and after mutation

of mutation of a gene is calculated using a random number function, where if the generated random number is less than a predetermined threshold value then that gene is selected for mutation.

4.3 Steady Phase

During the steady phase, each CH assigns a time slot using TDMA to its cluster members such that they transmit their sensed data to their corresponding CH in the assigned time slot only to avoid collision. The CH then aggregates the received packets and forward it to the RS following the routing tree constructed in the network setup phase. This completes one round of network operation. The steady phase runs for a fixed number of rounds then the information sharing and the network setup phase are executed again to form new routing tree.

5 Simulation Results

5.1 Simulation Setup

The simulation is carried out using MATLAB 2012b and C programing language on Intel i7 processor, 3.40 GHz CPU, 8 GB RAM running on Windows 7 operating system. We assume a two-dimensional plane of (100×100 m^2) where 200 ordinary sensor nodes and 50 advanced sensor nodes are deployed randomly. The initial energy of ordinary and advanced sensor nodes is 0.05 J and 0.1 J, respectively. A node is considered dead once its remaining energy reaches 0 J or it can not able to communicate with other nodes due to any hardware failure. In our simulation, the probability of failure of a node follows Weibull distribution. The energy model parameters discussed in Sect. 3 are $\varepsilon_{\text{fs}} = 10\,\text{pJ/m}^2/\text{bit}$ and $\varepsilon_{\text{mp}} = 0.0013\,\text{pJ/bit/m}^4$.

To evaluate the performance of the proposed algorithm, the simulation results were compared with MHRM and EHE-LEACH.

5.2 Evaluation of Experimental Results

Here, we compare the performance of our proposed algorithm with other existing algorithms in terms of energy efficiency, number of alive nodes, and packet delivery ratio.

In Fig. 8, we have shown the average remaining energy of CHs per round in FAR, MHRM, and EHE-LEACH. From Fig. 8, we can observe that in FAR CHs have conserved more energy compared to MHRM and EHE-LEACH. This is due

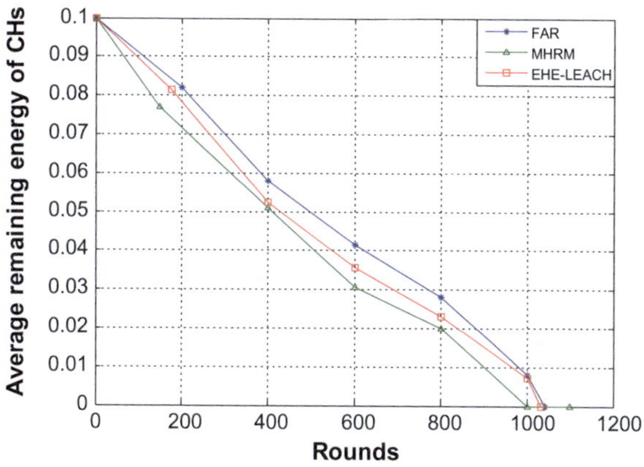

Fig. 8 Average remaining energy of CHs per round

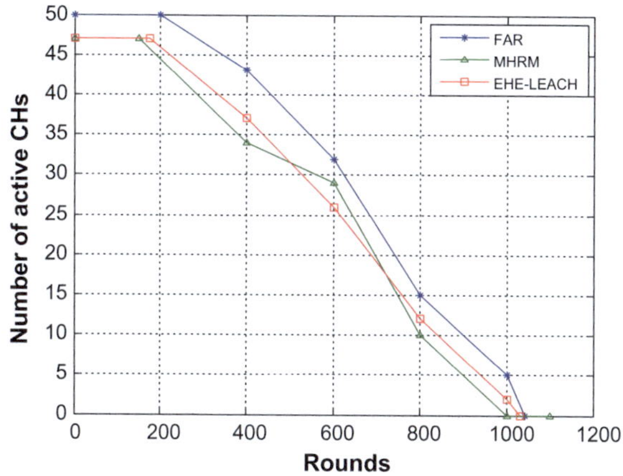

Fig. 9 Number of active CHs per round

to the fact that in FAR fitness function is designed in such a way that the load is balanced among the CHs. In MHRM, the next-hop relay node selection objective of CHs is to minimize the hop count to RS. So, CHs may select a next-hop CH which is farthest from it to reduce the hop count. Since, energy consumption during data communication is directly proportional to the communication distance so, MHRM consumes much higher energy while reducing the hop count. Again, in EHE-LEACH a greedy approach is used by CHs to route their data to RS. But, in large scale WSN it is very difficult to find optimal routing path using greedy method, whereas during routing path construction, FAR used GA which is widely used for solving problems in large solution space.

In Fig. 9, we have shown the number of active CHs per round. A CH is considered active if it has connectivity to RS and has sufficient energy to route its cluster data.

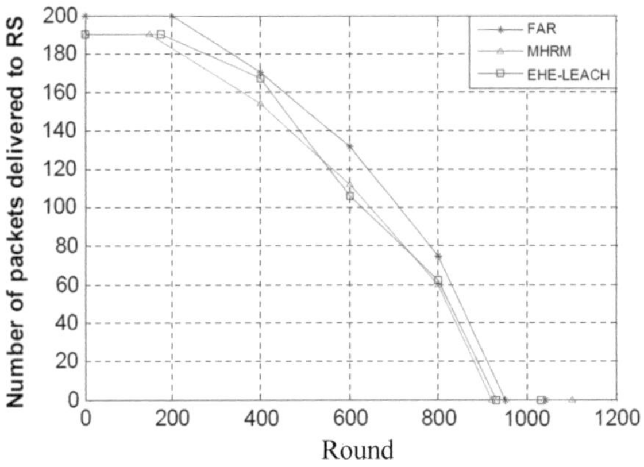

Fig. 10 Packets delivered to RS per round

From Fig. 9, we can notice that in MHRM and EHE-LEACH all CHs are not active from the beginning because these algorithms do not consider any routing technique for unconnected CHs. So, such CHs will have no connectivity to RS as a result they are considered inactive. Whereas in FAR, a cleaver routing strategy is adopted so that all CHs have connectivity to RS. Again, FAR balanced the loads of CHs as a result, it has more active CHs per round compared to MHRM and EHE-LEACH.

Although energy efficiency is the primary objective of most routing algorithms designed for WSNs. But, the successful packet delivery is an equally important measure to evaluate any routing algorithm. During our simulation run, we assume that each ordinary sensor node sends one packet per round to its CH and then the CH forwards it to RS. Figure 10 shows the number of packets delivered to RS in FAR, MHRM, and EHE-LEACH. Here, FAR has delivered more packets compared to MHRM and EHE-LEACH because FAR has established a routing path for each CH to RS which ensure all cluster data packets will reach the RS. Furthermore, if any CH failure occurs then FAR provides an intelligent routing strategy to recover from the path failure.

6 Conclusion

In this paper, we have employed a fault-aware routing algorithm (FAR) to route each CH data to RS by adopting a genetic algorithm approach such that load among various CHs are balanced. For CHs that does not have any next-hop CH within its reach to forward their data, our algorithm considers cluster members of such CHs to establish connectivity with other CHs. FAR includes a novel chromosome generation algorithm which ensures each CH has a routing path to RS. We have also derived a fitness function to evaluate the load balancing among CHs. Furthermore, the crossover and mutation operation are also properly described. To evaluate the performance of FAR, we have compared the experimental results with two related algorithms MHRM and EHE-LEACH. The experimental results show that FAR has outperformed MHRM and EHE-LEACH in terms of energy efficiency, number of active CHs, and data packet delivery. In the future, we would like to extend the algorithm for mobile scenarios of WSN.

References

1. Dey N, Hassanien AE, Bhatt C, Ashour AS, Satapathy SC (eds) (2018) Internet of things and big data analytics toward next-generation intelligence. Springer, Berlin
2. Elhabyan RS, Yagoub MC (2015) Two-tier particle swarm optimization protocol for clustering and routing in wireless sensor network. J Netw Comput Appl 52:116–128
3. Mazumdar N, Om H (2017) A distributed fault-tolerant multi-objective clustering algorithm for wireless sensor networks. In: Proceedings of the international conference on nano-electronics, circuits & communication systems. Springer, Singapore
4. Mazumdar N, Om H (2016) An energy efficient GA-based algorithm for clustering in wireless sensor networks. In: 2016 international conference on emerging trends in engineering, technology and science (ICETETS). IEEE
5. Mazumdar N, Om H (2017) A distributed fault-tolerant multi-objective clustering algorithm for wireless sensor networks. In: Proceedings of the international conference on nano-electronics, circuits & communication systems, Springer, Singapore, pp 125–137
6. Das SK, Tripathi S (2018) Adaptive and intelligent energy efficient routing for transparent heterogeneous ad-hoc network by fusion of game theory and linear programming. Appl Intell 48(7):1825–1845
7. Das SK, Tripathi S (2017) Energy efficient routing formation technique for hybrid ad hoc network using fusion of artificial intelligence techniques. Int J Commun Syst 30(16):e334
8. Mukherjee A, Dey N, Kausar N, Ashour AS, Taiar R, Hassanien AE (2019) A disaster management specific mobility model for flying ad-hoc network. In: Emergency and disaster management: concepts, methodologies, tools, and applications. IGI Global, pp 279–311
9. Das SK, Tripathi S (2018) Intelligent energy-aware efficient routing for MANET. Wirel Netw 24(4):1139–1159
10. Akyildiz IF, Weilian Su, Sankarasubramaniam Y, Cayirci E (2002) A survey on sensor networks. IEEE Commun Mag 40:102–114. https://doi.org/10.1109/mcom.2002.1024422
11. Anastasi G, Conti M, Di Francesco M, Passarella A (2009) Energy conservation in wireless sensor networks: a survey. Ad Hoc Netw 7(3):537–568
12. Abbasi AA, Younis M (2007) A survey on clustering algorithms for wireless sensor networks. Comput Commun 30:2826–2841. https://doi.org/10.1016/j.comcom.2007.05.024
13. Heinzelman WB, Chandrakasan A, Balakrishnan H (2000) Energy-efficient communication protocols for wireless microsensor networks. In: Proceedings of Hawaii international conference on system sciences. https://doi.org/10.1109/hicss.2000.926982
14. Younis O, Fahmy S (2004) HEED: a hybrid, energy-efficient, distributed clustering approach for ad hoc sensor networks. IEEE Trans Mob Comput 3:366–379. https://doi.org/10.1109/TMC.2004.41
15. Bandhopadhyay S, Coyle E (2003) An energy efficient hierarchical clustering algorithm for wireless sensor networks. In: Proceedings of IEEE INFOCOM, vol 3, pp 1713–1723. https://doi.org/10.1109/infcom.2003.1209194
16. Manjeshwar A, Agarwal D (2001) TEEN: a protocol for enhanced efficiency in wireless sensor networks. In: Proceedings of 15th parallel and distributed processing symposium San Francisco. IEEE Computer Society, pp 2009–2015

17. Lindsey S, Raghavenda CS (2002) PEGASIS: power efficient gathering in sensor information systems. In: Proceedings of the IEEE aerospace conference, Big Sky, Montana. https://doi.org/10.1109/aero.2002.1035242
18. Tyagi S, Gupta SK, Tanwar S, Kumar N (2013) EHE-LEACH: enhanced heterogeneous LEACH protocol for lifetime enhancement of wireless SNs. In: 2013 international conference on advances in computing, communications and informatics (ICACCI). IEEE, pp 1485–1490
19. Kumar D, Aseri TC, Patel RB (2009) EEHC: energy efficient heterogeneous clustered scheme for wireless sensor networks. Comput Commun 32(4):662–667
20. Kuila P, Gupta SK, Jana PK (2013) A novel evolutionary approach for load balanced clustering problem for wireless sensor networks. Swarm Evol Comput 12:48–56
21. Azharuddin M, Jana PK (2015) A distributed algorithm for energy efficient and fault tolerant routing in wireless sensor networks. Wirel Netw 21(1):251–267
22. Dey N (ed) (2017) Advancements in applied metaheuristic computing. IGI Global
23. Powell O, Leone P, Rolim J (2007) Energy optimal data propagation in wireless sensor networks. J Parallel Distrib Comput 67(3):302–317. https://doi.org/10.1016/j.jpdc.2006.10.007
24. Chiang S, Huang C, Chang K (2007) A minimum hop routing protocol for home security systems using wireless sensor networks. IEEE Trans Consum Electron 53:1483–1489. https://doi.org/10.1109/TCE.2007.4429241
25. Tarachand A, Jana PK (2015) Energy-aware routing algorithm for wireless sensor networks. Comput Electr Eng 41:357–367. https://doi.org/10.1016/j.compeleceng.2014.07.010
26. Heinzelman WB, Chandrakasan AP, Balakrishnan H (2002) An application-specific protocol architecture for wireless microsensor networks. IEEE Trans Wirel Commun 1(4):660–670
27. Mazumdar N, Om H (2015) Coverageaware unequal clustering algorithm for wireless sensor networks. Procedia Comput Sci 57:660–669
28. Rahmanian A, Omranpour H, Akbari M, Raahemifar K (2011) A novel genetic algorithm in LEACH-C routing protocol for sensor networks. In: 2011 24th Canadian conference on electrical and computer engineering (CCECE). IEEE, pp 001096–001100
29. Safa H, Moussa M, Artail H (2014) An energy efficient Genetic Algorithm based approach for sensor-to-sink binding in multi-sink wireless sensor networks. Wirel Netw 20(2):177–196
30. Bhatia T et al (2016) A genetic algorithm based distance-aware routing protocol for wireless sensor networks. Comput Electr Eng 56:441–455
31. Azharuddin Md, Jana PK (2017) PSO-based approach for energy-efficient and energy-balanced routing and clustering in wireless sensor networks. Soft Comput 21(22):6825–6839
32. Gupta G, Younis M (2003) Fault-tolerant clustering of wireless sensor networks. In: 2003 IEEE wireless communications and networking, WCNC 2003, vol 3. IEEE, pp 1579–1584
33. Haseeb K et al (2016) A dynamic energy-aware fault tolerant routing protocol for wireless sensor networks. Comput Electr Eng 56:557–575
34. Boukerche A, Martirosyan A, Pazzi R (2008) An inter-cluster communication based energy aware and fault tolerant protocol for wireless sensor networks. Mob Netw Appl 13(6):614–626
35. Azharuddin M, Jana PK (2015) A PSO based fault tolerant routing algorithm for wireless sensor networks. In: Information systems design and intelligent applications. Springer, New Delhi, pp 329–336
36. Lee JJ, Krishnamachari B, Kuo CCJ (2008) Aging analysis in large-scale wireless sensor networks. Ad Hoc Netw 6(7):1117–1133
37. Rausand M, Hoyland A (2004) System reliability theory: models, statistical methods, and applications, vol 396. Wiley

Chapter 3
GA-Based Fault Diagnosis Technique for Enhancing Network Lifetime of Wireless Sensor Network

Ruchika Padhi and Bhabani Sankar Gouda

1 Introduction

Wireless sensor network contained group of sensor node which monitor the network and record the condition. It collects the information as data and sent that to a central location known as sink node. It measures the temperature, sound, pollution level, humidity and so on as information. There is a different type of sensor network present, i.e. terrestrial network, underground network and underwater network so on. The wireless sensor is mostly used in continuous communication area like ecological habitat monitoring, health monitoring, environmental pollution detection, industry process control and military target tracking, etc.

1.1 Issues

Wireless sensors have different issues as design issues, topology issues and others which affect the performance of sensor node.

(a) **Design issues**: The issues are like fault tolerant, low latency, scalability, transmission media and coverage problem. Sensor nodes are deployed at uncontrolled area so the node can faulty and unreliable. For this reason, nodes cannot be communicated with sink node directly within time. So sensors vary in numbers; if

R. Padhi (✉) · B. S. Gouda
National Institute of Science and Technology (Autonomous), Institute Park, Pallur Hills, Berhampur 761008, Odisha, India
e-mail: ruchika.padhi@gmail.com

B. S. Gouda
e-mail: bhabani012@gmail.com

D. De et al. (eds.), *Nature Inspired Computing for Wireless Sensor Networks*,
Springer Tracts in Nature-Inspired Computing,
https://doi.org/10.1007/978-981-15-2125-6_3

some addition required, it is an issue known as scalable issue. In the communication path of sensor node, many interruptions may arise and that need to verify the transmission media and every node is not being in a same network to communicate.

(b) **Topology issues**: Some topological issues like geographic routing, sensor holes and coverage topology. In geographical routing, the sender node sends the data to the geographical location of network instead of proper network address. In routing, a number of holes are present and that are cannot be participated in routing process. So there is a requirement to find the holes which have light weight, low capability and not aware of geographical position. Every network covered at least k-sensors which has parameter as coverage area. Some other issues are like wireless radio communication, hardware and operating system of each node, network layer like different layer, localization, etc.

1.2 Challenges

WSN nodes have large potential. It has the ability to connect with other nodes present in the physical world. It collects information where it impractically deploys. It improves the potential by finding the limitation and technical issues of network. The nodes are synchronized for data fusion. Wireless sensor nodes are number of challenges which need to overcome.

1.2.1 Energy

Most design challenge for nodes is energy efficiency. Power consumption is more in case of nodes, which are under three well-designed domains, i.e. sensing, communicating and data processing, all require optimization. Nodes use battery power, so lifetime is limit. That battery power can be recharged or replaced. For non-rechargeable batteries, the nodes can be used up to transmission then it replaced.

1.2.2 Bandwidth Limited

For transmission of data less amount of energy consumed. But very limited data transmit in one time, i.e. 10–100 kB/s. Data transmission between two sensor nodes required message exchange and synchronization is also impossible without message exchange. Sensor nodes operated in bandwidth and performance of node controlled in multi-hop communication medium. That communication links operate or active in the radio, infrared and visual range.

1.2.3 Node Costs

Wireless network contains a large amount of sensor nodes. Nodes are critical to overall financial metric, so the cost of every one node has reserved low for global metric. If the cost of nodes appropriate for sensor node then it used more as required.

1.2.4 Deployment

Nodes need to deploy in proper place otherwise it create more complexity. Some special techniques implement to reduce the complexity like in limited energy data can transmit, accurate data transmit, etc. There are two different models for deployment, i.e. (i) static deployment (ii) dynamic deployment. In static deployment need to choose the best location according to optimization strategy and the location cannot be changed throughout the lifetime. In dynamic deployment, nodes throw randomly for optimization and can change the place as requirement.

1.2.5 Design Constraints

The main goal is to design a node smaller in size so it can carry anywhere, cheaper for decrease the cost and more efficient electronic devices. It has challenges for both software and hardware design with restricted constraints.

1.2.6 Security

High security required for resources to collect sensitive information. The remote and unattended operations of sensor nodes increase their exposures to malicious intrusion and attacks. Node authentication and data confidentiality are the main security of nodes. Nodes are passing through the authentication exam at manager node and cluster node for identifying dependable and unpredictable nodes. Sensor nodes have some key establishment, distribution and node authentication for security [5].

2 Fault, Errors and Failures

In sensor nodes, the fault may occur due to problem in position of hardware or program since a continuous crash in components of node. Fault arises due to many cases like energy depletion, loss of connectivity and delay. Errors present in nodes evaluated and collected in an error log by accepting and detecting errors. Errors in node found as fault that can define by diagnosis tests, report the location, report the cause and manipulating the database information. When fault found notification send to the cluster head or manager node. There are different types of nodes like network fault, hardware fault and software fault.

2.1 Types of Fault

(a) **Network fault**: This type of fault found in connection failure, channel blockage, asynchronous clocks, illegal interference and address partiality. Unreadable data found in the fault area.
(b) **Hardware fault**: Sensor node damage found in this case. Damage found due to memory, power supply, processor and wireless communication, etc.
(c) **Software fault**: It consists of drift, correctness decline, fixed unfairness and complete bias. The collected data processed to manager node or canter node which contain abnormal value created by sensor node known as software fault. The abnormal values found because of degradation in the sensor node. Faults in sensor node also classified according to data sent by sensor. Those are given as:

    i. **Offset fault**: A constant value added to expected data which found fault node.
    ii. **Gain fault**: The expected collected data change in a period of time and evaluate the fault node.
    iii. **Stuck-at fault**: When sensed data series varied as zero, this type of fault found.
    iv. **Out of bound**: When sensed data values found out of bound of normal running values, this type of fault found.

2.2 Fault Diagnosis

Fault diagnosis is the method to improve the network, lifetime of nodes, increase the performance, service type and security of the data. It reduced the network problem, control the cost and extend service life of nodes. It acted as a type of pattern recognition for the diversity and complexity of network equipment and fault. In this case, calculate the fault symptoms, diagnose the state, and define the network status and running status of equipments to evaluate the complete fault diagnosis.

Basically, the fault tolerance method is used to analyze the variation amongst several terms like failure, errors and fault, where fault is a main cause of error. Failure indicates a notification about fault.

Let us assume there is a situation that sensor node A sends the measurement of its sensors or the nearer sensors to another running sensor node B. Condition is node A crash due to lose connection with neared sensor node and it could not send the data to the aggregation provision within the specific time interval. This is called breakdown of node A observed by neighbour node B. In the condition, two states found one loose connection of the sensor node, second does not send sensor information within specific time interval. To provide suppleness in faulty situations two main actions should be performed.

(d) **Fault detection**: This is the first step in a scheme that must be carry-out to sense a specific functionality of node is faulty or not.
(e) **Fault recovery**: This is the next step to make sure or recover from it. In this case, replicate the mechanism of the system which is important for its correct process.

3 Related Work

Wireless sensor networks [6] should be energy constraints, flexible for different network topology scalable and its changes. All the sensor nodes in the network works in a data-centric manner and all nodes connect to a central node. In the sensor network nodes have an address that is specific and address, which can define dynamically and statically [7]. This is the process in which, if we found a dead node, it is discarded and useless. So we need an efficient protocol. In the wireless sensor network, we have used different varieties of routing [8] processes and implemented from that by using direct diffusion (DD) algorithm to analysis a specific routing in which sensor fault node can be avoided and discarded and form a new path as their requirement. In this survey, in WSN is presented the different fault management [9]. This paper specially addresses a different challenge of the existing network fault management process that concluded the existing network management structure and architectures. In this paper presented [10], different types of fault-tolerant topology which control of mobile sensor network as well as in static [11]. This paper represents the distributed process to all the network nodes by assigning the minimum possible power. In this paper, the work [12] represents the asymmetric links in distributed network control for sensor networks. It considers the network topology control problems in a heterogeneous network. In this method, different wireless sensor devices are with a different maximum transmission range within a time period of alive nodes. In this paper [13] presents a decision model by using Bayesian for intelligent routing. To implement the learning patterns approach which is based on energy-aware routing? In this model to find out the energy of the entire neighbour nodes by using estimation based on probabilistic decision model. The work presently in [14] addresses the faulty control in a heterogeneous wan. It represents the k-degree network topology control problem for different routing. The work presented in [15] addresses the new Bayesian fusion approach to combine the trust component of more than one to infer the whole trust nodes in between. The results shown the node is highly trustworthy, and both the trust components simultaneously confirm its trust and entrust by these components. This work is [16] made by algorithms based on dynamic routing. In the papers, it describes how management is flexible for dynamic changes in broadband networks. Here, it is shown how the path is made inside both the nodes and how the new path is started. Doing this, get a good solution or improved the speed of the genetic algorithm (GA). In [17] shown, how the routing process works in wireless sensor network. How to further encouraged the network's efficiency and lifetime from the new schemes. In [18], these algorithms are made and processed for a whole lot of ways. Already a lot of concepts have been designed, which require a new application for network speed; here, the concept is very good because of the genetic-based optimal scheme. A new approach (FNR) has been given on how sensor networks work [19]. Here, a new approach is designed which shows that its performance is the best because it is based on grade diffusion algorithms. By doing this, the data loss is also a little bit, the life of the node is also good and all the nodes of the network are active too. Here, it is shown [20] that whatever fault is being done with regard to cluster head, it is removed. Here, the energy from both of them was shown with regard to distance. How did the cluster head get good benefit from doing it based on the cluster head recovery algorithms [21–26]. In [27], these algorithms are made and processed for optimal solutions and converge to a global network one. Here, the concept is based on the optimization of optimal path which is represented on genetic algorithm and support vector model. In this paper presented [28], the different types of machine learning techniques which are described the models. By using techniques to consider the minimization of data loss, reduced the error and increase the lifetime of network. In this paper presented [29], the different types of IoT controlling system techniques which are described and analyzed the data stream methods. By using techniques to consider the minimization of data loss, reduced the error and increase the lifetime of network. WSN is one type of wireless network which increases rapidly due to its flexible frameworks [30]. Wireless network has different variations in terms of applications, which is totally based on the requirements of the users. Some of the authors proposed several effective works in wireless ad hoc network as well as WSN to enhance the routing with network lifetime of the network [31–35].

4 Proposed Method

This section highlights the main proposed work, which is a blend of two existing techniques such as direct diffusion algorithm and genetic algorithm. Direct diffusion algorithm [6] uses four types of messages, i.e. interest, exploratory data, reinforcement and data messages. Initially, the base node forward an interest message for source node that means sink node sends signal to the source node to collect information from network then send it back to sink. After this, there is a four-way message transmission begins. Then exploratory data message send to the senders because the sender does not know their destinations. Interest, data aggregation and data propagation are determined by localized interactions. It has four main features, i.e. interest, data, gradient and reinforcement.

(a) **Interest**: It means query or interrogation which specifies what a user wants.
(b) **Data**: It indicates collected or processed information.
(c) **Gradient**: It defines direction state created in each node that receives interest. Gradient direction is towards the neighbouring node which the interest is received. Events start flowing from originators of interest along multiple gradient paths.
(d) **Reinforcement**: It is an optimization technique which is used for learning purpose.
(e) **Attribute**: It indicates value pairs that describe a task.

In Fig. 1, the interest is usually injected into the network from sink. For each active task, the sink broadcasts an interest message to each of its neighbour time to time. The first interest contains the specified neighbour and duration attributes, but larger interval attributes which are in the given range. The interest tries to find the active sensor node in specified range, which specifies what a user wants by naming the data. Sink periodically broadcasts the interest message to each neighbour. All nodes maintain the interest cache. Sensor node receives interest packet. Nodes are must be in given range and the task for the sensor system to generate samples at the highest rate of all the gradients. At data propagation, the sender node forwards data message to base node with the initially setup gradient direction. Base node forwards a reinforced message to the nearest neighbour node.

The neighbour node which receives the reinforced message can also forward this to the selected next neighbour node which can receive the new data first. A path

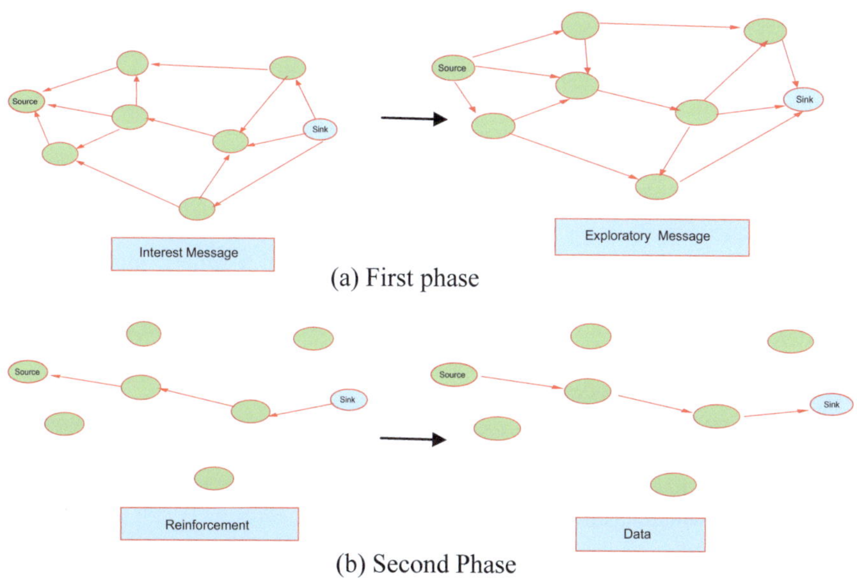

Fig. 1 Implementing DD algorithm

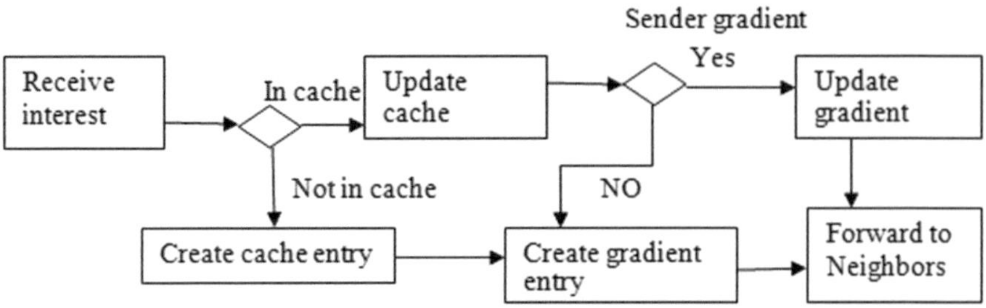

Fig. 2 Procedure of interest in cache of every active sensor node

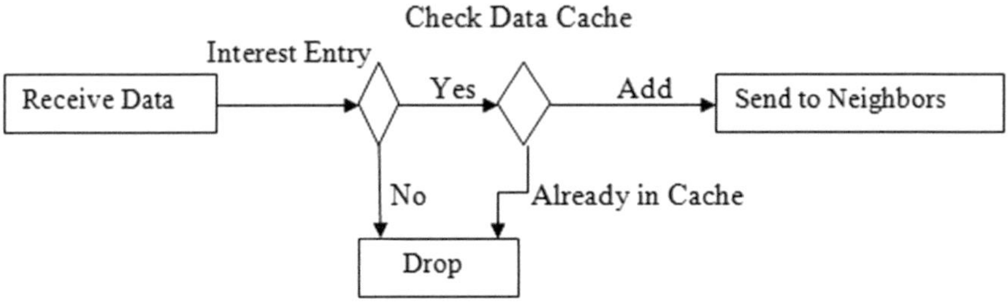

Fig. 3 Procedure to send interest messages to neighbours

determines when a maximum gradient is formed, so that the future arriving data message can be transmitted along the best unbreakable path (Figs. 2 and 3).

DD algorithm is planned for healthiness, scale and efficient energy. It reduces the data pass on transmission, which count for power administration. It is a query-driven communication protocol. The collected data are communicated if it matches the query from the base node which sends as interest message. All the sensor nodes are bound to a path, when broadcasting the interested queries and the path is also a new route. Routes are in a circular form, i.e. defined at the time of broadcasting the query message (Fig. 4).

There is fault node recovery concept implemented in a genetic algorithm for increase the lifetime of sensor network. If some nodes have no battery lifetime or they reach to their threshold value, then few replacements can be done and recycled the routing path using this fault node recovery algorithm. So it reduces the cost and increase the lifetime of nodes.

Genetic algorithm implemented as fault node recovery algorithm. This is based on set of solutions called population transformed into new population by reproducing [36]. It does not have efficient fault detection concept. This is based on the fitness value of nodes which has five steps as follows.

a. Check the expected solution and replaced, if it not functioning.
b. Check the replaced node information through expected solution and calculate a fitness function.
c. Selection is done based on high fitness value.

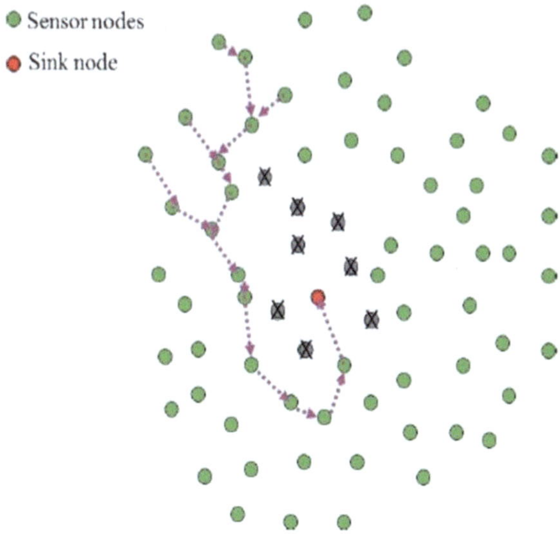

Fig. 4 Implementing DD algorithm for finding path where some nodes are dead

d. Crossover step used to implement single crossover strategy.
e. Mutation step used to flip node randomly to reuse the most direction-finding paths and maximize the WSN lifetime.

For these above five steps implemented using a different process like initialization, evaluation, selection, etc.

(i) **Initialization**: In genetic algorithm, chromosomes generate in the first step. Each chromosome is an in need solution. The number of chromosome (gens) depends on number of nodes in a network. Each chromosome initialized with a value known as fitness value, i.e. 0 or 1.

Ex:

Node no.	5	4	17	20	35
Fitness value	0	1	0	1	0

Here, the chromosome length is 5. The generated value is 0 or 1, which is randomly generated. Chromosome length is the number of sensor nodes that are not functioning (deactivated node). The node should replace if the value is 1.

(ii) **Evaluation**: The generated fitness value is evaluated as a fitness function and by given parameters, which are chromosome genes. The generated value cannot be put directly into the fitness function in the energy-efficient distributed dynamic diffusion-based algorithm. Mainly this fault node recovery technique is to reuse the most direction-finding paths and to replace the least sensor nodes, which are non-functioning.

The criteria for path finding is based on non-functioning sensor nodes and calculated fitness function shown in given equation:

$$f_n = \sum (P \times \mathrm{TN})/(N \times \mathrm{TP} \times i) \quad \text{at } 1 < i < n \tag{1}$$

In Eq. (1) some notation given as:

N No. of replaced sensor nodes and their grade value at i.
P No. of reusable routing paths from sensor nodes with their grade value at i.
TN Total no. of sensor nodes in the novel network.
TP Total no. of routing path in the novel network (Fig. 5).

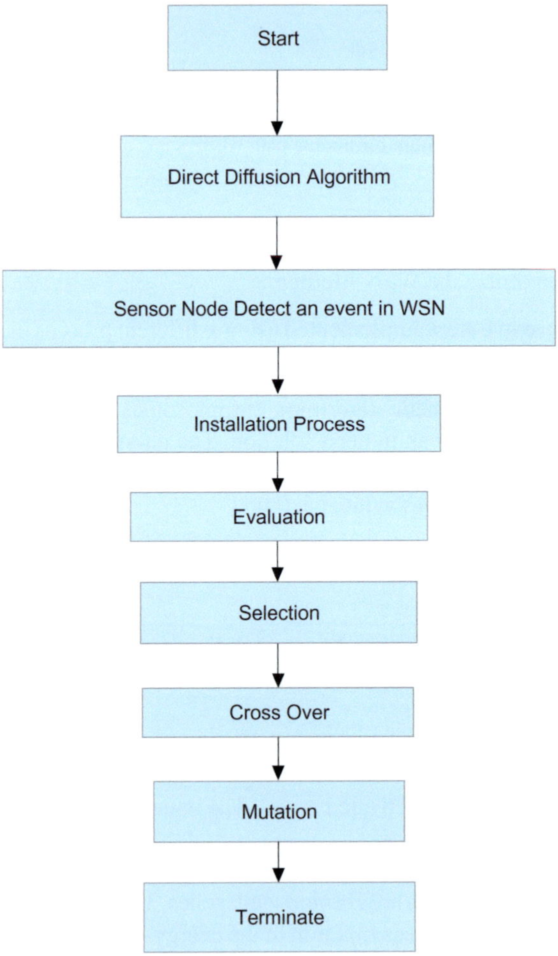

Fig. 5 Procedure of implementing genetic algorithm

(ii) **Selection**: This process implemented to select higher fitness value sensor node. It will eliminate the lowest fitness value genes. It selects the higher fitness value genes into a mating pool and the other worse genes are deleted. New genes required to replace the poorer genes after crossover step.
(iii) **Crossover**: This is implemented to change the individual genes. Here, the single crossover plan used to create new chromosomes and two different chromosomes are chosen for the mating pool to found two new offspring. Crossover end is selected between the initial and last genes of the parent individual genes. Division of each individual genes on both side of the crossover end is exchanged and determined.

5 Performance Evaluation

This proposed plan has been simulated in the variety of network scenarios. The simulation result is considered lengthily for random 500 iterations. In this section, simulation results represented as different model, which contain details of procedure, performance parameters and results. The projected simulation model consists of two parameters (i) N number of nodes, and (ii) deployed as arbitrarily in a distributed atmosphere. The simulation results are prepared for 500 nodes and calculate the probability of fault (P), fault alarm ratio (FAR) and data accuracy (DA) of fault tolerance and fitness values of GA for the proposed scheme. The simulation atmosphere comprises with the different models that are network, battery, channel and energy model. Some of the models are used in this performance evaluation which are follows as:

(a) **Network model**: Considered area as $L * B$ square metres for distributed sensor network (DSN) environment. Each try out corresponds to a random deployment of sensors in a fixed or variable network area and performs the fault alarm ratio (FAR) and data accuracy(DA) of given network using GA by using lowest fitness value of the chromosomes, it satisfy the fault status.
(b) **Channel model**: In sensors network, data packets generate with different sizes. Sensor network protocol as S-MAC protocol is used for media (channel) access. The packet transmission in a sensor network is said to occur in discrete nature. At the destination side or sink node receives all packets by bearing the sender info in an interval, when the sender node is in off state. Intended for cleanness, it has been well thought out that initially an error-free channel present.
(c) **Battery model**: In the sensor network, each of sensors has some specific battery with finite and non-replicable. The node was located with an initial energy in terms of joules. When a sensor transmits to the specified location or it receives a data packet from the sink, it consumes and produced some amount of energy level.
(d) **Energy model**: The radio transmission can perform the power control. So it uses the minimum required energy to reach the fault tolerance criteria. It is

assumed that at any given time, the energy required for transmitted (ET) and sensing (ES) for k' bit packet to another node at distance d' metres is EN Joules for node. So total energy calculated as Eq. (2) given as:

$$\text{Total Energy (TE)} = (\text{ES} \times Pk) + (\text{ET} \times Pk) * di \quad (2)$$

where Pk = Packets size in terms of bits.

In this paper, different simulation parameters are used, which are shown in Table 1. There are different performance metrics are illustrated that highlight the performance of the projected method in Figs. 6, 7, 8, 9 and 10.

Table 1 Simulation parameters

Parameters	Notations	Data
Simulation area	L	1000 m
Width	B	1000 m
Number of used nodes	NN	500 nodes
Message transmission range	CR	250 m
Type of selection	S_r	Selection of wheel (Roulette)
Type of crossover	C_r	Two cross points
Crossover possibility	C_P	0.07–1.10
Mutation possibility	M_P	0.20–0.40
Population density	PS	500
Number of repetitions	I_N	2000
Energy of each nodes	EN	2 Joules
Hop bandwidth	HW	1 mbps
Threshold energy	TH_{EN}	0.05 Joules
Threshold link	THoff	0.287 mbps
Energy of each sensor nodes	E_s	50 nJ/bit level
Used energy for communication	E_r	50 nJ/bit level

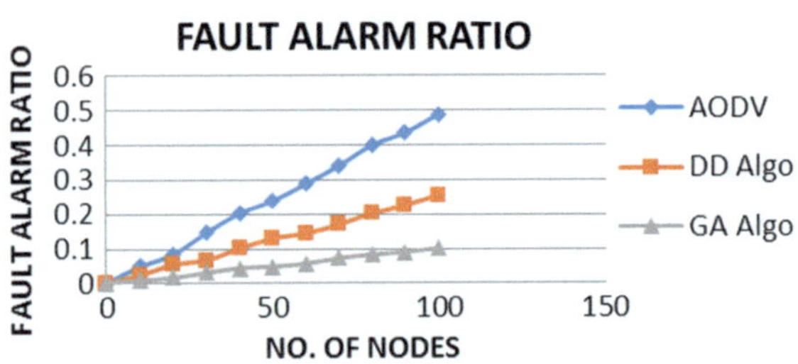

Fig. 6 Fault alarm ratio verses no. of nodes

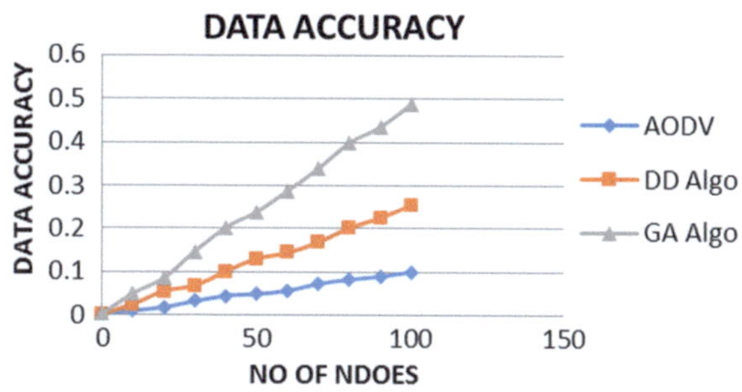

Fig. 7 Data accuracy verses no. of nodes

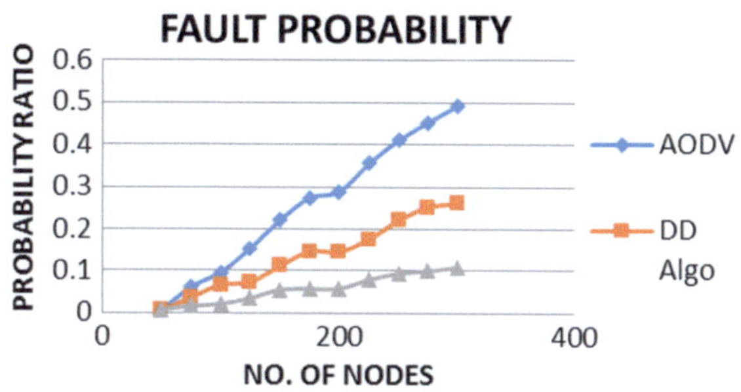

Fig. 8 Fault probability verses no. of nodes

(a) **Fault alarm ratio**: Fig. 6 shows the FAR is approximating same due to less no. of nodes and that all are scattered at distance position so FAR cannot be so high. So FAR is slightly more in it. In DD, the faulty nodes are avoided and routing method is implemented to change the path from source to destination. According to AODV, FAR value is less in DD. But in GA, the faulty nodes avoid by implementing chromosome concept in which the faulty node avoid and choose the nearer active node by implementing crossover technique.

(b) **Data accuracy**: Fig. 7 shows the DA value is more in GA compared to DD and AODV. DA is very less in AODV because faulty node cannot be detecting in it. In DD, DA is high in comparison with AODV because fault node is detected but not avoided.

(c) **Fault probability**: Fig. 8 indicates the fault probability is approximating same due to less no. of nodes and that all are scattered at distance position so FAR cannot be so high. In AODV, the routing concept is implemented for transfer the packets from source to destination without avoiding the fault. So fault probability is slightly more in it. In DD, the faulty nodes are avoided and routing method is implemented to change the path from source to destination. According to AODV, fault probability value is less in DD. But in GA, the faulty nodes avoid by implementing chromosome concept in which the faulty nodes avoid and

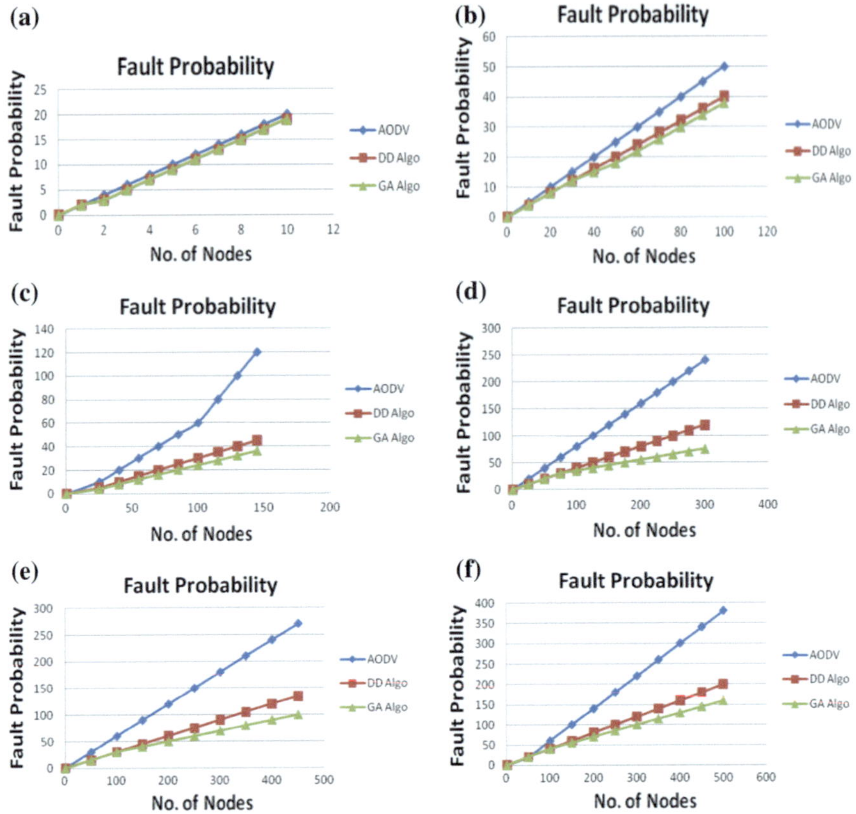

Fig. 9 Fault probability versus no. of nodes

choose the nearer active node by implementing crossover technique, so the fault probability value is less as compared to other two.

In Fig. 9, fault probability nearly same in less number of node, i.e. 0–10 and 0–100 but in case of more number of nodes 0–150, 0–300 and 0–500 fault probability is as compared to DD and GA algorithm.

In Fig. 10, as like as fault probability is same in less number of node and that will in more number of node, FAR is same performance as it in less number of node and more number of node.

In Fig. 11 shows the step response of GA algorithm system, which is better optimized than AODV and DD algorithms. Number of nodes increased with data accuracy. Initially, all the shown response is same but increment of node after a certain stage the other two goes down.

In the GA, the faulty nodes avoid by implementing chromosome concept in which the faulty node avoid and choose the nearer active node by implementing crossover technique, so the nodes are increased but response of data accuracy is maintained and as compare to other two.

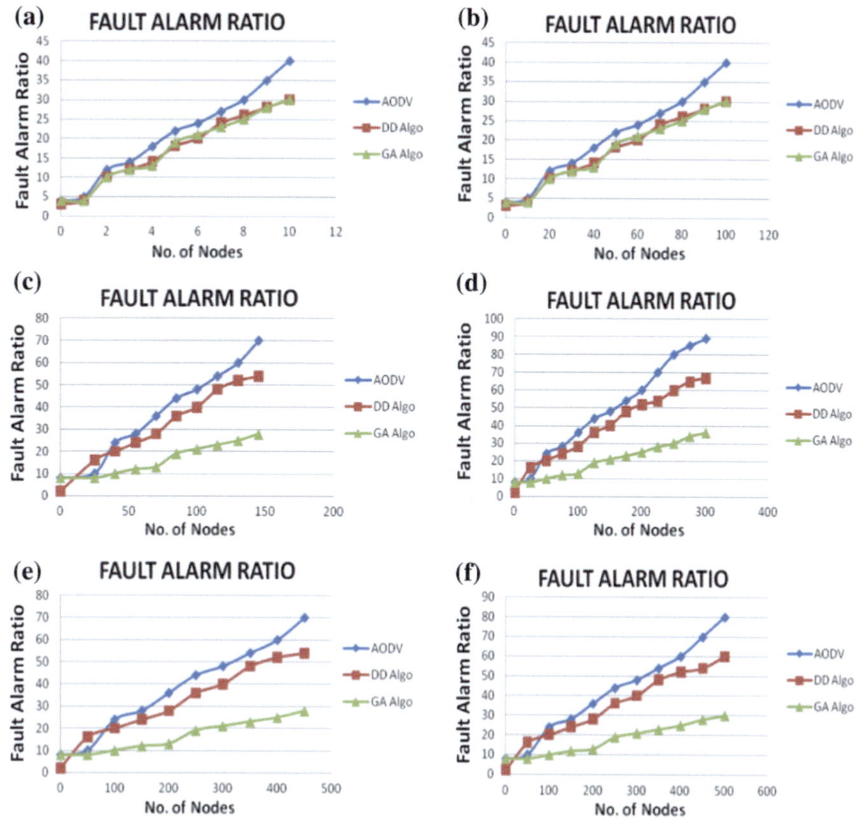

Fig. 10 Fault alarm ratio versus no. of nodes

6 Conclusion

In our paper, our work projected as based on GA. The GA concept implemented in DSN network. Its efficiency is more in case of present of fault node. This is a concept of fault tolerance by which get fault probability, FAR, data accuracy efficiently from DD algorithm and AODV. This GA implementation determines fault nodes avoid by chromosome technique and increase the network lifetime. In our imitation results we proposed, our system is more capable than other different networks. The constitution parameters are analyzed by fault possibility for tolerance, time complexity, fitness values, energy optimization, lifespan of network and possibility of fault detection in rate of the DSN environment.

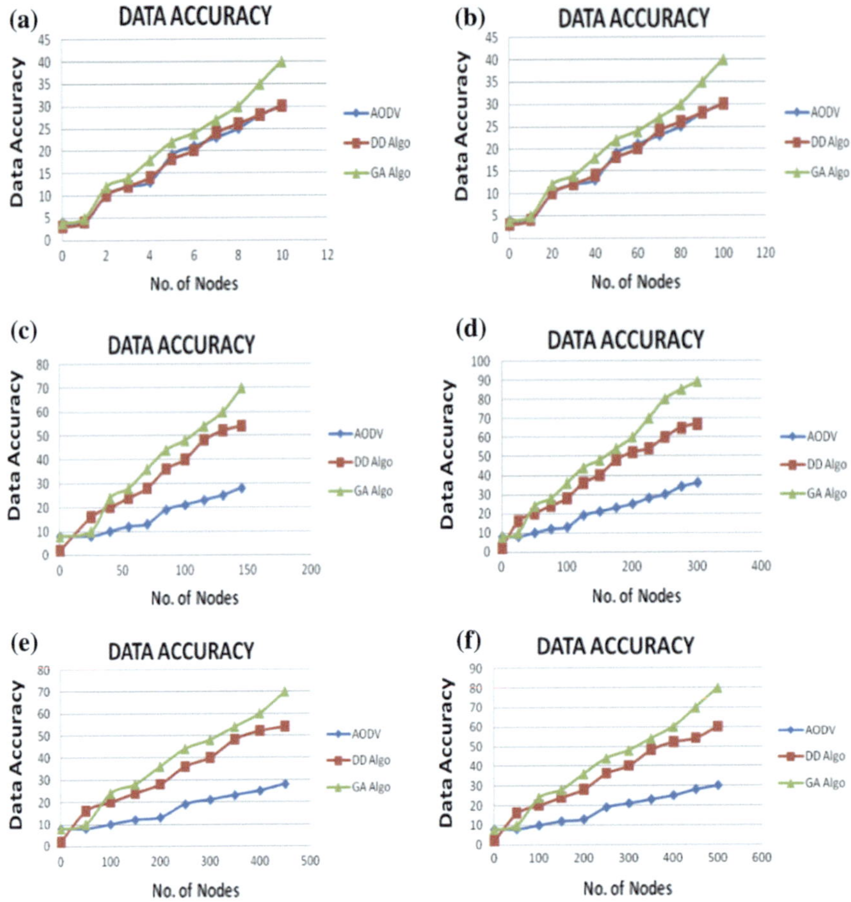

Fig. 11 Data accuracy for no. of nodes

References

1. Rongbo Z (2010) Efficient fault-tolerant event query algorithm in distributed wireless sensor networks. Int J Distrib Sens Netw 2010(1155):1–7
2. Herbert T, Donald L (1986) Schilling principles of communication systems. McGraw-Hill, New York
3. Heinzelman W, Chandrakasan A, Balakrishnan H (2000) Energy efficient communication protocol for wireless micro sensor networks. In: Proceedings of the IEEE Hawaii international conference on system sciences, vol 8, pp 8020–8030
4. Bhajantri LB, Nalini N (2014) Genetic algorithm based node fault detection and recovery in distributed sensor networks. Int J Comput Netw Inf Secur 6(12):37–46
5. Raza HA, Sayeed G, Sajjad H (2010) Selection of cluster heads in wireless sensor networks using bayesian network. In: Proceedings of international conference on computer, electrical, systems, science and engineering, pp 1–7

6. Iyengar SS, Ankit T, Brooks RR (2004) An overview of distributed sensors network. Chapman and Hall/CRC, London, pp 3–10. http://books.google.com/books/about/Distributed−sensor−networks.html?id=Nff5
7. Al Karaki JN, Kamal AE (2004) Routing techniques in wireless sensor networks: a survey. J IEEE Wirel Commun 11:6–28
8. Sitharam SI, Mohan BS, Kashyap RL (1992) Information routing and reliability issues in distributed sensor networks. IEEE Trans Signal Process 40(12):3012–3021
9. Lilia P, Qi H (2007) A survey of fault management in wireless sensor networks. J Netw Syst Manage 15(2):171–190
10. Mihaela C, Shuhui Y, Jie W (2007) Fault-tolerant topology control for heterogeneous wireless sensor networks. In: Proceedings of the IEEE international conference on mobile adhoc and sensor systems, pp 1–9
11. Bhaskar K, Sitharama I (2004) Distributed bayesian algorithms for fault-tolerant event region detection in wireless sensor networks. IEEE Trans Comput 53(3):241–250
12. Jilei L, Baochun L (2003) Distributed topology control in wireless sensor networks with asymmetric links. In: Proceedings of the IEEE globecom, wireless communications symposium, vol 3, pp 1257–1262
13. Arroyo VR, Marques AG, Vinagre-Diaz J, Cid-Sueiro JA (2006) Bayesian decision model for intelligent routing in sensor networks. In: Proceedings of 3rd international symposium on wireless communication systems, pp 103–107
14. Mihaela C, Shuhui Y, Jie W (2008) Algorithms for fault-tolerant topology control for heterogeneous wireless sensor networks. IEEE Trans Parallel Distrib Syst 19(4):545–558
15. Mohammad M, Subhash C, Rami A (2010) Bayesian fusion algorithm for inferring trust in wireless sensor networks. J Netw 5(7):815–822
16. Shimamoto N, Hiramatsu A, Yamasaki K (1993) A dynamic routing control based on a genetic algorithm. In: Proceedings of IEEE international conference on neural networks, vol 2, pp 1123–1128
17. Ayon C, Swarup Kumar M, Mrinal Kanti N (2011) A genetic algorithm inspired routing protocol for wireless sensor networks. Int J Comput Intell Theory Pract 6(1):1–10
18. Bhattacharya R, Venkateswaran P, Sanyal SK, Nandi R (2005) Genetic algorithm based efficient routing scheme for multicast networks. In: Proceedings of international conference on personal wireless communications, pp 500–504
19. Hong-Chi S, Jiun-Huei H, Bin-Yih L (2013) Fault node recovery algorithm for wireless sensor network. IEEE Sens J 13(7):2683–2689
20. Elmira MK, Sanam H (2012) Recovery of faulty cluster head sensor by using genetic algorithm. Int J Comput Sci Issues 9(1):141–145
21. Lokesh BB, Nalini N (2012) Energy aware based fault tolerance approach for topology control in distributed sensor networks. Int J High Speed Netw 18(3):197–210
22. Alaa FO, Mohammed Al (2012) Improving the performance of the networks using genetic algorithm. In: Proceedings of international conference of advances in computer networks and its security, vol 2, no 3, pp 117–120
23. Myeong HL, Yoon HC (2008) Fault detection of wireless sensor networks. J Comput Commun 31:3469–3475
24. Xiaofeng H, Xiang C, Lloyd LE, Chien-Chug S (2010) Fault-tolerant relay node placement in heterogeneous wireless sensor networks. IEEE Trans Mob Comput 9(5):643–656
25. Biao C, Ruixiang J, Kasetkasem T, Varshney PK (2004) Channel aware decision fusion in wireless sensor networks. IEEE Trans Signal Process 52(12):3454–3458
26. Darrell W (1994) A genetic algorithm tutorial. J Stat Comput 4:65–85
27. Karaa WBA, Ashour AS, Sassi DB et al (2016) Medline text mining: an enhancement genetic algorithm based approach for document clustering. In: Applications of intelligent optimization in biology and medicine. Springer, Cham, pp 267–287
28. Chatterjee S, Sarkar S, Hore S, Dey N, Ashour AS, Shi F, Le DN (2017) Structural failure classification for reinforced concrete buildings using trained neural network based multi-objective genetic algorithm. Struct Eng Mech 63(4):429–438

29. Fong S, Li J, Song W, Tian Y, Wong RK, Dey N (2018) Predicting unusual energy consumption events from smart home sensor network by data stream mining with misclassified recall. J Ambient Intell Humaniz Comput 9(4):1197–1221
30. Das, SK, Samanta S, Dey N et al (2019) Design frameworks for wireless networks. Lecture Notes in Networks and System, ISBN: 978-981-13-9573-4, Springer, pp 1–439
31. Das SK, Yadav AK, Tripathi S (2017) IE2M: design of intellectual energy efficient multicast routing protocol for ad-hoc network. Peer-to-Peer Netw Appl 10(3):670–687
32. Das SK, Tripathi S, Burnwal AP (2015, February) Fuzzy based energy efficient multicast routing for ad-hoc network. In: Proceedings of the 2015 third international conference on computer, communication, control and information technology (C3IT), IEEE, pp 1–5
33. Das SK, Tripathi S (2018) Adaptive and intelligent energy efficient routing for transparent heterogeneous ad-hoc network by fusion of game theory and linear programming. Appl Intell 48(7):1825–1845
34. Das SK, Tripathi S (2016). Energy efficient routing protocol for manet using vague set. In: Proceedings of fifth international conference on soft computing for problem solving, Springer, Singapore, pp 235–245
35. Dey N, Ashour AS, Shi F, Fong SJ, Sherratt RS (2017) Developing residential wireless sensor networks for ECG healthcare monitoring. IEEE Trans Consum Electron 63(4):442–449
36. Sajid H, Abdul WM, Obidul I (2007) Genetic algorithm for hierarchical wireless sensor networks. J Netw 2(5):87–97

Chapter 4
A GA-Based Intelligent Traffic Management Technique for Wireless Body Area Sensor Networks

Kanhu Charan Gouda, Santosh Kumar Das, Om Prakash Dubey and Efrén Mezura Montes

1 Introduction

Wireless Body Area Network (WBAN) or Wireless Body Area Sensor Network (WBASN) is a collection of different sensor nodes [1, 2]. These sensor nodes may be homogeneous or heterogeneous depending on the medical requirements. These nodes are known as biological sensors. The nodes are placed into the body as wearable or fixed modes. The function of each node is different. It helps to measure imprecise vital signs of a patient and also helps to detect conflicting emotions such as happiness, fear, stress, human positions, etc. as distributed system [3]. Each sensor node is directly or indirectly connected with the main coordinator node which has low energy and high processing capacities [4]. The purpose of this controller node is to send biological signals of patient to the doctor. So, that doctor takes right decision in the medical diagnosis. One of the modern facilities is Internet of Things (IoT) based WBAN in healthcare system. It paves the gap of traditional system like to visit hospital. IoT allows facilities like communicating, sensing, processing with biomedical and physical parameters [5, 6]. Cloud computing also provides some advantages to the

K. C. Gouda · S. K. Das (✉)
School of Computer Science and Engineering, National Institute of Science and Technology (Autonomous), Institute Park, Pallur Hills, Berhampur, Odisha 761008, India
e-mail: sunsantosh2014@gmail.com

K. C. Gouda
e-mail: kanhu367@gmail.com

O. P. Dubey
Department of Mathematics, Jag Jiwan College, Unit of Veer Kunwan Singh University, Arrah, Bhojpur, Bihar 802312, India
e-mail: omprakashdubeymaths@gmail.com

E. M. Montes
Artificial Intelligence Research Center, University of Veracruz, Xalapa, Mexico
e-mail: emezura@uv.mx

D. De et al. (eds.), *Nature Inspired Computing for Wireless Sensor Networks*, Springer Tracts in Nature-Inspired Computing,
https://doi.org/10.1007/978-981-15-2125-6_4

WBAN because it has large processing and storage infrastructure. It helps to process the data and information as offline as well as online by body sensor streams [7–10]. There are several applications of WBAN, which are as follows.

(i) Fitness monitoring
(ii) Interaction between Body Area Networks (BANs)
(iii) Mobile device centric
(iv) Remote control
(v) Social applications
(vi) Video stream
(vii) Wearable audio.

Due to the limited capacity of sensor nodes and different type of uncertainty causes several challenges in terms of privacy and security. Security means safeguard the data or data set or different related information and privacy indicates safeguarding of user information or identity [11]. To solve different type problems such as outlier detection or anomaly detection, fault diagnosis, intrusion detection, mobility prediction several works have been proposed [12, 13]. Most of the outlier detection and prediction algorithms are outperform with the help of machine learning algorithms [14, 15]. Because machine learning algorithm helps to reduce the critical situation of the outlier detection in order to optimize way [16–18]. There are some basic steps that follow by machine learning algorithms:

(i) Feature selection and output labeling
(ii) Sample collection
(iii) Offline training
(iv) Online classification.

The basic aim of this proposal to design an intelligent traffic management technique for WBAN using Genetic Algorithm (GA). The proposed technique is based on Maxone technique of GA. It is used to model the network traffic with the help of crossover and mutation.

The rest of this paper is organized as follows: Sect. 2 provides related work done in WBAN with sensor and ad hoc networks. Details of the proposed method are described in Sect. 3. Section 4 illustrates the performance evaluation of the proposed method and compares it with existing protocols. Finally, Sect. 5 concludes the paper.

2 Literature Review

In the last decade, several routing and traffic management techniques are proposed in WSN, as well as WBAN. A details survey is given in the proposed article by Curry and Smith [19]. In this survey, each literature is described an intelligent algorithm that directly or indirectly uses one or more techniques for enhancing network lifetime or network metrics efficiently. This survey also illustrates impractical situations and their optimal solution with respect to network lifetime and network metrics. Yan et al.

[20] proposed a low energy based node positioning system in WSN. The proposed work is based on PSO optimization technique. The basic aim of this proposal to optimize the dynamic position of the sensor nodes efficiently in terms of minimum energy consumption based on the particle of optical sensor nodes. Lai et al. [21] design a Protocol for Traffic Safety (PTS) for WSN. This protocol is based on energy-efficient routing protocol that helps to manage traffic by reducing collisions of the network. It collectively analyze the traffic and handles it efficiently. Srivastava and Sudarshan [22] design an Intelligent Traffic Management (ITM) system for WSN. Basically, this work proposed for vehicular travel for handling surveillance cameras, wired sensors, inductive loops, etc. that support vehicular travel with the help of WSN. It consists of an intelligent method that helps to reduce Average Waiting Time (AWT). This time indicates average waiting time at each junction point of the network. Besides the reduction time, it also helps regulate traffic management system. Jiménez and García [23] proposed a technique for avoidance of jams in WSN named as Traffic Jams Avoidance (TJA). The main aim of this avoidance is to manage traffic of the network. The proposed technique is used for surveillance as well as traffic monitoring. So, it is an energy-efficient technique that also reduces the hardware cost of the network. Yu et al. [24] proposed a hybrid localization method for WSN. The basic keyword of this method is Chicken Swarm Optimization (CSO). The nature optimization technique CSO is used for deep mining based on the wheel graph. This method helps in transformation of the clusters and improves the precision of the location for the sensor nodes. Phoemphon et al. [25] proposed a hybrid method for localization system in WSN. This method is the fusion of fuzzy logic, machine learning, and vector particle swarm optimization. It helps to improve the traditional localization system, i.e., centroid by using the hybrid mechanism. Finally, it overcomes the limitation of estimation precision. Sun et al. [26] proposed an attack localization system in WSN. The basic aim of this method allows task allocation in the network using binary POS mechanism. The complete process is based on the three objective functions: (i) maximization of load balancing system, (ii) minimization of the energy consumption of the network, and (iii) minimization of the cost of execution time. The combination of the stated optimization helps to construct the constraints of the received signal strength and enhance the network lifetime. Cao et al. [27] proposed a technique for deployment in WSN using PSO in distributed environment. In this method, there are two types nodes are used to sensor and relay nodes for prolong the lifetime and maximize the coverage. The stated method is basically used for 3D industrial WSN. Finally, it helps to reduce computational costs and enhance the network lifetime. Das and Tripathi [28] proposed a routing technique for Transparent Heterogeneous Ad hoc Network (THANET). This technique basically used non-cooperating game theory optimization for managing several conflicting strategies of the network. This non-cooperative game theory optimization is used fuzzy logic for making rigid goals to fuzzy goals. So, that dynamic environment of THANET manages efficiently. Das et al. [29] proposed a routing protocol for multicast ad hoc networks in order to create an energy effective path from origin to every multicast set built on two vague parameters such as distance and energy. Where, as other parameters were not been considered by this proposed work which leads to its limitation. Hence, Yadav et al. [30] stretched the

effort placed on multi constraints method. They have considered three parameters that are delay, bandwidth, and energy. It supports to elect the best route with the help of fuzzy cost. Here too limitations occurred as it is a point-based membership function and fails to hold the fuzziness information. Henceforth, Das and Tripathi [31] proposed an energy-conscious based routing protocol by considering five parameters such as distance, energy, delay, packets, and hop-count. The objective of this routing is to search for an optimal route by considering multi-criteria decision making and intuitionistic fuzzy soft set. As it doesn't use any optimization method leads to its limitation. So, by used above techniques several contradictory objectives may not be optimized. Das and Tripathi [32] brought up a routing technique based on non-linear optimization technique. This non-linear optimization technique is based on geometric programming which works with polynomial environment instead of polynomial environment. It helps to determine non-linear parameters efficiently and enhance the network lifetime. Das and Tripathi [33] design a fusion algorithm for managing dynamic and conflicting environment of the Hybrid Ad Hoc Network (HANET). This fusion is based on several Artificial Intelligence (AI) techniques such non-linear geometric programming, fuzzy logic, multi-objective optimization, aspiration level, and optimset. The basic aim of this fusion to manages multiple non-linear conflicting objectives of the network efficiently. So, that network lifetime as well as several network metrics are increases simultaneously in several scenarios of passes. Zahedi et al. [34] proposed an intelligent routing protocol for clustered WSN. This is fuzzy based routing protocol that used swarm intelligence to manage all cluster head nodes of WSN. In this routing protocol Mamdani inference system is used for making decision making of the fuzzy logic. Finally, it helps to prolong the network lifetime and balanced the cluster using firefly swarm algorithm. Shankar et al. [35] proposed a hybrid algorithm for energy efficiency in WSN. The aim of this algorithm is selection of cluster head and reduce energy consumption of the network. The both stated operation is performed with the fusion of Harmony Search Algorithm (HAS) and PSO. The dynamic capability of the algorithm is more with respect to topology of the network. It helps to judge alive and dead nodes, increase throughput, and reduce residual energy of the network. Azharuddin and Jana [36] proposed an enhanced algorithm for WSN with the help of PSO. The PSO is a machine intelligence technique that used here to select cluster head and cluster members. The proposed technique helps to manage distributed traffic load. It possesses network lifetime enhancement as well as network metrics enhancement. Ouchitachen et al. [37] proposed a multi-objective optimization technique in WSN for weighted clustering algorithm. In this algorithm, a Genetic Algorithm (GA) based Base Station (BS) is used to manage consumed energy of the sensor nodes. The proposed algorithm helps to satisfy and fulfill the requirement of the sensor nodes. It also helps to improve the communication between sender and received nodes by crossing neighbor nodes. Gholipour et al. [38] proposed a technique for congestion control in WSN using fusion of genetic algorithm and support vector machine. In this technique, SVM parameters are tuned using GA and match actual data with current in different phases. The purpose of this algorithm is to increase energy efficiency and throughput and decrease packet loss. Bhatia et al. [39]

designed a distance aware routing protocol for WSN using GA. The proposed routing protocol is based on the LEACH protocol. This protocol stands for Low Energy Adaptive Clustering Hierarchy. It is used to enhance the network lifetime by reducing energy consumption. In this protocol, random probability is added into the GA for improving the CH selection and establish efficient communication between CH and BS. Finally, it enhances the network lifetime of the network based on different metrics. Ray and De [40] designed an algorithm for WSN using swarm optimization. The main purpose of this algorithm is to solve two issues like energy conservation and coverage. The main keyword of this algorithm is Glowworm Swarm Optimization (GSO) which is a bio-inspired algorithm. It helps to reduce redundant coverage by sensor traveling one place to another place. Finally, it helps to optimize distance traversal and reduce energy consumption of the sensor nodes. Taherian et al. [41] proposed a secure and optimal routing protocol for WSN using PSO nature-inspired algorithm. In this algorithm, an efficient technique is used to divide each sensor as clustering method and apply there PSO optimization for efficient and safe routing systems in WSN. Barekatain et al. [42] designed a routing protocol for WSN using fusion of k-means and GA. In this routing protocol, the CH collects data from all the cluster members i.e. simple sensor nodes and sends into BS time to time. This system helps to aggregate the necessary information into single place. The main aim of this routing protocol is to reduce energy consumption and extends the network lifetime. Dhivya and Sundarambal [43] proposed a Tabu Swarm Optimization (TSO) technique for network lifetime maximization in WSN. This is a QoS based routing optimization which is design by the fusion of PSO and tabu search. This fusion technique helps to enhance the network lifetime and reduce the energy consumption of the WSN. Das et al. [44] provide a details book for wireless networks. It consists of several frameworks of wireless networks along with WSN and ad hoc networks. Some of works are [45, 46] describe intelligent routing technique using fuzzy petrinet and strategy management using non-linear formulation technique. Both works are helping to estimate uncertainty parameters of the network. WBASN is a part of wireless network that has several benefits in terms of real-life applications. It works with ad hoc as well as sensor networks to achieve the purpose of the users. The different applications and uses are described in [47–50]. Although there are several advantages to this network, it has also some limitations in terms of their parameters.

The above-mentioned literature describes several types of issues and problems with respect to the mentioned domains and parameters. But most of the literature is describe the issue based on solution point of view. The proposed problem of this article indicates the traffic management technique using meta-heuristic method.

3 Proposed Method

In this section, the proposed methodology is illustrated with respect to domain and its related attributes. The basics of WBAN and its related functionalities and applications are already highlighted in the first section of the proposed. Some concepts are

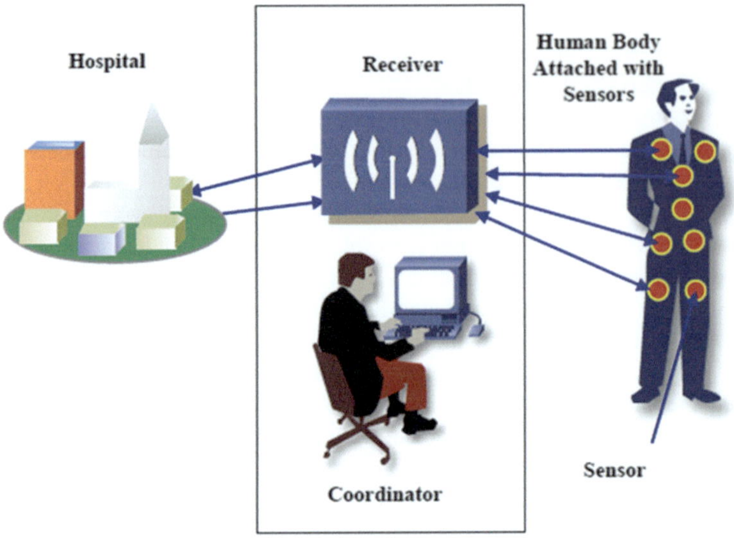

Fig. 1 Basic architecture of WBAN

described here to understand the proposed problem efficiently. Hence, the basic architecture of WBAN shown in Fig. 1 where coordinator is attached with a receiver. This coordinator is responsible to handle and manage all sending and receiving information of the sensor nodes. This combined illustration is known as Receiver Controller (RC). Left-hand side of the RC is a hospital and right-hand side of the RC is a human body that attached with some sensor nodes. These sensor nodes are responsible to receive different variation signal from the human body and send it to the RC. The complete procedure is attached to hospital for diagnosis and treatment purposes. But here is a limitation that sometimes RC is in sleep mode that time received signal from the sensor nodes is loss due to inactivation mode of RC. This situation is illustrated in Fig. 2.

In Fig. 2 different sensor nodes are illustrated by Transmitter i where $1 \leq i \leq n$ and different signals are illustrated by Si. During data transmission of Transmitter 1, i.e., signal S1 so on for S2 to S*n* of Transmitter 2 to Transmitter *n*, the RC mode is sleep mode. It means when RC is in active mode that time only data send by Transmitter i. But during sleep mode of RC data is loss. This problem causes different type faults in medical diagnosis. This issue is overcome in Fig. 3.

Figure 3 illustrates the solution of above-mentioned problem. In this situation, the RC sends a beacon signal to all Transmitter *i* as *i* lies from 1 to *n*. Transmitter i receive that signal from the RC. If any Transmitter i wants to send data to the RC then it requires to send ACK packet as ACKi. The RC receives that particular ACKi from Transmitter *i*. This receiving indicates a "green signal" for data transmission. Finally, Transmitter i starts data transmission.

The proposed work is based on intelligent traffic management techniques for WBASN. In this work, GA plays the main role of intelligence. GA is a nature-inspired

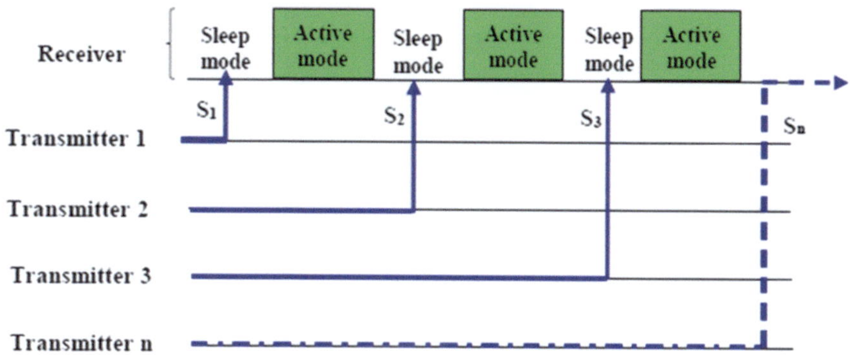

Fig. 2 Data loss problem in WBAN

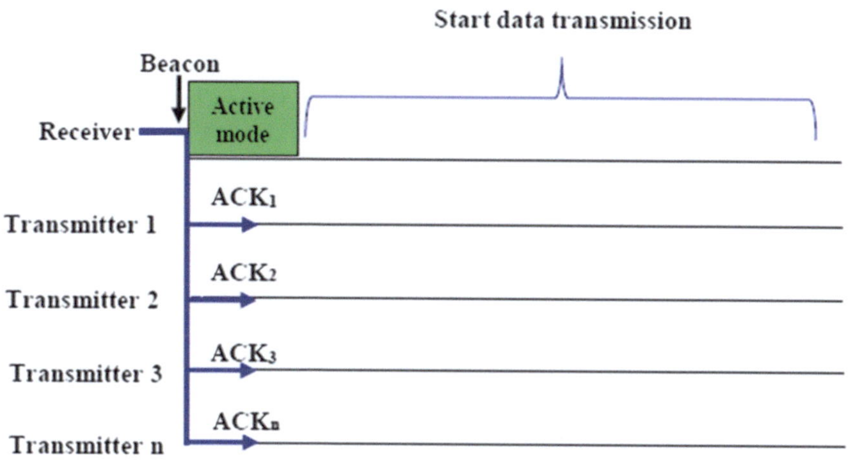

Fig. 3 Solution of data loss problem in WBAN

technique that is based on meta-heuristic approach [51–53]. It is also known as bio-inspired algorithm. We know that basic structure of the GA based on its operations shown in Fig. 4. The complete structure consists of some steps. Short descriptions are as follows.

(a) **Initialization**: This is first step that indicates population initialization.
(b) **Selection**: This is second step that indicates derive one feasible category within-population like in Fig. 5, in first figure the set of population is shown wherein second figure of Fig. 5 the set of population is divided into two categories (i) Good people, (ii) Bad people. So, here permission is given to only good people to increase their generation, but this permission not given to bad people, so after one or two generations the whole population becomes good people. But, practically, it is not possible.

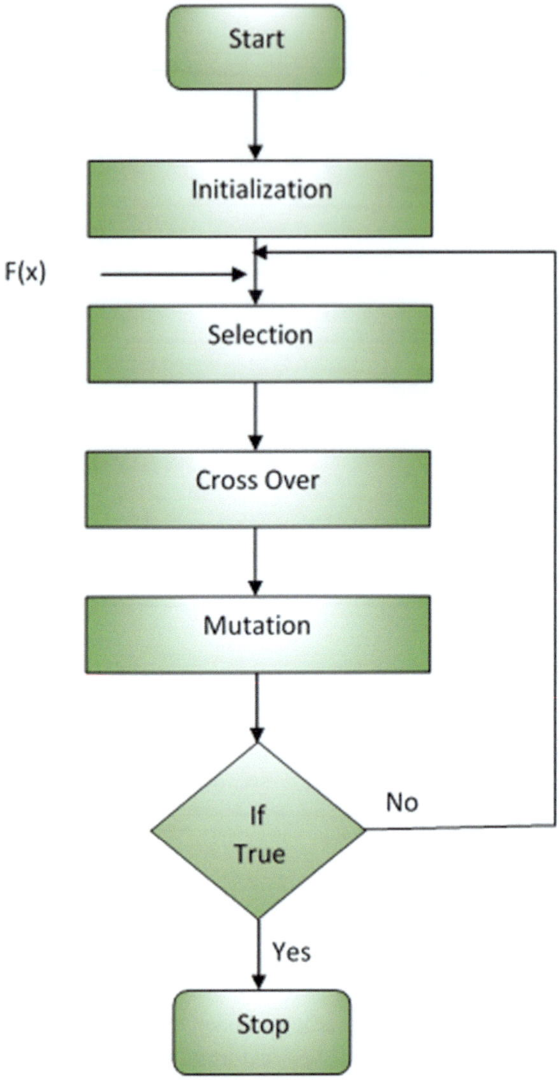

Fig. 4 Basic structure of genetic algorithm

(c) **Crossover**: This is the third step of the operation that indicates to increase the generation.
(d) **Mutation**: This is a genetic operator which is used to get new solution.

All the above-mentioned phases or steps are work on the fitness function i.e. $F(x)$, this is the main objective function of the whole problem. Finally, check desire solution is arrived or not if arrived then stop otherwise repeat selection, cross over,

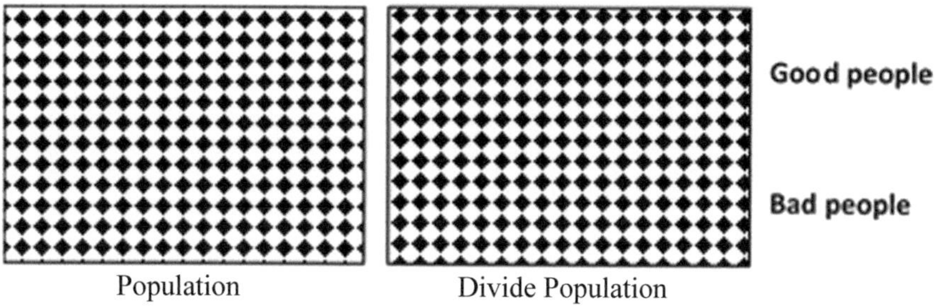

Fig. 5 Example of set of population

and mutation steps. Stopping condition is decides by the used. Sometimes, it defines a number of iteration as 4 which means after 4 iteration processes will be stopped.

Figure 6 shows the chromosome within the population in terms of "green signal" and "non-green signal", i.e., also known as sleep mode. In this proposed model green signal denoted by 1 and non-green signal denoted by 0. Signals within the population are known as chromosomes. In the proposed method we used GA as an intelligent technique, and we select the 0 and 1, i.e., non-green and green signals based on their fitness function (F(x)) which resides between the steps initialization and selection in Fig. 4. Each cell is also known as gene or particle in terms of GA.

Let us select two good chromosomes with the help of fitness function shown in Fig. 7.

Chromosomes →

Population

0	0	1	0	1	1	1	1
0	0	1	1	0	1	1	1
0	0	0	0	0	1	1	1
0	1	0	1	0	0	0	0

Caption:
1: Green Signal
0: Non-green Signal

Fig. 6 Chromosome within population

0	0	1	0	1	1	1	1
0	0	1	1	0	1	1	1

Fig. 7 Selected chromosomes before crossover

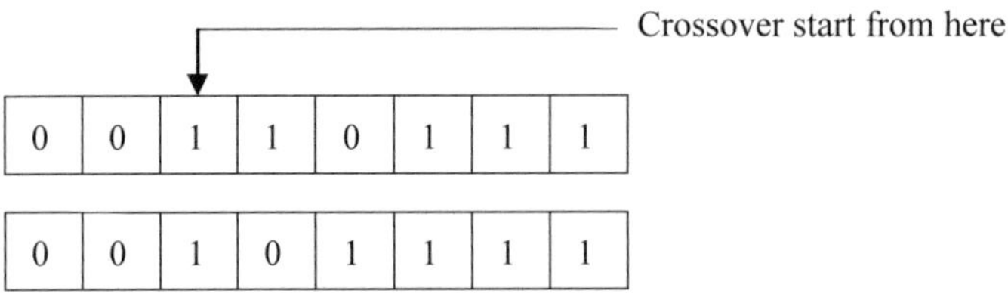

Fig. 8 Selected chromosomes after crossover

After the selection of the chromosomes two operations are performed as crossover and mutation. Here, crossover means to mixed both selected chromosomes and performed some exchange. Like we want to perform crossover after point 2 then resultant chromosomes after crossover is shown in Fig. 8.

Mutation is a genetic operator that is used to get a new solution. Means already selected good signal as green in terms of chromosomes and now we want to add some extra good quality to the green signal which is previously not available in the signal. So, adding this extra quality is known as mutation. It just reverses the bit of particular gene. By applying mutation operation, fitness of the chromosome is enhanced. Figure 9 shows the mutation operation where change is occurred after bit four, i.e., bit number five, just exchanging between 0 and 1 and after mutation of crossover bits the population becomes a new population.

Based on the above-mentioned crossover and mutation operations, the basic operational structure of the GA is changed as shown in Fig. 10. In Fig. 10, after mutation two steps are added as "evaluate" and "generate" that indicate after mutation again calculate fitness of the updated chromosomes and generate new population. This process is repeated continuously and checks how result is improved in each generation.

Maxone technique for proposed method: This Maxone technique is a GA-based technique that is used to maximize the desired output, i.e., green signal instead of non-green signal.

In the proposed method, a total number of signal is 50 which is divided into five slots. Each slot consists 10 signals. Slot denoted by si where $i \in 1$ to 5 given in Fig. 11.

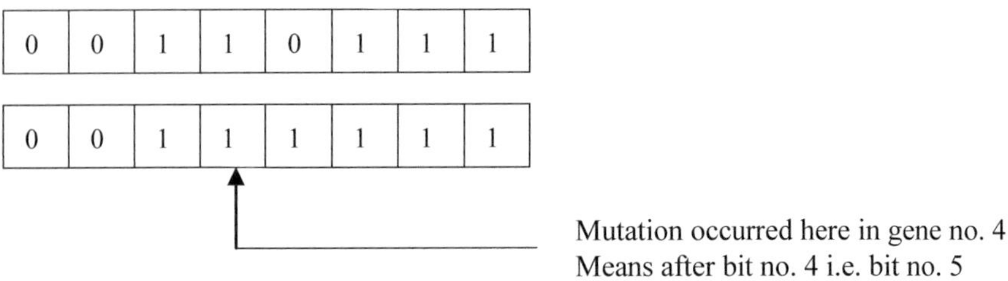

Fig. 9 Mutation operation

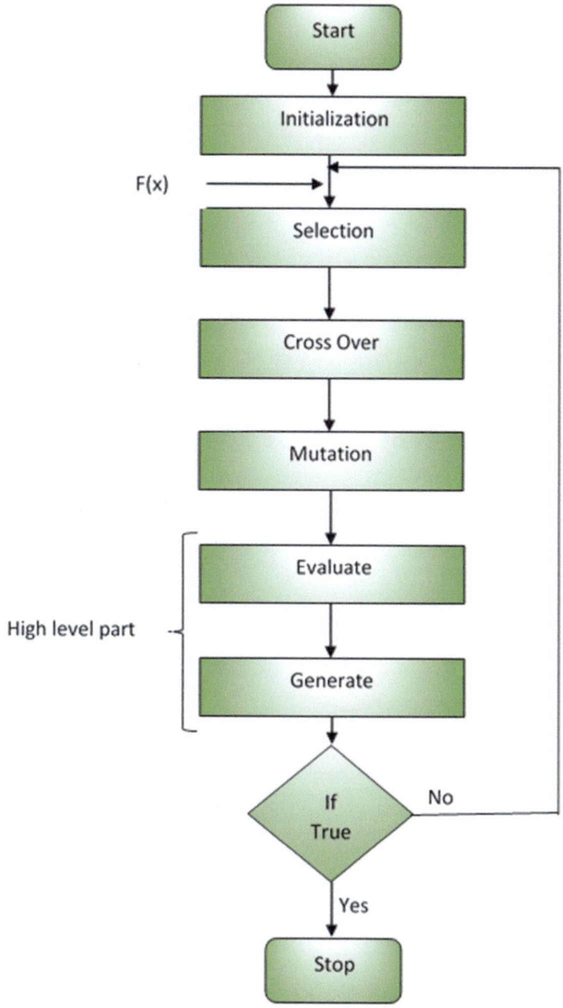

Fig. 10 Extended structure of genetic algorithm

Figure 11 indicates initialization of the population of the signals. Then calculate fitness of each chromosome by calculating number of desired output in each chromosome shown as:

$f(s1) = 4$, $f(s2) = 3$, $f(s3) = 4$, $f(s4) = 3$, $f(s5) = 6$. So, total fitness is $= 4 + 3 + 4 + 3 + 6 = 20$.

Now, select the chromosomes by roulette wheel method. Roulette wheel design by circle and we know total angle of circle is 360°. Here, 20 is equal to 360°, so maximum value is s5 so, total percentage of s5 is = (6/20)*100 = 30%, in this way $s1$ to $s4$ are 20%, 15%, 20%, 15%. Hence, roulette wheel is shown in Fig. 12.

s1	1	0	0	1	0	0	0	1	0	1
s2	0	1	0	1	0	0	0	0	0	1
s3	0	1	1	0	0	1	0	0	0	1
s4	0	0	1	0	0	0	0	0	1	1
s5	1	0	0	0	1	1	1	1	0	1

Fig. 11 Population of signals of the proposed method

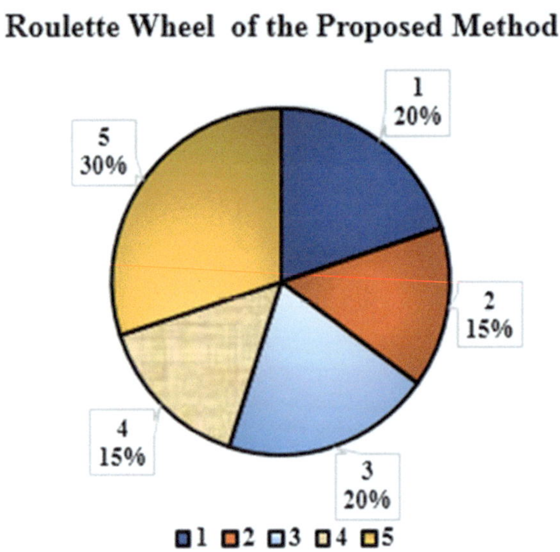

Fig. 12 Roulette wheel of the proposed method

Arrange all chromosomes based on the fitness function by roulette wheel in descending order and assign new name for better understanding:

s1′ = 1000111101
s2′ = 1001000101
s3′ = 0110010001
s4′ = 0101000001
s5′ = 0010000011

Now, performed selection process by selecting any two pairs randomly and in this model the selected pairs are Part 1(s1′–s2′) and Part 2 (s4′–s5′) and perform crossover between members (i.e. s1′ with s2′ and s4′ with s5′).

Part 1:

s1'	1	0	0	0	1	1	1	1	0	1
s2'	1	0	0	1	0	0	0	1	0	1

Part 2:

s4'	0	1	0	1	0	0	0	0	0	1
s5'	0	0	1	0	0	0	0	0	1	1

Now, the modified population after crossover as follows.
Modified Part 1:

s1'	1	0	0	1	0	0	0	1	0	1
s2'	1	0	0	0	1	1	1	1	0	1

Modified Part 2:

s4'	0	0	1	0	0	0	0	0	1	1
s5'	0	1	0	1	0	0	0	0	0	1

The modified population as follows.

s1″ = 1001000101
s2″ = 1000111101
s3″ = 0110010001
s4″ = 0010000011
s5″ = 0101000001

Now, for more improving again we apply crossover between s2′ and s3′ as Part 3 and between s1′ and s5′ as Part 4 as follows.

Part 3:

s2''	1	0	0	0	1	1	1	1	0	1
s3''	0	1	1	0	0	1	0	0	0	1

Part 4:

s1''	1	0	0	1	0	0	0	1	0	1
s5''	0	1	0	1	0	0	0	0	0	1

Now, modified population after crossover as follows.

Modified Part 3:

s2'''	0	1	1	0	0	1	0	0	0	1
s3'''	1	0	0	0	1	1	1	1	0	1

Modified Part 4:

s1'''	0	1	0	1	0	0	0	0	0	1
s5'''	1	0	0	1	0	0	0	1	0	1

The modified population as follows.

$s1''' = 0101000001$
$s2''' = 0110010001$
$s3''' = 1000111101$
$s4''' = 0010000011$
$s5''' = 1001000101$

Now, apply mutation for particular bits that are underlined as given population.

$s1''' = 0101\underline{0}00\underline{0}01$
$s2''' = 0110010001$
$s3''' = 1000111101$
$s4''' = 0010\underline{0}0\underline{0}011$
$s5''' = 10010\underline{0}0101$

And, again the modified population is as follows.

s1'''' = 0101100101
s2'''' = 0110010001
s3'''' = 1000111101
s4'''' = 0010101011
s5'''' = 1001010101

Now, again calculate fitness value of each slot.

s1'''' = 0101100101 = 5
s2'''' = 0110010001 = 4
s3'''' = 1000111101 = 6
s4'''' = 0010101011 = 5
s5'''' = 1001010101 = 5

Total fitness value is = 5 + 4 + 6 + 5 + 5=25.

So, the previous fitness value is 20 and new fitness value is 25, its means number of 1 increases. So, number of green signals also increases. It is the main goal of the problem. This process repeats continuously until we get the desired output, maybe desired output defined by the administrator controller.

4 Performance Evaluation

The proposed method is a short communication that is compared theoretical instead of simulation. It compared with some existing methods (e.g. ITM [22], TJA [23], and PTS [21]) and shown in Table 1. The proposed method along with all three existing methods are based on routing loop avoidance and source initiated. But all these methods are not based on receiver-initiated. The proposed method is based on GA technique which is a meta-heuristic technique. This technique supports multi-objective and its optimization is good in "noisy" environment. Hence, based on different features the proposed method is much better than existing methods. The existing method ITM [22] is based on average waiting time which is used in each junction point of the network. So, it is used to regulate network traffic. Hence, different parameters of network such as QoS, network lifetime, delay packet delivery ratio, communication overhead, throughput, packet loss, scalability, bandwidth, robustness, and handling high mobility, traffic load, and mutual interference is more better than other two existing methods like TJA [23] and PTS [21] but less better than the proposed method. The existing method TJA is based on traffic avoidance as well as energy efficiency. It is used both network surveillance and monitoring. So, this is less good than the proposed method and ITM but better than existing method PTS. The existing method PTS is an energy-efficient routing protocol that handles only traffic. So, its performance is poor than other existing methods as well as the proposed method.

Table 1 Feature comparison of the proposed method with some existing methods

Features	Proposed method	ITM [22]	TJA [23]	PTS [21]
1	√	√	√	√
2	√	√	√	√
3	×	×	×	×
4	4	3	2	1
5	4	3	2	1
6	4	3	2	1
7	1	2	3	4
8	4	3	2	1
9	1	2	3	4
10	4	3	2	1
11	1	2	3	4
12	4	3	2	1
13	4	3	2	1
14	4	3	2	1
15	4	3	2	1
16	4	3	2	1
17	4	3	2	1
18	4	3	2	1

Caption:

1: Routing loop avoidance		2: Source initiated		3: Receiver initiated	
4: QoS support		5: Residual energy		6: Network lifetime	
7: Delay		8: Packet delivery ratio		9: Communication overhead	
10: Throughput		11: Packet loss		12: Scalability	
13: Bandwidth		14: Robustness		15: Connectivity status	
16: Handling high mobility		17: Handling traffic load		18: Handling mutual interference	
Yes: √	No: ×	Very high: 4	High: 3	Medium: 2	Low: 1

5 Conclusion

In this paper, GA-based traffic management technique is proposed with the help of Maxone technique. The proposed method is used to maximize the green signal of the network and minimize the non-green signal of the network. It takes several steps of GA such as crossover and mutation to enhance the network parameters efficiently. The final results conclude that the proposed method outperformed the three existing methods in terms of several characteristics.

References

1. Negra R, Jemili I, Belghith A (2016) Wireless body area networks: applications and technologies. Procedia Comput Sci 83:1274–1281
2. Masdari M, Ahmadzadeh S, Bidaki M (2017) Key management in wireless body area network: challenges and issues. J Netw Comput Appl 91:36–51
3. Panda SK, Naik S (2018) An efficient data replication algorithm for distributed systems. Int J Cloud Appl Comput (IJCAC) 8(3):60–77
4. Jain PK, Quamer W, Pamula R (2018, April). Electricity consumption forecasting using time series analysis. In: International conference on advances in computing and data sciences. Springer, Singapore, pp 327–335
5. Karati A, Biswas GP (2019) Provably secure and authenticated data sharing protocol for IoT-based crowdsensing network. Trans Emerg Telecommun Technol 30(4), e3315:1–22
6. Karati A, Islam SH, Karuppiah M (2018) Provably secure and lightweight certificateless signature scheme for IIoT environments. IEEE Trans Industr Inf 14(8):3701–3711
7. Panda SK, Jana PK (2019) An energy-efficient task scheduling algorithm for heterogeneous cloud computing systems. Cluster Comput 22(2):509–527
8. Panda SK, Jana PK (2018) Normalization-based task scheduling algorithms for heterogeneous multi-cloud environment. Inf Syst Frontiers 20(2):373–399
9. Panda SK, Pande SK, Das S (2018) Task partitioning scheduling algorithms for heterogeneous multi-cloud environment. Arab J Sci Eng 43(2):913–933
10. Karati A, Amin R, Islam SH et al (2018) Provably secure and lightweight identity-based authenticated data sharing protocol for cyber-physical cloud environment. IEEE Trans Cloud Comput 1–14
11. Karati A, Islam SH, Biswas GP (2018) A pairing-free and provably secure certificateless signature scheme. Inf Sci 450:378–391
12. Jain PK, Pamula R (2019) Two-step anomaly detection approach using clustering algorithm. international conference on advanced computing networking and informatics. Springer, Singapore, pp 513–520
13. Mishra G, Agarwal S, Jain PK, Pamula R (2019) outlier detection using subset formation of clustering based method. International conference on advanced computing networking and informatics. Springer, Singapore, pp 521–528
14. Kumari P, Jain PK, Pamula R (2018, March) An efficient use of ensemble methods to predict students academic performance. In: 2018 4th international conference on recent advances in information technology (RAIT), IEEE, pp 1–6
15. Punam K, Pamula R, Jain PK (2018, September) A two-level statistical model for big mart sales prediction. In: 2018 international conference on computing, power and communication technologies (GUCON), IEEE, pp 617–620
16. Das SP, Padhy S (2018) A novel hybrid model using teaching–learning-based optimization and a support vector machine for commodity futures index forecasting. Int J Mach Learn Cybernet 9(1):97–111
17. Das SP, Padhy S (2017) Unsupervised extreme learning machine and support vector regression hybrid model for predicting energy commodity futures index. Memetic Comput 9(4):333–346
18. Das SP, Padhy S (2017) A new hybrid parametric and machine learning model with homogeneity hint for European-style index option pricing. Neural Comput Appl 28(12):4061–4077
19. Curry RM, Smith JC (2016) A survey of optimization algorithms for wireless sensor network lifetime maximization. Comput Ind Eng 101:145–166
20. Yan Z, Goswami P, Mukherjee A, Yang L, Routray S, Palai G (2019) Low-energy PSO-based node positioning in optical wireless sensor networks. Optik 181:378–382
21. Lai Y, Zheng Y, Cao J (2007, June) Protocols for traffic safety using wireless sensor network. In: International conference on algorithms and architectures for parallel processing. Springer, Berlin, Heidelberg, pp 37–48

22. Srivastava JR, Sudarshan TSB (2013, May) Intelligent traffic management with wireless sensor networks. In: 2013 ACS international conference on computer systems and applications (AICCSA), IEEE, pp 1–4
23. Gil Jiménez VP, Fernández-Getino García MJ (2015) Simple design of wireless sensor networks for traffic jams avoidance. J Sens 2015:1–7
24. Yu X, Zhou L, Li X (2019) A novel hybrid localization scheme for deep mine based on wheel graph and chicken swarm optimization. Comput Netw 154:73–78
25. Phoemphon S, So-In C, Niyato DT (2018) A hybrid model using fuzzy logic and an extreme learning machine with vector particle swarm optimization for wireless sensor network localization. Appl Soft Comput 65:101–120
26. Sun Z, Liu Y, Tao L (2018) Attack localization task allocation in wireless sensor networks based on multi-objective binary particle swarm optimization. J Netw Comput Appl 112:29–40
27. Cao B, Zhao J, Lv Z, Liu X, Kang X, Yang S (2018) Deployment optimization for 3D industrial wireless sensor networks based on particle swarm optimizers with distributed parallelism. J Netw Comput Appl 103:225–238
28. Das SK, Tripathi S (2018) Adaptive and intelligent energy efficient routing for transparent heterogeneous ad-hoc network by fusion of game theory and linear programming. Appl Intell 48(7):1825–1845
29. Das SK, Yadav AK, Tripathi S (2017) IE2M: Design of intellectual energy efficient multicast routing protocol for ad-hoc network. Peer-to-Peer Netw Appl 10(3):670–687. https://doi.org/10.1007/s12083-016-0532-6
30. Yadav AK, Das SK, Tripathi S (2017) EFMMRP: design of efficient fuzzy based multi-constraint multicast routing protocol for wireless ad-hoc network. Comput Netw 118:15–23
31. Das SK, Tripathi S (2018) Intelligent energy-aware efficient routing for MANET. Wireless networks, Springer, https://doi.org/10.1007/s11276-016-1388-7, May 2018, vol 24, no 4, pp 1139–1159
32. Das SK, Tripathi S (2019) Energy efficient routing formation algorithm for hybrid ad-hoc network: a geometric programming approach. Peer-to-Peer Netw Appl 12(1):102–128
33. Das SK, Tripathi S (2017) Energy efficient routing formation technique for hybrid ad hoc network using fusion of artificial intelligence techniques. Int Journal Commun Syst 30(16), e3340:1–16
34. Zahedi ZM, Akbari R, Shokouhifar M, Safaei F, Jalali A (2016) Swarm intelligence based fuzzy routing protocol for clustered wireless sensor networks. Expert Syst Appl 55:313–328
35. Shankar T, Shanmugavel S, Rajesh A (2016) Hybrid HSA and PSO algorithm for energy efficient cluster head selection in wireless sensor networks. Swarm Evol Comput 30:1–10
36. Azharuddin M, Jana PK (2016) Particle swarm optimization for maximizing lifetime of wireless sensor networks. Comput Electr Eng 51:26–42
37. Ouchitachen H, Hair A, Idrissi N (2017) Improved multi-objective weighted clustering algorithm in Wireless Sensor Network. Egypt Inf J 18(1):45–54
38. Gholipour M, Haghighat AT, Meybodi MR (2017) Hop-by-hop congestion avoidance in wireless sensor networks based on genetic support vector machine. Neurocomputing 223:63–76
39. Bhatia T, Kansal S, Goel S, Verma AK (2016) A genetic algorithm based distance-aware routing protocol for wireless sensor networks. Comput Electr Eng 56:441–455
40. Ray A, De D (2016) An energy efficient sensor movement approach using multi-parameter reverse glowworm swarm optimization algorithm in mobile wireless sensor network. Simul Model Pract Theory 62:117–136
41. Taherian M, Karimi H, Kashkooli AM, Esfahanimehr A, Jafta T, Jafarabad M (2015) The design of an optimal and secure routing model in wireless sensor networks by using PSO algorithm. Procedia Comput Sci 73:468–473
42. Barekatain B, Dehghani S, Pourzaferani M (2015) An energy-aware routing protocol for wireless sensor networks based on new combination of genetic algorithm & k-means. Procedia Comput Sci 72:552–560
43. Dhivya M, Sundarambal M (2012) Lifetime maximization in wireless sensor networks using tabu swarm optimization. Procedia Eng 38:511–516

44. Santosh Kumar D, Sourav S, Nilanjan D et al Design frameworks for wireless networks. Lecture Notes in Networks and Systems, Springer, ISBN: 978-981-13-9573-4, pp 1–439
45. Samantra A, Panda A, Das SK et al (2020) Fuzzy petri nets-based intelligent routing protocol for ad hoc network. In: Design frameworks for wireless networks. Springer, Singapore, pp 417–433
46. Santosh Kumar D, Tripathi S (2020) A nonlinear strategy management approach in software-defined ad hoc network. Design frameworks for wireless networks. Springer, Singapore, pp 321–346
47. Fong S, Li J, Song W, Tian Y, Wong RK, Dey N (2018) Predicting unusual energy consumption events from smart home sensor network by data stream mining with misclassified recall. J Ambient Intell Humaniz Comput 9(4):1197–1221
48. Mukherjee A, Dey N, Kausar N et al (2019) A disaster management specific mobility model for flying ad-hoc network. In: Emergency and disaster management: concepts, methodologies, tools, and applications, IGI Global, pp 279–311
49. Das SK, Tripathi S, Burnwal AP (2015, February) Fuzzy based energy efficient multicast routing for ad-hoc network. In: Proceedings of the 2015 third international conference on computer, communication, control and information technology (C3IT), IEEE, pp 1–5
50. Das SK, Tripathi S (2015) Energy efficient routing protocol for manet based on vague set measurement technique. Procedia Comput Sci 58:348–355
51. Dey N, Ashour AS, Bhattacharyya S (2019). Applied nature-inspired computing: algorithms and case studies, pp 1–275, ISBN 978-981-13-9263-4
52. Dey N, Ashour A, Beagum S, Pistola D, Gospodinov M, Gospodinova E, Tavares J (2015) Parameter optimization for local polynomial approximation based intersection confidence interval filter using genetic algorithm: an application for brain MRI image de-noising. J Imaging 1(1):60–84
53. Chatterjee S, Sarkar S, Hore S, Dey N, Ashour AS, Shi F, Le DN (2017) Structural failure classification for reinforced concrete buildings using trained neural network based multi-objective genetic algorithm. Struct Eng Mech 63(4):429–438

Chapter 5
Fault Diagnosis in Wireless Sensor Networks Using a Neural Network Constructed by Deep Learning Technique

Meenakshi Panda, Bhabani Sankar Gouda and Trilochan Panigrahi

1 Introduction

Wireless sensor networks (WSNs) consist of tiny inexpensive low power sensor nodes with computing, processing and wireless communication capability. WSNs are popular due to their ease of deployment in any kind of environment and wide range of remote sensing applications starting from defence, terrestrial to underground water application [1, 2]. Now, WSN-based applications also developed in domains like healthcare, medical, industrial and home automation [3, 4]. Most recent design framework for WSN is provided in [5]. A sensor node equipped with one or more sensors (such as mechanical, thermal, biological, chemical, optical and magnetic sensors based on the application area on which sensor nodes are deployed), high power processors, memory, battery, radio for wireless transmission and an actuator. Each module performs a different task. WSN is running with limited battery power, short communication range to save power, low bandwidth because of limited channel in industrial scientific and medical (ISM) band of frequency and limited storage in each sensor node [6].

As sensor nodes are deployed in unattended and hostile environments. The coverage area of the sensor network can be improved by employing optimization algorithms like genetic algorithm, swarm optimization and cuckoo search [7]. Sensor nodes are supposed to be operated autonomously, robust and adaptive to the change

M. Panda · T. Panigrahi
National Institute of Technology Goa, Farmagudi, Goa, India
e-mail: meenakshi.nitrkl@gmail.com

T. Panigrahi
e-mail: tpanigrahi80@gmail.com

B. S. Gouda (✉)
National Institute of Science and Technology, Brahmapur, Odisha, India
e-mail: bhabani012@gmail.com

D. De et al. (eds.), *Nature Inspired Computing for Wireless Sensor Networks*, Springer Tracts in Nature-Inspired Computing,
https://doi.org/10.1007/978-981-15-2125-6_5

in environment. But the sensors are prone to have faults due to internal (problems in electrical circuits inside the sensor node, battery depletion or hostile tampering, etc.) or external (environmental degradation) reason. In fact, ageing is another problem for arbitrary behaviour of sensor node during operation [8].

Sometimes, the sensor nodes are unable to communicate when the transceiver module becomes faulty or the battery is completely drained. Such fault is known as hard fault [9, 10]. Sometimes, the behaviour of the sensor node is random which provides suspicious reading. The kind of fault is known as soft fault. But in real time, the sensor nodes generate faulty and fault free reading at different time instants arbitrarily [11, 12]. This intermittent behaviour in the sensor node is because of loose battery contacts, overheating of ICs, an impulsive noise from the sensors and environments. Different types of faults and fault diagnosis algorithms are described in [13].

In literature, statistical methods are used to find the soft or data faults in WSNs [14]. Intermittent faults of sensor nodes have been diagnosed by calculating the number of instants the sensor nodes produced faulty measurement in a specified period [15]. Recently, in [16], the authors have explained how the robust statistical method can be used to detect the intermittent faulty sensor nodes with a minimum number of repetitions. Distributed approach also used to do this. The disadvantage here is that each time the sensors to be diagnosed using the statistical methods which need more computational time and delay. The accuracy of the algorithm depends upon the methods used to detect the fault in each time and also the number of times the fault is diagnosed. Many authors have used hit and trial method to decide how many times the fault diagnosis to be done.

There are some disadvantages in the algorithms which use statistical method to detect the intermittent fault in WSNs. These algorithms follow repeated tests to find the intermittent faults, but not clear how many times the detection to be repeated. There some ambiguity in choosing the threshold values in the statistical methods. To overcome these issues in statistical methods to detect the intermittent faults in WSNs, soft computing [15] and machine learning approaches are used in literature [17, 18]. But the authors have used data from the sensors as an input to the neural network to detect the intermittent fault. This increases the size of neural network which leads to high computational complexity and delay in the network. Whereas, in this chapter, the features are extracted from the measured data of the sensor nodes. The features are and given as input to the neural network. This reduces the size of the network and can be used in a sensor node for the self-fault diagnosis. Further, deep learning approach has been used to train the neural network to improve performance.

Finally, the chapter is organized as follows. The work done for diagnosing sensor nodes using various machine learning approaches is given in Sect. 2. The system and data model is developed for intermittent fault diagnosis algorithm in Sect. 3. Problem formulation is described in Sect. 4. Feature selection is described in Sect. 5. Neural network for fault classification is given in Sect. 6. Performance of various neural network algorithms for different parameters is discussed in Sect. 7. In Sect. 8, the chapter is concluded.

2 Related Work

When sensor nodes become faulty, they cannot communicate with other members in the wireless network if they die. Sometimes, they may be still alive but produce incorrect data, and they may be unstable jumping between normal state and faulty state which is in general known as intermittent behaviour. To improve quality of the data and to increase the lifetime of WSN, many studies have focused on fault diagnosis. Fault diagnosis algorithms are categorized as centralized approaches, distributed approaches and hybrid approaches. Details survey on fault diagnosis is provided in [13].

An overview of the work done for intermittent fault diagnosis based on statistical test is briefly introduced first. Then soft computing and neural network approaches for fault diagnosis in WSN are presented.

2.1 Statistical Test-Based Intermittent Fault Diagnosis

In the early days, fault diagnosis of system and software is available. The similar kinds of algorithm are extended to the fault diagnosis of sensor nodes in WSN. Detection of faults in digital circuits where the concept of a fault pattern as intermittent is presented in [19] . In [11], a diagnosis procedure is presented which is to be repeated in discrete event systems [20]. A method for modelling intermittent faults and their resets in the context of discrete event system modes is proposed in [21]. However, these approaches are mainly used for the diagnosis of either a system or software. These methods are may not be feasible to apply to diagnose the nodes in WSNs as sensors are suffering from resource constraints.

They diagnose the intermittently faulty processor by using the comparison model as discussed in [22]. A threshold and count-based intermittent fault diagnosis algorithm is presented in [23] where a clear distinction between transient and intermittent faulty processor is described. Probabilistic-based fault diagnosis approach based on the remaining energy of the sensor node is given in [24]. Each sensor node exchanges message related to their remaining energy. This method does not give how to select minimum number of tests required detecting the fault which is essential for intermittent fault diagnosis. Choi et al. [25] proposed an adaptive fault detection algorithm to identify the transient and intermittently faulty sensors over the static network which closely follows [26]. Time redundancy method is proposed in [27]. It is assumed that every node has minimum three neighbouring nodes. This may not be always guaranteed for strongly sparse sensor networks. Later, soft computing methods are used for the intermittent fault diagnosis algorithm.

In the recent past, the extended Mobile Ad-hoc Network architecture is enhanced to smart phone and open source unmanned aerial vehicle (UAV) technology. Disaster aware mobility modelling for a Flying Ad-hoc Network infrastructure, where the UAV group is considered as nodes, is discussed in [28]. The impact of various parameters like UAV node attitude, geometric dilution precision of satellite, Global Positioning System (GPS) visibility and real-life atmospheric upon the mobility model is analyzed by the authors.

2.2 Soft Computing and Neural Network Approach for Fault Diagnosis

The use of machine learning is one of the most convenient approaches to detect faulty sensor nodes in WSNs. Support vector machines (SVMs) classification method is used for this purpose in [29]. SVM is used to define a decision function which is based on statistical learning theory. The disadvantage is, the decision function is executed at cluster heads to detect faulty sensor nodes in WSN. In [30], the authors have extended the SVM-based fault detection algorithm into distributed scenario. But these methods are not for detecting intermittent faulty sensor nodes in WSNs.

The most prominent evolutionary or heuristic approaches like neural networks [31, 32], perceptron neural network [33], multi-objective particle swarm optimization [15], genetic algorithm [34], back propagation neural network [35], support vector machines [36], etc. are applied to fault diagnosis in WSNs.

Particle swarm optimization algorithm is used to find the optimum parameters to diagnose the intermittent faults in sensor network. Multi-objective particle swarm optimization (MOPSO) algorithm is used to select parameters for intermittent fault detection [15]. Evolutionary algorithms are not computationally efficient if each of the sensor nodes runs the MOPSO algorithm and also not feasible for dynamic sensor network. Clustering-based intermittent fault diagnosis approach is proposed in [18] which is suitable for both sparse and dense sensor networks.

Regressional learning-based fault tolerance technique where hard, soft, intermittent and transient fault are identified [17]. To identify the hard fault, the neighbour coordination-based time out concept is used. Neighbour majority voting method is applied to detect the permanent soft, intermittent and transient faults. Then regressional learning method is used to calculate how many times a faulty sensor persists in the network. This method is neither self-diagnosis nor distributed in nature.

The major demerits of present approaches are the optimum threshold selection and diagnosing the system more accurately when both testing and tested nodes are faulty using comparison model. To overcome this situation, in this chapter, a robust statistical-based self-intermittent fault diagnosis protocol is established. This method is capable to generate an optimum threshold for testing the intermittently faulty sensor node which enhances the detection accuracy.

A neural network-based fault diagnosis approach has proposed for diagnosing the multi-processor system in [32]. This is a centralized approach in which they consider only two hidden layers. The activation function of hidden layer is Sigmoid function. Gaussian weights are used to connect the input units to the second layer units, which are used to classify inputs internally. The third-layer units are used to classify outputs internally, and the associations between output units and them are set up through implement supervised learning.

Elhadef et al. [33] have proposed a fault diagnosis approach using perceptron-based neural network. In this approach, asymmetric comparison model is used for giving input to the neural network. The disadvantage of this approach over [31] is the accuracy for detecting faulty node is less compared to [33].

Ji et al. [37] exploit the redundant or complementary information of multi-sensor in space or time to detect and isolate the faulty sensor nodes in WSN. The authors present a structure of three layers to detect and isolate multiple faults. The first layer is a state recognition network. It is composed of some modularity radial basis function neural network (RBFNN). The belief assignment of a sensor state is obtained by RBFNN with two inputs and one output. The two inputs are the data provided by sensor v_i and v_j. The output is $m_{ij}(\{OK_i, OK_j\})$. Here, $m_{ij}(OK)$ means both v_i and v_j are fault free. Each trained RBFNN is used as one model. The second layer is merging of the different frames of discernment. These frames of discernment are merged into a common frame of discernment by refinement operation. The third layer is evidence fusion and state decision.

Jabbari et al. [38] present the fault detection and isolation technique based on artificial neural network (ANN). This approach follows two phases namely residual generation and residual verification phases. Two separate ANN algorithms are considered for these phases respectively. This approach compares the measured data with network prediction and generates fault residuals. All the residuals are evaluated and analyzed. A residual is a signal that is used as a fault detector. Normally, the residual is considered to be zero (or small in a realistic case where the process is subjected to noise and the model is uncertain) in the fault free case and deviates significantly from zero when a fault occurs. For generating residuals, it considers generalized regression neural network architecture data approximation. In this phase, measurement residuals are generated by comparing measured data with network prediction. In the second phase, it uses a probabilistic neural network (PNN) for analyzing probable fault/failure conditions and fault/failure classification.

Moustapha and Selmic [39] introduced a neural network modelling approach for sensor node identification and fault detection in WSNs. The recurrent neural networks (RRNs) have the ability to capture and model the dynamic properties of nonlinear systems. In this approach, RRNs are used to model the sensor node, the node's dynamics and interconnecting with other sensor nodes. The RRN nodes have their own dynamics with interconnecting weights between the nodes similar to WSNs, and each sensor node has its own dynamics. The dynamic RRNs consist of a set of dynamic nodes that provide internal feedback to their own inputs. This is used to simulate a network of sensors. This approach assumes that there is one sensor per sensor node where the sensor nodes are viewed as small dynamic systems with memory-like features. The introduced ad-hoc RRN is analogous to WSN systems with confidence factors ($0 < CF_{ij} < 1$) between sensor nodes v_i and v_j. The confidence factor depends on the signal strength and the data quality in communication links between the nodes. The overall modelling process is divided into two phases such as the learning phase and the production phase. In the learning phase, the neural network adjusts its weights that correspond to the healthy and *N* faulty models. The production phase compares the current output of the sensor node with the output of the neural network. The difference between these two signals is the basis to detect a sensor's health status. Barron et al. [40] implement this approach on Moteiv's Tmote Sky platform with TinyOS operating system.

Swain et al. proposed a probabilistic neural network-based heterogeneous fault diagnosis algorithm [18]. This approach diagnoses the hard, soft, intermittent, transient faulty nodes present in the WSN. To identify the soft, intermittent and transient faults, analysis of variance method is used and neighbour coordination-based time out status register mechanism is used to identify the hard faulty nodes. After finding the different faulty nodes, they are classified into different categories based on the feed-forward probabilistic neural network. A fuzzy inference-based investigation scheme using kernelized feature space to diagnose the intermittent, transient and permanent soft faulty nodes proposed in WSN [41].

Many researchers have proposed neural network-based intermittent fault diagnosis in WSNs. Most of the authors have used sensor data for certain number of time instants as an input data vector to the neural network. By doing this, the size of the network is more which required more memory and computation time. In this book chapter, the statistical feature vector has been derived from the input data of the sensor and then used as input vector to the neural network. This process improves the performance of the diagnosis method and reduces the size of the network.

3 System Model

Description of wireless sensor network and intermittent fault model is given in this section. In network model, the topology of the sensor network and communication methods among the sensors is described. In fault model, the behaviour of the various kinds of faulty sensor nodes is presented.

3.1 Assumptions

The following assumptions are used for the development of the WSNs. These are as follows [14]

1. Sensor nodes are homogeneous in nature having uniform energy.
2. In WSNs, sensor nodes are able to send and receive the sensed data.
3. If a sensor node fails to communicate with neighbours, refereed as hard fault.
4. Network is static, i.e. the position of sensor nodes and the network topology remains same during the fault detection period.
5. The error in the communication link is taken care by the MAC layer.
6. Sensors communicate by using UDP/IP communication protocol.

3.2 Sensor Network Model

Let us consider N number of sensor nodes are randomly deployed in a square terrain of length R. In a network, each sensor node s_i, $i = 1, 2, 3 \ldots, N$ has known unique identifier and position $P_i(xc_i, yc_i)$, where $0 \leq xc_i \leq R$, $0 \leq yc_i \leq R$. In sensor network, s_i interacts with the neighbours and employs a one-to-many broadcast primitive in their basic transmission mode. Let T_r be the transmission range of s_i and which is assumed to be uniform for all the sensors. Sensor node lies within T_r of s_i at nth time instant is assumed to be connected. A sensor network is generally considered as a graph $G(S, C)$ in which S represents a set of sensors and C is the set of communication link among the sensors.

All the sensor nodes in WSNs are connected by a wireless link. Each sensor sends and receives message from the neighbours within a bounded time period in a synchronous WSNs. The MAC layer protocol used here for data communication of sensor nodes is IEEE 802.15.4. The sensor nodes are sending the data to the base station through multi-hop communication which is feasible through the immediate neighbours. Nodes can communicate with the neighbours by using multicast routing protocols presented in the literature [42]. In fact, routing is an important task in WSN. Because of the current demands, wired and wireless networks are combined together to fulfil the basic need of the customers. In this hybrid network, a geometric and linear programming approach based routing protocol is proposed in literature[43, 44]. Energy-efficient routes in Mobile Ad-hoc Networks (MANETs) are challenging tasks. It is duet to dynamic and mobile nodes are fitted with limited capacity of batteries in MANET. An energy-aware efficient routing protocol for MANET is proposed in [45]. Artificial intelligence-based routing is proposed in [46]. These routing algorithms are adaptive to the changes in the system.

3.3 Fault Model

When sensor nodes send their sensed data to central processor through multi-hop communication, error may occur due to faulty link. But as per the assumption, the links are fault free in WSNs. The error is taken care by the MAC layer of underlying sensor networks. But the sensor nodes are subjected to faults due to various reasons. In fact, the measured data of fault free sensors also noisy, but always lies within an acceptable range. When a sensor node becomes faulty, it gives arbitrary value where the error is beyond the acceptable range. As an example, the fault model is depicted in Fig. 1 where 50 sensors are deployed, and out of them, 15 (30% of total number of sensor nodes) sensor nodes are introduced as faulty. Let p be the probability of a sensor being intermittently faulty. The set of randomly chosen sensor nodes ($\lceil pN \rceil$ numbers of faulty sensor node) which are subject to failures is denoted as S_F. The set S_F contains both hard and soft faulty sensor nodes.

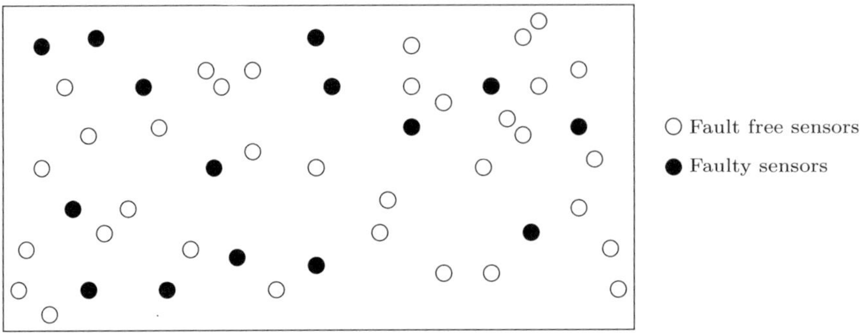

Fig. 1 A sensor network with fault free and faulty sensor nodes. There are 15 sensor nodes are introduced as faulty out of total 50 sensor nodes in the network

Each sensor node s_i observes the outcomes over some time period. A sequence of outcomes of a sensor node that satisfies the following assumptions:

- The measured data of a sensor node in each time instant has two possible outcomes either fault free or faulty.
- The measured data is temporarily and spatially independent, i.e. the outcome of a sensor node at one time instant has completely uncorrelated with the outcome at another time instant. Similarly, the data of each sensor node is independent to each other.
- Let us define, $\alpha \in [0, 1]$ is an intermittent fault probability of a sensor node. It can be described as, at each time instant, the probability of failure to provide actual data is α and the probability of sensor node able to provide data with acceptable noise is $1 - \alpha$.

Initially, it is assumed that the sensors in the network are fault free and supposed to be diagnosed with their fault status. Each of the sensors in the network is subjected to an intermittent fault during their course of action. Let p be the fault probability that a sensor is intermittently faulty which is same for every sensor.

3.4 Modelling of Sensor Data

Modelling of sensor data is an important task for fault diagnosis. The data measured by sensor nodes is typically erroneous or noisy. The error usually occurs due to the problems in hardware and data communication [47]. This becomes severe in adverse environmental conditions and also when battery is low. Here, the data transmission over ideal channels is assumed [48] which means no error occurs in the channel. In applications like parameter estimation [8], event boundary detection [49, 50] and most of the fault detection algorithms as well, the sensor observation is modelled as additive noise with the true value.

The outcome $x_k(n)$ of a senor node s_k at time instant n is the sum of actual value A of the measured physical parameter (e.g. the light intensity, temperature, humidity, etc.) and random error [47]. Thus, without losing the generality, the simplest data model of sensor nodes observation is given as [14]

$$x_k(n) = A + \upsilon_k(n), \quad n = 1, 2, \ldots, K \quad K = \frac{T}{\Delta T} \quad \text{and} \quad k = 1, 2, \ldots, N \tag{1}$$

where $\upsilon_k(n)$ is erroneous data at respective sensor nodes. The random error in the erroneous data is different that is having the different variance, but same mean [50, 51].

The sensor nodes are accumulating the observed data in a regular interval ΔT over the total time duration T. It is assumed that the erroneous data is temporally and spatially independent and has the same distribution function at each node. It follows that the observations $x_i(1), x_i(2), \ldots, x_i(K)$ are independent with common distribution function and can say that the $x_i(n)$'s are *i.i.d.*, i.e independent and identically distributed. A conventional way to represent well-behaved data, i.e. data without fault, is assumed F is a normal distribution with mean A and variance σ_i^2 which implies $F = \mathcal{N}(A, \sigma_i^2)$ [14].

In fact, the measurement is erroneous whether the sensor is faulty or fault free. But the only difference is its variance of the measurement error for a fault free sensor node is very less (nearly 10,000 times) compared to that faulty sensor node [50]. However, the scenario is completely different when a sensor node suffered from intermittent fault. Whereas the intermittent fault sensor nodes provide an arbitrary data for some time duration and behave as good in another time. Let α is the intermittent fault probability of a sensor node and N be the number of observations measured. Then the number of data are faulty is $N\alpha$ for a intermittent faulty sensor node. Data from sensor nodes with and without intermittent fault is given in Sect. 5.5.

4 Problem Formulation

Let us consider N number of sensor nodes are distributed in a geographical area for any remote sensing application. Let the data from a kth sensor s_k is represented as $x_k(t)$ defined in the closed interval $[0, T]$ where t is the time variable and T is the total duration of the signal. The signal $x_k(t)$ is sampled with frequency F_s. The sampling period $T_s = \frac{1}{f_s}$, that is the signals from the sensor are measured at a regular interval of time T_s. The discrete time signal is

$$x_s[n] = x(nT_s) \tag{2}$$

where $x_k[n]$ is sampled signal and n is the sampling index. The discrete time signal $x_k[n]$, where $n = 1, 2, \ldots, L$. The value of L is defined as

$$L = \left\lfloor \frac{T}{T_s} \right\rfloor \tag{3}$$

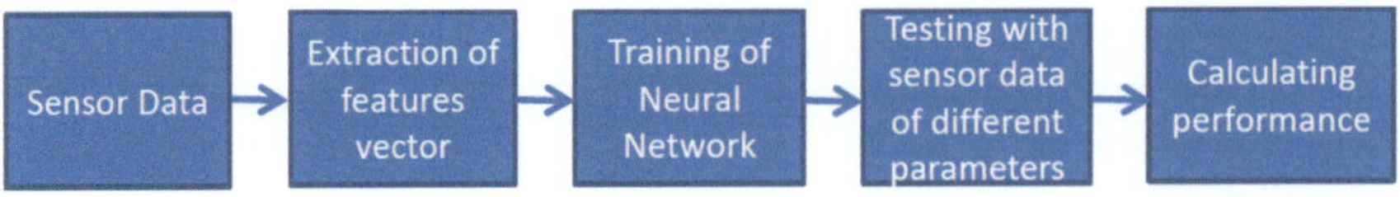

Fig. 2 Block diagram for intermittent fault diagnosis in sensor network by using neural network

The sensors data $x_k[n]$ is modelled as

$$x_k[n] = A + w_k[n], \quad k = 1, 2, \ldots, N \tag{4}$$

Now, the data $\{x_k[n]\}_{n=1}^{L}$ from each sensor k is used to find the kind of faults, a sensor suffers. The step by step process is given in Fig. 2.

5 Feature Selection

In machine learning and pattern recognition problems, features play a key role in extracting dominant component of input sequence data. Features represent the component that may be least or frequently occurring or can be derived from sensor's data. The importance of feature extraction is

- Reduce redundancy present in input data.
- Provides an accurate representation of input data.
- Reduce input data to neural network which leads to the reduction of network size and computational complexity.

On the other hand, feature extraction always adds extra computational burden to extract them; in fact, this can reduce the size of the neural network. We should also care the features which are distinct for different fault classes of sensor nodes. The features used in our problem of sensor node classification are given in the next section. Here, time domain features are used to classify the type of sensor node instead of frequency domain feature.

5.1 Mean

Statistical mean is the most common and easy implemented feature in the time domain. The mean $\mu(k)$ of the kth sensor node is defined as

$$\mu(k) = \frac{1}{L}\sum_{n=1}^{L} x_k[n], \quad k = 1, 2, \ldots, N \tag{5}$$

5.2 Standard Deviation (SD)

It is one of the features measured in time domain. The SD can be determined at each sensor node k as

$$\sigma(k) = \sqrt{\frac{1}{L-1}\sum_{n=1}^{L}(x_k[n] - \mu(k))^2}, \quad k = 1, 2, \ldots, N \tag{6}$$

5.3 Skewness and Kurtosis

Statistical analysis of sensor data is to characterize the location and variability of a data set available over time. This can be done by measuring skewness and kurtosis. A data set from the sensor is symmetric if it looks the same to the left and right of the centre point (usually mean). Skewness is a measure of symmetry, in other words, the lack of symmetry. It is also the measure of third-order cumulative, whereas kurtosis is a measure of whether the data is heavy-tailed or light-tailed relative to a normal distribution. That is, data sets with high kurtosis tend to have heavy tails or outliers. Data sets with low kurtosis tend to have light tails or lack of outliers [52].

5.3.1 Skewness

The skewness can be redetermined by using the following formula

$$skew(k) = \frac{\frac{1}{L}\sum_{n=1}^{L}(x_k[n] - \mu(k))^3}{\sigma_k^3}, \quad k = 1, 2, \ldots, N \tag{7}$$

The formula for skewness defined in (7) is referred to as the Fisher-Pearson coefficient of skewness. The adjusted Fisher-Pearson coefficient of skewness is given as

$$SKEW(k) = \frac{\sqrt{L(L-1)}}{L-2}\frac{\frac{1}{L}\sum_{n=1}^{L}(x_k[n] - \mu(k))^3}{\sigma_k^3}, \quad k = 1, 2, \ldots, N \tag{8}$$

This is an adjustment for inadequate sample size getting from each sensor node. The adjustment factor approaches 1 as the sample size L gets large.

If the data from a sensor node falls into a normal distribution which is most common, then the skewness is zero. In fact, for any symmetric data, the skewness is near zero. Negative values for the skewness indicate data that is skewed left which means that the left tail is long relative to the right tail. Similarly, positive values for the skewness indicate data that is skewed right. It means that the right tail is long relative to the left tail.

5.3.2 Kurtosis

The kurtosis can be defined as

$$KURT(k) = \frac{\frac{1}{L}\sum_{n=1}^{L}(x_k[n] - \mu(k))^4}{\sigma_k^4}, \quad k = 1, 2, \ldots, N \tag{9}$$

As per the definition, the kurtosis for a standard normal distribution is three. Thus, the definition of kurtosis can be modified as by subtracting 3:

$$KURT(k) = \frac{\frac{1}{L}\sum_{n=1}^{L}(x_k[n] - \mu(k))^4}{\sigma_k^4} - 3, \quad k = 1, 2, \ldots, N \tag{10}$$

As per the definition in (10), the standard normal distribution has a kurtosis of zero now. Then the positive kurtosis indicates a "heavy-tailed" distribution and negative kurtosis indicates a "light-tailed" distribution.

5.4 Mean Absolute Deviation (MAD)

The average of the absolute deviations of data points from their mean is known as MAD. The definition of MAD is

$$MAD(k) = \frac{1}{L}\sum_{n=1}^{L}|x_k[n] - \mu(k)|, \quad k = 1, 2, \ldots, N \tag{11}$$

5.5 Extracting the Features From Sensor Data—An Example

Here, the features extracted from the data of a sensor node are given. Let us consider, a sensor does the noisy measurement of actual temperature of 25 °C. The noise is assumed to be normal distributed with zero mean and variance $\sigma^2 = 0.1$ when the node is fault free. In the case of intermittent faulty sensor node, some measurements are very noisy. The noise for the faulty measurement is modelled as the mixture of very normal noise and very high (10^5 times the variance of normal distribution). The data generated for fault free and intermittent faulty node is plotted in Figs. 3 and 4. It has been found from the figure that, when the node is fault free, all the measured values are very close to the actual value 25. Whereas, when the node suffers intermittent fault, few observations are very noisy. In Fig. 4, seven data are deviated from the normal distribution out of total 50 measured observations. All the features defined before are provided in Table 1.

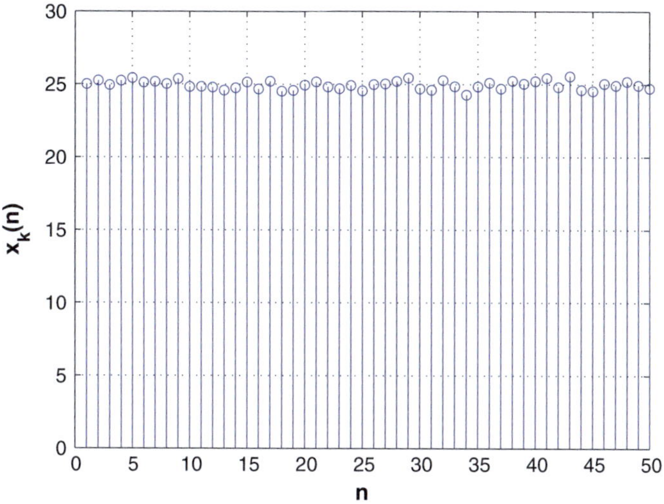

Fig. 3 Plot of data from a sensor node without intermittent fault

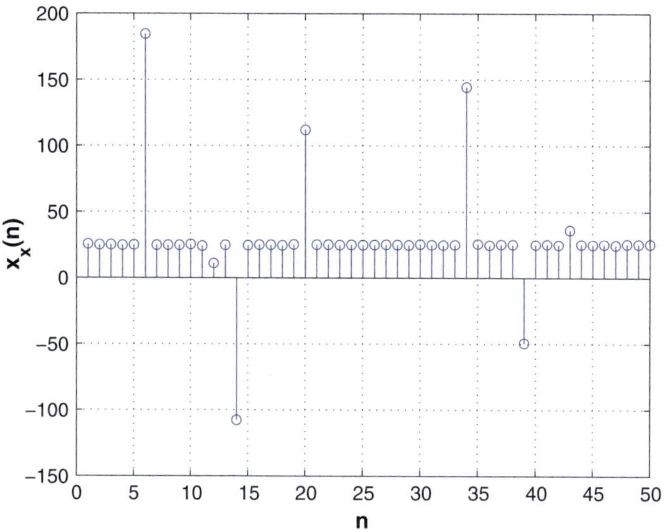

Fig. 4 Plot of data from a sensor node with intermittent fault

Table 1 Feature of the data shown in Figs. 3 and 4

Name of the feature	Feature value	
	Fault free	Faulty
Mean	24.955	28.18
MAD	0.23654	14.575
STD	0.28551	37.868
Skewness	−0.016009	1.1368
Kurtosis	−0.64994	9.2318

6 Neural Network with Deep Learning Algorithms For Intermittent Fault Detection of Sensor Nodes

Neural networks (NNs) are statistical learning algorithms in machine learning which are inspired by properties of the biological neural networks. They are used for a wide variety of tasks, from relatively simple classification problems to speech recognition and computer vision. NNs are implemented as a system of interconnected processing elements, sometimes called nodes (different from sensor nodes), which are functionally analogous with biological neurons. The connections between different nodes have numerical values, called weights, and by altering these values in a systematic way, the network is eventually able to approximate the desired function.

Each node in the neural network takes many inputs from other nodes and calculates a single output based on the inputs and the connection weights. This output is generally fed into another neuron and the process is repeating. In this way, one can easily envision the internal hierarchical structure of the artificial neural network, where neurons are organized into different layers. The input layer receives the inputs features vector and the output layer produces an output which is the prediction by the NN for the given input vector. The layers in between input and output layer are called hidden layers.

The neural network is able to approximate an arbitrary function by varying its weights in a systematic way which is known as training of the NN by using an algorithm. The weights are given random values initially and the network is trained in such a way to find the weight parameters that produce the desired output. First, the error is calculated by comparing the neural network output with the actual output and uses this error to adjust the weights of the network by following training algorithm.

Computationally expensive backpropagation phase has to be done for every input set in NN. Networks are usually very slow learners and need huge amount of computational power to produce desirable results. The computational complexity can be minimized by reducing the number of input data and also by using less complicated structure of the network. In our proposed method, we have used four statistical features as input instead of proving all sensor data to the NN.

6.1 Basic Neural Network Design

The basic neural networks such as the feed-forward net (FFNet), cascade forward net (CFNet), pattern net (Pnet) and fit net (FitNet) are very common in any computing tool like MATLAB or Python. Each of the networks can be trained by choosing one of the training algorithms. The architectures of neural networks are shown in Figs. 5, 6, 7 and 8. All the networks are designed with 1 input layer, 10 neurons in hidden layers and 1 output layer.

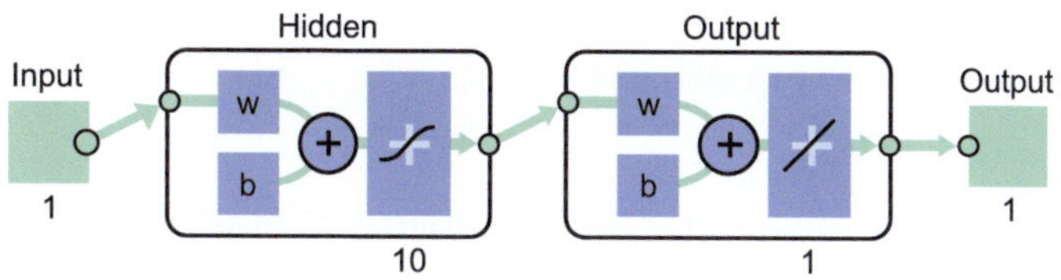

Fig. 5 A feed-forward network

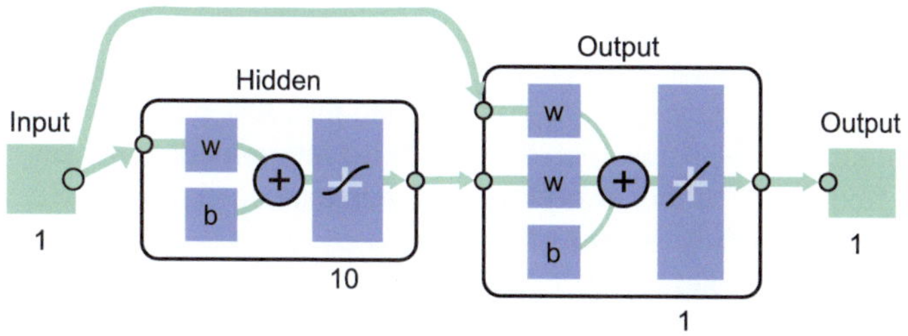

Fig. 6 A cascade forward network

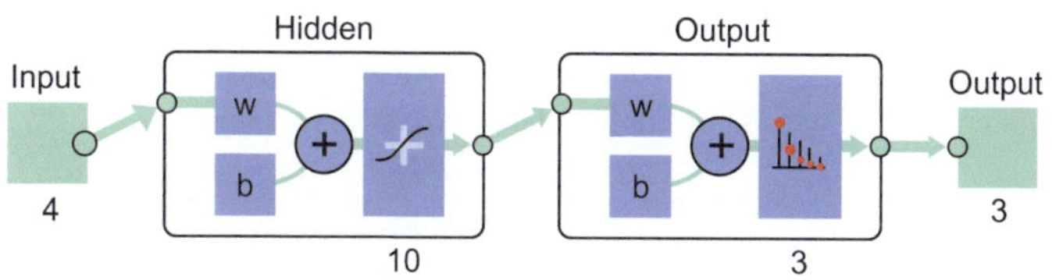

Fig. 7 A pattern network

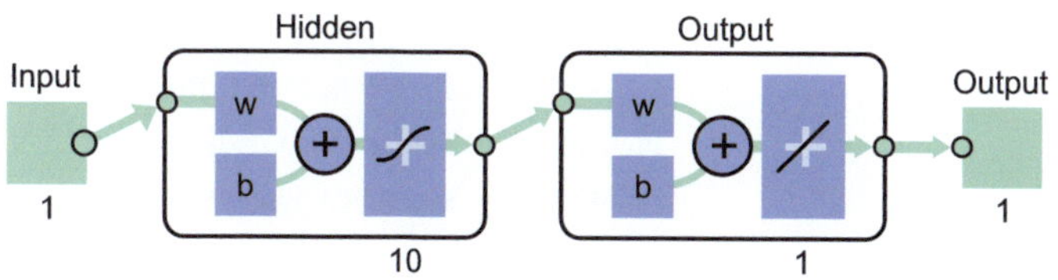

Fig. 8 A FitNet network similar to feed-forward network

An input is consisting of a column vector of four features such as mean, mean absolute deviation, standard deviation and skewness and produces output 1 if the node is faulty and 0 otherwise.

6.1.1 Choosing the Training Methods and Neural Networks

Different deep learning algorithms are available for training the neural network. The deep learning methods provided in MATLAB are listed in Table 2. All these deep learning algorithms are used to train three different kinds of neural networks for classification to know their corresponding mean square error performance. The available data set is used by employing the methods provided.

6.1.2 Performance Analysis for Different Neural Networks and Deep Learning Algorithms

The training methods as listed in Table 2 are used in different neural networks such as the FFNet, PNet and FitNet networks. The input to the network is four features of the recorded data from each sensor node in a sensor network of $N = 1024$ nodes. The number of data recorded is $L = 20$. It is assumed that 30% nodes are suffered from intermittent fault. As we know that the data from a intermittent faulty sensor node is suspicious frequently, but not always. For this comparison purpose, we have chosen the intermittent data fault is 10%. The input is common to all the networks. The performance of neural network is calculated as a mean squared error (MSE) of the difference between the training and validation set and detection accuracy (DA) during the validation. The number of epochs is also compared here. All these performances are listed in Table 3.

Table 2 Different training algorithms for multilayer shallow neural networks

Acronym	Function	Algorithm
LM	trainlm	Levenberg-Marquardt
BR	trainbr	Bayesian Regularization
BFG	trainbfg	BFGS Quasi-Newton
RP	trainrp	Resilient Backpropagation
SCG	trainscg	Scaled Conjugate Gradient
CGB	traincgb	Conjugate Gradient with Powell/Beale Restarts
CGF	traincgf	Fletcher-Powell Conjugate Gradient
CGP	traincgp	Polak-Ribire Conjugate Gradient
OSS	trainoss	One Step Secant
GDX	traingdx	Variable Learning Rate Gradient Descent
GDM	traingdm	Gradient Descent with Momentum
GD	traingd	Gradient Descent

Table 3 Performance comparison of neural networks with different training methods

Networks	Fitnet			Feedforward			Patternet		
Algorithm	DA	MSE	Epoch	DA	MSE	Epoch	DA	MSE	Epoch
BR	83.67	1.30E−13	158	81.33	1.41E−11	81	82.33	1.30E−09	32
LM	82.33	5.20E−11	70	82.67	5.41E−11	160	82	4.63E−09	18
BFG	87.33	4.80E−05	54	87.33	1.90E−05	51	81.67	1.00E−10	48
RP	86	4.17E−04	76	81.67	1.70E−04	70	81.67	1.00E−06	32
SCG	80.33	7.75E−05	84	83.33	2.49E−05	78	78.33	2.60E−07	23
CGB	80.33	7.01E−04	43	82.67	1.46E−04	45	81.00	2.22E−16	41
CGF	83.33	1.17E−04	57	84.32	7.19E−04	55	79.33	5.80E−16	37
CGP	81.67	2.14E−04	86	84	1.17E−04	40	81.67	1.73E−11	28
OSS	85.67	6.89E−04	62	87.33	4.83E−04	66	79.67	1.00E−13	28
GDX	81	4.75E−04	152	79	6.08E−04	161	82.33	2.76E−06	250
GDM	81.67	7.19E−03	1000	80	8.70E−03	1000	80.67	8.65E−03	1000
GD	82.67	4.17E−03	1000	81.33	2.84E−03	1000	80.67	1.75E−02	1000

Table 4 DA comparison of different training methods for FitNet neural networks and $L = 10$

Training	DA in percentage for different values of α							
Algorithm	0.05	0.10	0.15	0.20	0.25	0.30	0.35	0.40
BR	41.96	64.19	77.14	87.62	91.81	97.1	97.51	98.96
LM	41.55	64.19	75.88	86.32	91.22	96.13	97.15	98.31
OSS	43.66	66.45	79.74	89.58	93.24	97.42	97.51	99.65
RP	43.31	65.81	79.74	89.58	92.57	97.42	97.51	99.65
BFG	44.66	66.45	80.88	89.58	93.58	97.42	98.51	99.65

It is found while comparing all the neural networks with different deep learning algorithms in Tables 3 and 4 the FitNet and feed-forward neural networks provide comparable detection accuracy with the BFG, RP and OSS the training methods. Whereas the BR and LM provide best MSE performance as expected with little less DA. The patternet neural network does not provide good performance. Further, detection of intermittent faults of sensor nodes in sensor network has done by using FitNet neural network with LM, BR, RP, BFG and OSS training methods with different set of values to see which training algorithm better. Here, the number of data from the sensor nodes L is chosen 10. The detection accuracy for different values of intermittent fault probability is provided in Table 4.

In literature, a back propagation algorithm with gradient descent (GD) and geneticist algorithm (GA) training method is used to detect the faults [53]. It has been found that the DA is 85.5 and 87.77% when the training algorithms are GD and GA, respectively. But in this proposed neural network with deep learning algorithm, the DA can achieve more than 99%. The DA is increasing with increase in intermittent fault probability of sensor nodes which is unlikely happened in [53]. But the fact is the faulty node can easily detect faulty if, most of the time, the observation is

Table 5 DA comparison of different training methods for feed-forward neural networks and $L = 10$

Training	DA for different values of α network							
Algorithm	0.05	0.10	0.15	0.20	0.25	0.30	0.35	0.40
BR	42.96	65.48	78.78	88.27	92.23	97.42	97.51	99.65
LM	42.96	64.84	78.14	87.62	91.89	97.42	97.51	99.31
OSS	42.96	65.48	79.1	88.93	92.23	97.42	97.51	99.65
RP	42.96	65.16	78.14	88.27	92.23	97.1	97.51	99.65
BFG	42.96	64.19	77.17	87.3	91.89	96.77	97.15	99.39

suspicious. Thus, the neural network with deep learning algorithms is used and their results are provided here for the detection of intermittent faults in WSNs (Table 5).

The interesting fact is among the algorithms BR, LM, OSS, RP and BFG, the BFG provides a better detection accuracy performance over others. In fact, the LM and BR algorithms show less DA values over the other three algorithms. Therefore, the FitNet neural network with BFG training algorithm is chosen here for the fault diagnosis process over feed-forward and patternet networks and other training methods. The results are provided in the following section for different values of fault and network parameters.

7 Results and Discussions

In this section, the results of intermittent fault diagnosis algorithm by using neural network are provided. The performance is analyzed by using the most common fault diagnosis parameters in WSNs are detection accuracy, false positive rate and false alarm rate. These parameters are defined as [14].

- Detection accuracy (DA): It is defined as the ratio between number of intermittent faulty sensor nodes detected as faulty and the total number of intermittent faulty sensor nodes in the sensor network.
- False alarm rate (FAR): The FAR is defined as the ratio between number of fault free sensors detected as intermittent faulty and the total number of fault free sensor nodes in the network.
- False positive rate (FPR): The FPR is defined as the ratio of number of intermittent faulty sensor nodes detected as faulty free to total number of fault free sensors in the network. Simply, it is defined as $FPR = 1 - DA$.

Let us consider a sensor network with $N = 1024$ sensor nodes deployed in a geographical area randomly. The nodes are supposed to be prone to intermittent faults. Let us assume 30% sensor nodes are intermittent faulty with different intermittent fault probability α. The faulty nodes are introduced randomly. In the analysis, the α value varies from 0.05 to 0.5 with the step size of 0.05. The detection accuracy increases with the increase in α. Another parameter on which the performance depends is number of data recorded for testing the fault status of senor node L. The simulation results are provided for $L = 10{-}30$ with the step size of 5. In the results, it is reflected that the performance also improves with increase in the value of L.

The data is generated as per the data model given in (1). The actual value is considered as 25. But the measurement is always noisy. Noise is modelled as normalized Gaussian with variance of $\sigma^2 = 0.01$ which common to all the sensor nodes. Whereas when as sensor node suffers intermittent fault, the measurement noise is very hing and casing it suspicious that too not always. Few measurements which depend on intermittent faulty probability are supposed to be very high noisy with variance 10000 more than actual back ground Gaussian noise. Then the statistical features are determined from the data of each sensor node which will be the input to the neural network.

Initially, data is generated for training which contains data for all the combinations of L and α. While training the FitNet neural network with the FBG algorithm, 70% data are used for training, testing and validation uses 15% each randomly. The training performance is plotted in Fig. 9. While training the network with deep learning algorithm, target MSE is archived in 61 epochs. It is also found that the testing and validation results are matching, which shows that the network is trained properly. Therefore, the same network is used to generate the following results with different network and fault parameters.

After training the neural network properly, the performance is analyzed with different values of L and α. The overall performance of the diagnosis algorithm using FitNet neural network with BFG training method is plotted in Figs. 10, 11 and 12. It is found that the DA performance increases with the increase in the value of

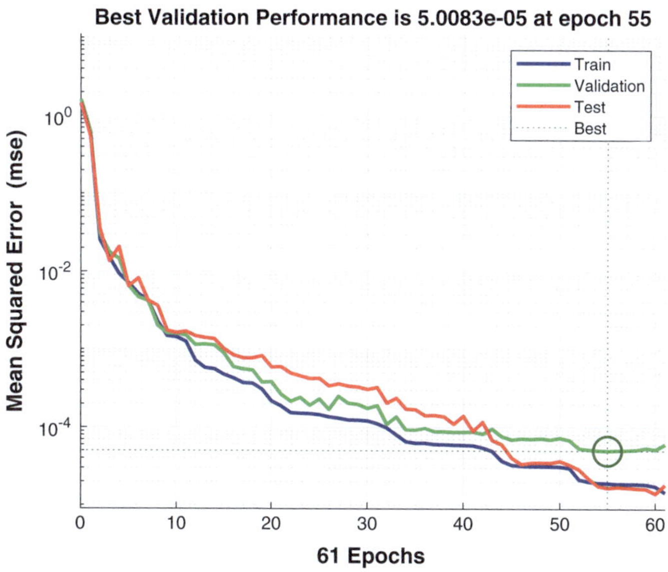

Fig. 9 Training performance of FitNet network with FBG algorithm

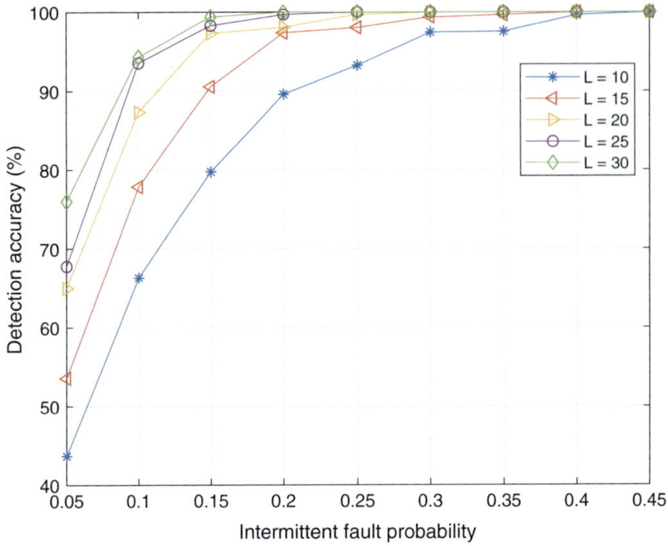

Fig. 10 DA versus α for different values of L for the neural network-based fault diagnosis method for different intermittent fault probabilities α

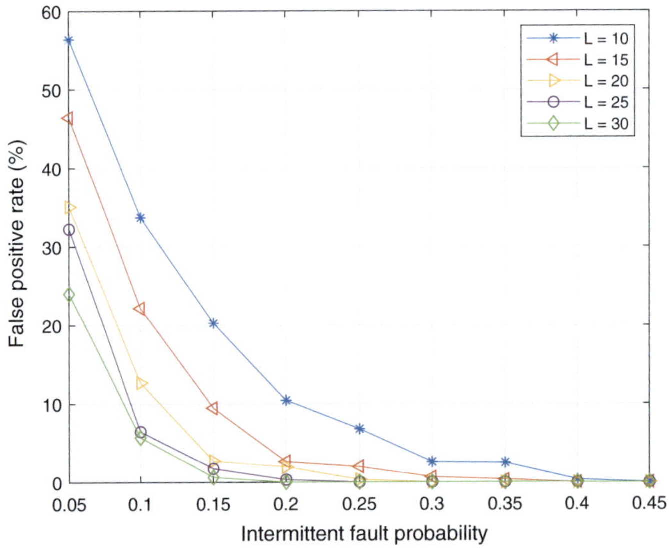

Fig. 11 FPR versus α for different values of L for the neural network-based fault diagnosis method for different intermittent fault probabilities α

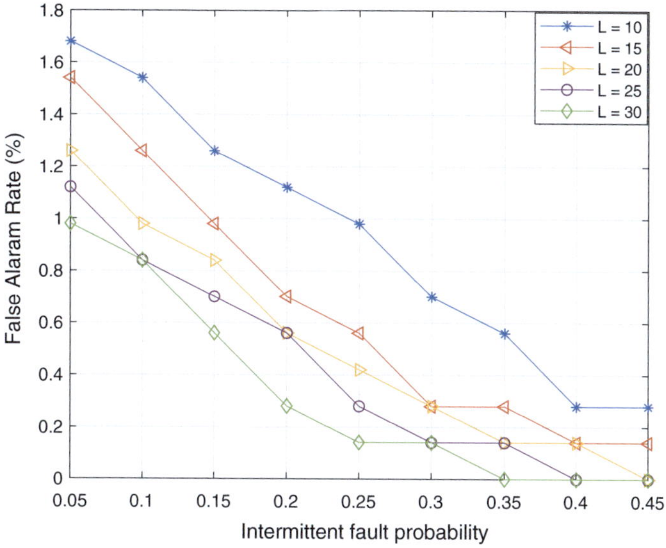

Fig. 12 FAR performance of the neural network-based fault diagnosis method for different intermittent fault probabilities α and data length L

intermittent fault probability α in Fig. 10. It is because, when the α value increases, the number of suspicious observation increases which leads to the deviation of its statistical parameters from the normal values (when the node is fault free). Then it is not difficult for the neural network to identify the intermittent fault of the sensor node in the network. When the α value is less than 0.2, the deep neural network is unable to detect the faulty node.
Even though when L value is large, the DA is less for small value of α. Thus, it is observed that the DA is quite good when the α value is more than 0.3 and becomes 1 when it is more than 0.45 (the L value is more than 10). But by using statistical method, even though the α value is more than 0.6, the DA is not achieved 99% [16]. The conventional neural network-based fault diagnosis algorithm provides very less DA performance which is given in [53].

It is quite obvious that when the DA increases, the FPR decreases which is observed in Fig. 11. When α is less, the deviation of the feature due to wrong observation may not be sufficient to detect the node is faulty. Thus, the faulty node may be detected as fault free. On the other hand, the FAR decreases with increase in the value of α for constant value for L. For small value of α, the feature from a good node may deviate from its normal value due to noisy observation. The FAR may decrease by two ways. One way is the minimization of the noise for good observation. This can be possible with high precision sensors. The noise introduced during the transmission of data from node to the sink or base station is controlled by the MAC layer. The second way of minimizing FAR is to pass the output of the neural network through proper threshold. The neural network may not provide exact 0 or 1 for the fault free and faulty node. Thus, appropriate thresholding is indeed for detection of faulty sensor nodes.

The value of L is also impacted on the performance of the algorithm. At constant α, the performance of diagnosis method improves with increase in L values. But the computational complexity and delay also increase. Therefore, the objective of any fault diagnosis algorithm is to detect the intermittent fault with as much as less number of data points. From Fig. 10, it has been observed that the DA performance is nearly 100% when $L = 30$ and the α is just 0.2.

8 Conclusion

A fault diagnosis method to diagnose the intermittent faulty sensor nodes in wireless sensor networks using neural network with deep learning is presented in this chapter. The data from sensor nodes is accumulated first and then four statistical features such as mean, standard deviation, mean absolute standard deviation and kurtosis are determined. These features are used as input data vector to the neural network. The performance of diagnosis method is determined by detection accuracy (DA), false alarm rate (FAR) and false positive rate (FPR). It is found in our study that the fit net neural network with BFG training algorithm provides best performance over other neural networks and training methods by providing best DA and less FAR and FPR as well. Simulation results show that, if 30 numbers of data from sensor used and the intermittent fault probability is more than 0.25, the faulty sensor node is detected with 100% detection accuracy. In future, different neural networks like convolutional neural network (CNN) can be used with recent developed deep learning algorithms. Features in time and frequency domain may be used to improve the performance of when intermittent fault probability and a number of accumulated data are less. Further, the method to correct the wrong data may be developed.

References

1. Akyildiz IF, Su W, Sankarasubramaniam Y, Cayirci E (2002) Wireless sensor networks: a survey. Sci Direct Trans Comput Netw 38(4):393–422
2. Yick J, Mukherjee B, Ghosal D (2008) Wireless sensor network survey. Comput Netw 52(12):2292–2330
3. Dey N, Ashour AS, Shi F, Fong SJ, Sherratt RS (2017) Developing residential wireless sensor networks for ECG healthcare monitoring. IEEE Trans Consumer Electron 63(4):442–449
4. Elhayatmy G, Dey N, Ashour AS (2018) Internet of things based wireless body area network in healthcare. In: Dey N, Hassanien AE, Bhatt C, Ashour AS, Satapathy SC (eds) Internet of things and big data analytics toward next-generation intelligence. Springer, Cham, pp 3–20
5. Das SK, Samanta S, Dey N, Kumar R (eds) (2020) Design frameworks for wireless networks. Lecture notes in networks and systems. Springer

6. Yuan H, Zhao X, Yu L (2015) A distributed Bayesian algorithm for data fault detection in wireless sensor networks. In: 2015 International conference on information networking (ICOIN). IEEE, pp 63–68
7. Binh HTT, Hanh NT, Quan LV, Dey N (2018) Improved cuckoo search and chaotic flower pollination optimization algorithm for maximizing area coverage in wireless sensor networks. Neural Comput Appl 30(7):2305–2317
8. Panigrahi T, Panda M, Panda G (2016) Fault tolerant distributed estimation in wireless sensor networks. J Netw Comput Appl 69:27–39
9. Nandi M, Dewanji A, Roy BK, Sarkar S (2014) Model selection approach for distributed fault detection in wireless sensor networks. Int J Distrib Sens Netw 2014(48234):1–12
10. Yu M, Mokhtar H, Merabti M (2007) Fault management in wireless sensor networks. IEEE Wirel Commun 14(6):13–19
11. Sampath M, Sengupta R, Lafortune S, Sinnamohideen K, Teneketzis D (1995) Diagnosability of discrete-event systems. IEEE Trans Autom Control 40(9):1555–1575
12. Ssu K-F, Chou C-H, Jiau HC, Hu WT (2006) Detection and diagnosis of data inconsistency failures in wireless sensor networks. Comput Netw 50:1247–1260
13. Zhang Z, Mehmood A, Shu L, Huo Z, Zhang Y, Mukherjee M (2018) A survey on fault diagnosis in wireless sensor networks. IEEE Access 6(2):11349–11364
14. Panda M, Khilar PM (2015) Distributed self fault diagnosis algorithm for large scale wireless sensor networks using modified three sigma edit test. Ad Hoc Netw 25, Part A(0):170–184
15. Mahapatro A, Khilar PM (2013) Detection and diagnosis of node failure in wireless sensor networks: a multi objective optimization approach. Swarm Evol Comput 13:74–84. https://doi.org/10.1016/j.swevo.2013.05.004
16. Panda M, Gouda B, Panigrahi T (2020) Distributed online fault diagnosis in wireless sensor networks. In: Das SK, Samanta S, Dey N, Kumar R (eds) Design frameworks for wireless networks. Lecture notes in networks and systems series. Springer, Singapore, pp 197–221
17. Swain RR, Khilar PM, Dash T (2018a) Fault diagnosis and its prediction in wireless sensor networks using regressional learning to achieve fault tolerance. Int J Commun Sys 31(14):e3769
18. Swain RR, Khilar PM, Bhoi S (2018b) Heterogeneous fault diagnosis for wireless sensor networks. Ad Hoc Netw 69:15–37
19. Breuer MA (1973) Testing for intermittent faults in digital circuits. IEEE Trans Comput 22(3):241–246
20. Jiang S, Kumar R (2006) Diagnosis of repeated failures for discrete event systems with linear-time temporal-logic specifications. IEEE Trans Autom Sci Eng 3(1):47–59
21. Contant O, Lafortune S, Teneketzis D (2004) Diagnosis of intermittent failures. Discrete Event Dyn Syst: Theory Appl 14(2):171–202
22. Malek M (1980) A comparison connection assignment for diagnosis of multiprocessor systems. In: Proceedings of the 7th annual symposium on computer architecture, ISCA'80, New York, NY, USA. ACM, pp 31–36
23. Bondavalli A, Chiaradonna S, di Giandomenico F, Grandoni F (2000) Threshold-based mechanisms to discriminate transient from intermittent faults. IEEE Trans Comput 49(3):230–245
24. Khilar PM, Mahapatra S (2007) Intermittent fault diagnosis in wireless sensor networks. In: 10th International conference on information technology (ICIT 2007), pp 145–147
25. Choi JY, Yim SJ, Huh JJ, Choi YH (2009) A distributed adaptive scheme for detecting faults in wireless sensor networks. WSEASE Trans Commun 8(2):269–278
26. Lee MH, Choi YH (2008) Fault detection of wireless sensor networks. Comput Commun 31(14):3469–3475
27. Xu X, Chen W, Wan J, Yu R (2008) Distributed fault diagnosis of wireless sensor networks. In: 11th IEEE international conference on communication technology, 2008. ICCT 2008, pp 148–151
28. Dey N, Mukherjee A, Kausar N, Ashour AS, Taiar R, Hassanien AF (2016) A disaster management specific mobility model for flying ad-hoc network. Int J Rough Sets Data Anal 3(3):72–103
29. Zidi S, Moulahi T, Alaya B (2018) Fault detection in wireless sensor networks through SVM classifier. IEEE Sens J 18(1):340–347

30. Yong C, Qiuyue L, Jun W, Shaohua W, Tariq U (2018) Distributed fault detection for wireless sensor networks based on support vector regression. Wirel Commun Mobile Comput
31. Mourad E, Nayak A (2012) Comparison-based system-level fault diagnosis: a neural network approach. IEEE Trans Parallel Distrib Syst 23(6):1047–1059
32. He JZ, Zhou ZH, Yin XR Chen SF (2000) Using neural networks for fault diagnosis. In: Proceedings of the IEEE-INNS-ENNS international joint conference on neural networks, 2000. IJCNN 2000, vol 5, pp 217–220
33. Elhadef M, Nayak A (2009a) Efficient symmetric comparison-based self-diagnosis using backpropagation artificial neural networks. In: 2009 IEEE 28th international performance computing and communications conference (IPCCC), pp 264–271
34. Elhadef M, Ayeb B (2001) Efficient comparison-based fault diagnosis of multiprocessor systems using an evolutionary approach. In: Proceedings 15th international parallel and distributed processing symposium, pp 1–6
35. Elhadef M, Nayak A (2009b) Efficient symmetric comparison-based self-diagnosis using backpropagation artificial neural networks. In: 2009 IEEE 28th international performance computing and communications conference (IPCCC), pp 264–271
36. Yuan S, Chu F (2007) Fault diagnosis based on support vector machines with parameter optimisation by artificial immunisation algorithm. Sci Direct J Mech Syst Sig Process 21(3):1318–1330
37. Ji Z, Bing-shu W, Yong-guang M, Rong-hua Z, Jian D (2006) Fault diagnosis of sensor network using information fusion defined on different reference sets. In: International conference on radar, pp 1–5
38. Jabbari A, Jedermann R, Lang W (2007) Application of computational intelligence for sensor fault detection and isolation. In: World academy of science, engineering and technology, pp 265–270
39. Moustapha AI, Selmic RR (2008) Wireless sensor network modeling using modified recurrent neural networks: application to fault detection. IEEE Trans Instrum Measur 57(5):981–988
40. Barron JW, Moustapha AI, Selmic RR (2008) Real-time implementation of fault detection in wireless sensor networks using neural networks. In: Fifth international conference on information technology: new generations, pp 378–383
41. Swain RR, Dash T, Khilar PM (2019) Investigation of RBF kernelized ANFIS for fault diagnosis in wireless sensor networks. In: Computational intelligence: theories, applications and future directions, vol II. Springer, pp 253–264
42. Das SK, Yadav AK, Tripathi S (2017) IE2M: design of intellectual energy efficient multicast routing protocol for ad-hoc network. Peer-to-Peer Netw Appl 10(3):670–687
43. Das SK, Tripathi S (2018a) Adaptive and intelligent energy efficient routing for transparent heterogeneous ad-hoc network by fusion of game theory and linear programming. Appl Intell 48(7):1825–1845
44. Dash SK, Tripathi S (2019) Energy efficient routing formation algorithm for hybrid ad-hoc network: a geometric programming approach. Peer-to-Peer Netw Appl 12(1):102–128 (Springer)
45. Das SK, Tripathi S (2018b) Intelligent energy-aware efficient routing for MANET. Wirel Netw 24(4):1139–1159 (Springer)
46. Das SK, Tripathi S (2017) Energy efficient routing formation technique for hybrid ad hoc network using fusion of artificial intelligence techniques. Int J Commun Syst 30(16):33–40 (Wiley)
47. Wang N, Wang J, Chen X (1916) A trust-based formal model for fault detection in wireless sensor networks. J Sens 19(8):2019
48. Tsang-Yi W, Li-Yuan C, Pei-Yin C (2009) A collaborative sensor-fault detection scheme for robust distributed estimation in sensor networks. IEEE Trans Commun 57(10):3045–3058
49. Tsang-Yi W, Qi C (2008) Collaborative event-region and boundary-region detections in wireless sensor networks. IEEE Trans Sig Process 56(6):2547–2561
50. Krishnamachari B, Iyenger S (2004) Distributed Bayesian algorithm for fault tolerant event region detection in wireless sensor networks. IEEE Trans Parallel Distrib Syst 24(8):1525–1534

51. Mahapatro A, Panda AK (2014) Choice of detection parameters on fault detection in wireless sensor networks: a multiobjective optimization approach. Wirel Pers Commun 78(1): 649–669. ISSN 0929-6212
52. Altn C, Er O (2016) Comparison of different time and frequency domain feature extraction methods on elbow gestures EMG. Eur J Interdiscip Stud 2(3):35–44
53. Swain RR, Khilar PM (2017) Composite fault diagnosis in wireless sensor networks using neural networks. Wirel Pers Commun 95(3):2507–2548

Chapter 6
Immune Inspired Fault Diagnosis in Wireless Sensor Network

Santoshinee Mohapatra and Pabitra Mohan Khilar

1 Introduction

As the name suggests WSN is a class of network, where the nodes are sensor n-odes which have the capability to sense the physical phenomena that occur around them. These sensing can be of different types like a particular sensor node can sense temperature, pressure or humidity. WSN has gained popularity over a decade now. They are very popular because of the types of application they can be used for tracking an object in a particular terrain, for agriculture, medical purposes [1–3], and so on. Sensor nodes have short transmission range due to which data is transmitted to the sink node through some intermediate nodes, which is also called multi-hop communication. Figure 1 illustrates the wireless sensor network.

A sensor node may fail due to various issues such as security, limited amount of energy, hardware problem, and so on [4]. To handle this, various energy-efficient routing protocols [5–8] and encryption technique for secure communication [9] have been developed. Various design and implementation of WSN is described in [10].

Since the sensor nodes are deployed in harsh environment they are subjected to different kinds of faults. Normal sensor nodes work perfectly in ideal environment condition. The behavioral fault of WSN is of two types, such as hard and soft fault. Hard faulty nodes are simply those unable to perform an operation, this could be because of power failure or environmental conditions. Soft faulty nodes exhibit features of failed nodes, but they send false data which will be a threat to the integrity of the network [11]. An overview of faulty and fault-free nodes are given in Fig. 2.

Different algorithms are there to diagnose the faulty nodes in the literature such as neural network [11], statistical method [12], neighboring co-ordination [13], and comparison [14]. In spite of the fact that the comparison-based approach is the most practical one, message complexity is too high. For example, PMC model [15], MM

S. Mohapatra (✉) · P. M. Khilar
Department of Computer Science and Engineering, National Institute of Technology Rourkela, Rourkela, Odisha 769008, India
e-mail: santoshinee88@gmail.com

P. M. Khilar
e-mail: pmkhilar@nitrkl.ac.in

D. De et al. (eds.), *Nature Inspired Computing for Wireless Sensor Networks*, Springer Tracts in Nature-Inspired Computing,
https://doi.org/10.1007/978-981-15-2125-6_6

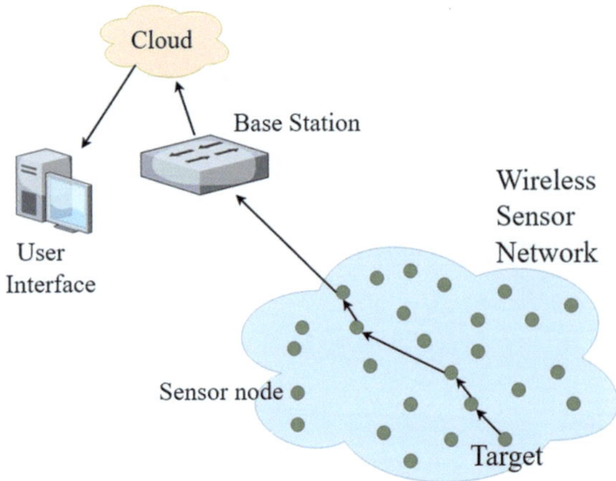

Fig. 1 Wireless sensor network

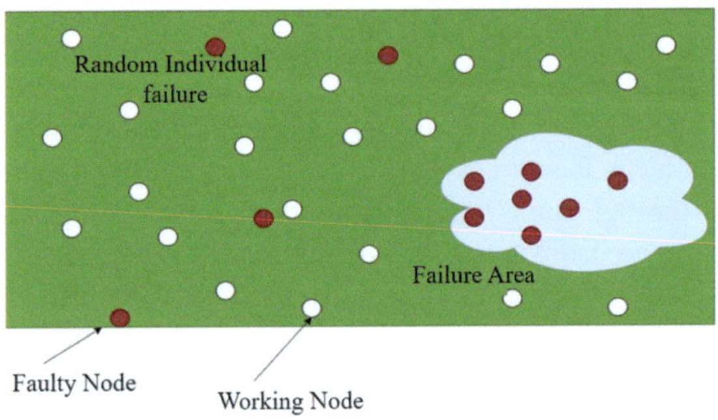

Fig. 2 An overview of faulty and fault-free nodes

model [16], and MM* model [17] need a test, in which the tasks are compared several times, so exchanges of messages is too high which ultimately consumed high energy. This method does not suit as, the WSN is battery powered followed by a wireless communication. The statistical method and neighboring co-ordination-based approaches depend on the number of nodes. The estimation of result is not proper if the network is sparse, as it has the least number of nodes. Neural network approaches mainly depend on data sets. It will give better results if the data set is good. Inspired by the principle and strategies of human immune system (HIS), we can relate the operation of the immune system of human being in the diagnosis process of WSN.

There is a greater interest seen among scientists and researchers in developing biologically inspired algorithms in the past. Artificial immune system (AIS) is considered as one of the most popular approaches due to its principle [18]. AIS is influenced from the principle of HIS, which can expertly save our bodies from bacteria and viruses [19]. The AIS is widely used in anomaly detection [20], pattern recognition [21], computer security [22], and fault detection [23]. In this chapter, we

Table 1 List of notations and their meaning

Notation	Meaning
WSN	Wireless sensor network
AIS	Artificial immune system
ACO	Ant colony optimization
PSO	Particle swarm optimization
ABC	Artificial bee colony
GA	Genetic algorithm
HIS	Human immune system
APC	Antigen-presenting cell
WBC	White blood cell
MHC	Major histocompatibility complex
CSA	Clonal selection algorithm
NSA	Negative selection algorithm
DCA	Dendritic cell algorithm
INA	Immune network algorithm
DA	Detection accuracy
FAR	False alarm rate
FDI	Fault detection and isolation
RBF	Radial basis function
AINC	Artificial immune network classification

have discussed the overall view of biological immune system and various approaches of AIS that can be applied to the fault diagnosis of WSN. Different notations used in this chapter are listed in Table 1.

1.1 Motivation

The major drawbacks of traditional methods are high message complexity and high energy consumption. In order to overcome these drawbacks, we have used AIS because of its adaptive nature. It has the capacity to recognize specific faults and memorize them for future response. Artificial immune system uses the concept of human immune system and utilizes them for the computational problems. The primary role of the immune system is to protect the body from the foreign pathogens such as bacteria and viruses. In the same way, this concept can be used in the fault diagnosis problem in wireless sensor network.

1.2 Contribution

In this chapter, we have discussed the overall view of biological immune system and various approaches of AIS that can be applied to the fault diagnosis of WSN. We have also discussed various applications of AIS.

The organization of the chapter is as follows. Section 2 presents a brief overview of the biological immune system. Various approaches of AIS for fault diagnosis are discussed in Sect. 3. The applications of AIS are described in Sect. 4. Finally, the chapter concludes in Sect. 5.

2 Biological Immune System: An Overview

The immune system's function is to protect our body from being attacked by various pathogens such as viruses and bacteria. There are two types of immune system, i.e., (1) innate immune system and (2) adaptive immune system. Our body is protected and completely covered with so many regions so that the invaders cannot enter at first and those are known as barriers [24]. For example, skin, which is the largest organ of our body and it covers the whole body. So, if any of the bacteria or virus enters the body, they need to evade our skin by means of any injury or any vector like a mosquito. Figure 3 depicts the biological immune system.

At the primary level, it is the most defensive part and besides that we have epithelial cells in the gut, intestine, and mucus layers. The gut is filled with so many good bacteria which resides in our body and produce vitamin k which helps us to fight against those pathogens. The pathogens can take entry only after breaking the barrier. There are different cells present throughout our body that are present in our bloodstream and different tissue layers, and the job of those cells is that if there is any foreign pathogen invades, they will directly kill them, which is known as innate immune system.

Some part is always present in our body to fight against infection, e.g., barriers, and another thing is complement system, which is not cells, but there are chemical

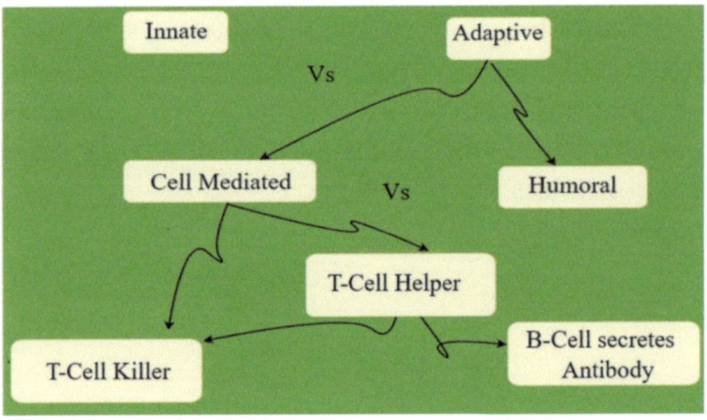

Fig. 3 Biological immune system

factors that are present in our body. They will target specific pathogen, and as a result of this, they will create pores in bacteria which ultimately kill them. These barriers and complement systems are always present in our body no matter if there are any pathogens present or not. There is another mode of the immune system that is not always on, but when the bacteria enter, it will develop very rapidly. For example, neutrophils and antigen-presenting cells (APCs) are cells which are present, but they will activate when there is any invading agent present in our body. Neutrophil is a part of white blood cell (WBC) and all those APCs are from white blood cells and some lymphocyte cells. Dendritic cells and macrophages are antigen-presenting cells that are always present, but they will become activated when there is any pathogen enters in our body. Neutrophils, dendritic cells, and macrophages are phagocytes which engulf other cells or bacteria and destroy them. They can present the antigen by destroying it that is why it is called antigen-presenting cells. Figure 4 shows the multiple layers of the immune system.

The innate immune system does not know the type of bacteria that enter into the body. They only concerned about the destruction of the bacteria or pathogen. Another type of immune system is an adaptive immune system which takes longer time to develop but more specific. When bacteria will start producing different pathogenic materials, which will break down the immune system and the innate immune system is not enough to fight against the bacteria then there is a need to have a more specific response to kill that bacteria. In case of adaptive immune system, immunity is developed once a pathogen enters and showcases its properties.

Pathogens have different types of antigens and once our body sees those antigens they will start producing specific antibodies against those antigens to specifically destroy those cells. The antibody response will very specific though it will take longer time to develop that is why it is also called delayed response. This delayed response produces some cells known as lymphocytes. There are two types of cells that are produced in lymphocytes, they are T cells and B cells. T cells are helping other cells, for example, T cells activate B cells. Once B cells are active they will produce antibodies. T cells can divide into two types one is a helper T cell which helps other cells to develop and the other is a killer T cell which kills after finding specific target. There are signaling processes between innate and adaptive immune system. There are two factors that are involved in the immune system one is cellular type factors

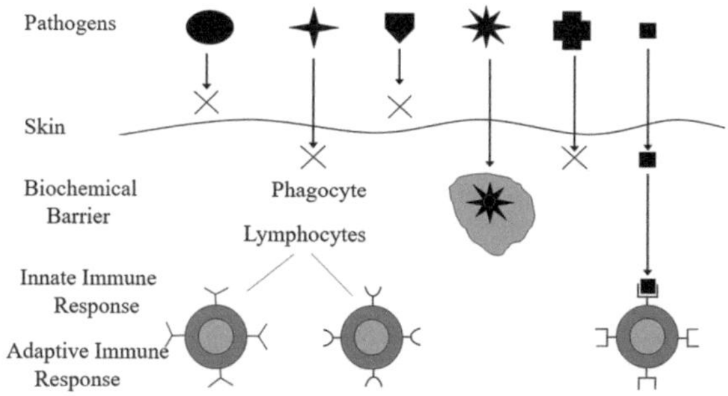

Fig. 4 Multiple layers of immune system (adapted from [25])

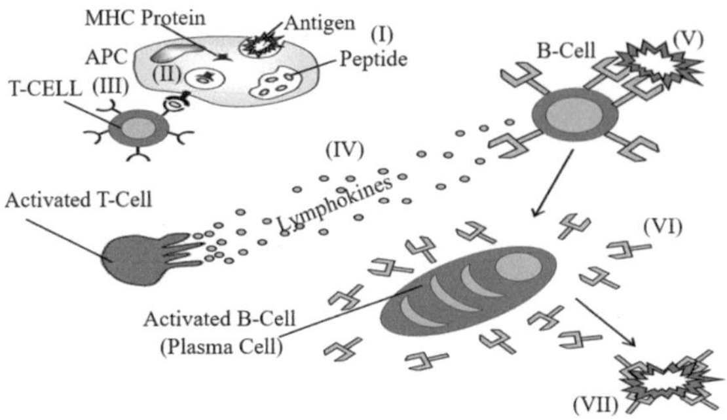

Fig. 5 Mechanism of immune system (taken from [25])

and another one is chemical or protein type factors. The cell type factors mediated system of immunity is known as cell-mediated immunity, e.g., APCs, natural killer cells, T lymphocytes and neutrophils. The complement system, antibodies is protein factors which are known as humoral immune system, because they are not the part of the cell in our bloodstream rather they are found in blood plasma.

The mechanism of the immune system is shown in Fig. 5. I-II show pathogens are entered into the body and activation of T cell occurs in III, which activates B cells in IV, antigens are matched in V, antibodies are produced in VI and in VII antigen destroy it [25].

The cells of our immune system protect us against infection. The most important thing of the immune system is not only to fight against infection, but also to distinguish between what is foreign and what is self. First, they need to identify fragments of the bacterial body or the fragment of a virus that can be detected by cells of the immune system. Once they detect those regions as foreign, then they can go against them to develop immune system, but if it fails, then it can hamper our body. There are some cells present in our tissue all the time and they are circulating in the bloodstream so that if any infection occurs they can find out. For example, there is a damage in our skin, and through that, bacteria enter into the tissue associated with skin. Macrophage and dendritic cells are phagocytic cells that stay in the nearby adjacent tissue where the bacteria enter. At this stage, this macrophage will engulf these bacteria inside by the process known as phagocytosis. Once they engulf these bacteria by phagocytosis, they will start releasing certain chemical factors in this nearby tissue which is known as cytokines. As the cytokine releases to the nearby tissue, those chemical factors further signal other cells, lymphocytes and white blood cells to come and join. For recognizing the pathogenic factors, there are specific and unique sequences that are not present in our body cells.

The immune system works in two different fashions. One is if there is any problem with a cell or the cell is damaged and this damage can not be fixed in that case immune system will signal that cell to be killed by apoptosis or programmed cell death. Another one is if there is any chance of improvement or if any pathogen found by any of the immune system cells they will recall all the other cells.

Major histocompatibility complex (MHC) is important for the compatibility of transplantation of tissue graft. MHC are protein molecules that are present in the surface of the membrane also called membrane-embedded proteins. MHC-I and MHC-II are nothing but glycoproteins that are found on the surface of eucaryotic cells. All the nucleated cells in our body contain MHC-I and it is different for different tissues. MHC-II is only found in APCs such as macrophage, dendritic cells, and B cells. Because of a specific function, if there is a pathogen, they will engulf that pathogen with phagocytosis; then, they will break the pathogen into small fragments that are known as antigens and they will showcase the antigens by loading them with a receptor and show them to rest of the immune system cells. The receptors where they load antigens are MHC-I and MHC-II.

MHC-II will help to activate more T helper cells that will help in bringing more macrophages and dendritic cells so as more APCs [26]. The interaction between the APCs and T helper cell with the help of MHC-II molecule can also help in activating the B cells and start producing antibodies. The function of MHC-I is to designate specific cells of our body as weak and killed that cells to prevent damage. The functional difference between MHC-I and MHC-II is that MHC-I is associated with T killer cell, whereas MHC-II is associated with the T helper cell.

The B cell is responsible for the humeral side of immunity, because B cell has the capability of converting itself into plasma cells. When a B cell matures, it produces plasma cells and these are antibody-producing factories of our body. These antibodies are very specific against specific pathogen [27]. B cells normally present in an inactive form and it has B cell receptors on the surface known as BCR. B cell has a unique capability that it can produce antibodies which are found to be attached tightly to the B cell receptors. So, the antibodies present on the surface of the B cell can bind specifically to antigens, and B cell also has another antigen-presenting cell which means it can also engulf pathogens and break it into different fragments. Structure of B cell is shown in Fig. 6.

The interaction between B cell and T helper cell is very important to get the B cell activated. There are varieties of B cells and T cells present in our body which can find out specific antigens and bind with that antigen. Antigens are foreign particles that are coming from outside of our body, so our body should not always recognize them. By the help of variable region, different types of modification can occur in

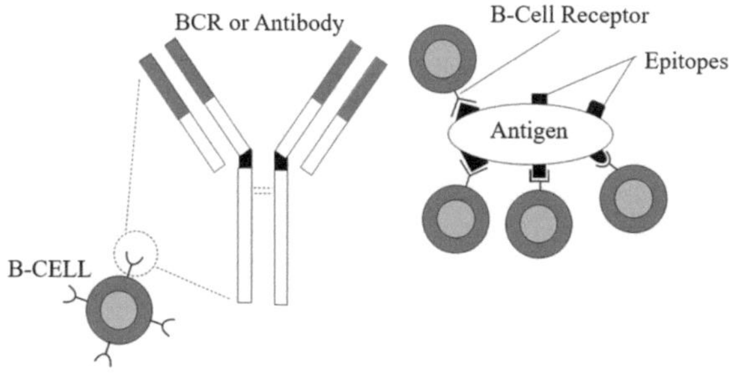

Fig. 6 Structure of B cell

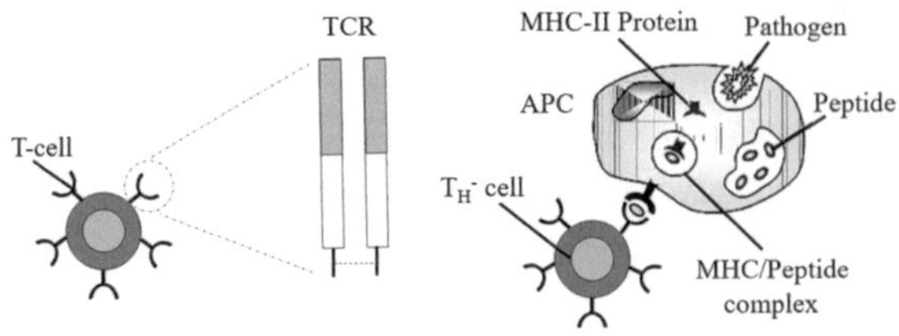

Fig. 7 Structure of T cell

different antibodies and mutation that occurs in variable region is called somatic hypermutation. Due to mutation they have different types of structures so they can bind with many varieties of antigens. Every time when an antigen enters, there is at least one or two cells capable of binding to that antigen and identify it. Then, that proportion of the cell starts producing the same type of cells so that immunity gets boosted. Those cells that can bind to those antigens are selected, and the rest are not required.

Once the T helper cell gets activated, so many responses start to happen inside our body, such as the cell can signal more T cells to get activated, can signal macrophages or dendritic cells or APCs to get activated. They can signal B cell to mature into plasma cells and also signal to natural killer cell to get activated and kill the disease cells. The cell-mediated and humoral-mediated immunity are balanced due to this activation of T cell. There are two types of T cells: T helper cells and T killer cells. Activation of T helper cell is more important for the response and beginning of cell-mediated immunity, and in that case, they need APCs for the activation process. The structure of T cell is shown in Fig. 7.

The maturation of T cell takes place in thymus. There are two processes to distinguish between T helper cell and T killer cell. There are subcapsular epithelial cells present in subcapsular region. If some of the thymocyte cells can interact with MHC-II T helper cell receptor will keep while T killer cell receptor will be degraded. Similarly, if some of the thymocyte cells can properly interact with MHC-I, T killer cell receptor will be selected and the T helper cell receptor will be destroyed. Some of the cells can not bind with either MHC-I or MHC-II or they can bind with both. In both of the cases, they fail to engage in proper interaction and will be signaled for apoptosis or program cell death.

3 AIS Approaches for Fault Diagnosis in WSN

AIS is a subfield of biological inspired computing and is a part of artificial intelligence. There are four algorithms of the AIS which are typically modeled for problem-solving based on the features of immune system. The AIS field is concerned with abstracting the role of the immune system and studying the application of the system toward solving computational problems. The algorithms can explain the behavioral function of HIS which can also be applied to fault diagnosis of WSN.

Table 2 Comparison of AIS approaches

Characteristics	CSA (clonal selection algorithm) [28]	NSA (negative selection algorithm) [29]	DCA (dendritic cell algorithm) [30]	INA (immune network algorithm) [31]
Affinity	Required	Not required	Not required	Required
Data set	Not required	Required	Not required	Not required
Computational time	Less	More	Less	Less
Message exchanged	More	Less	Less	More
Detection accuracy	Low	High	High	Low
False alarm rate	High	Low	Low	High

- **Clonal selection algorithm (CSA)**: Clonal selection algorithms are inspired by the clonal selection theory of the adaptive immune system which explains the affinity maturation. These algorithms focus on antigen–antibody interaction, reproduction, and variation by somatic hypermutation. These are mostly applied to pattern recognition or optimization problem [28].
- **Negative selection algorithm (NSA)**: The main purpose of the negative selection algorithm is to categorize the cells into self and non-self. This algorithm states that the body prevents its self pattern (known pattern) from being attacked by various pathogens such as bacteria and viruses [29]. One of the advantages of this method is that no prior knowledge of non-self is required.
- **Dendritic cell algorithm (DCA)**: Dendritic cells are antigen-presenting cells of the immune system which are always present, but will activate when an invading agent enters to our body. These are phagocytes that engulf other cells or bacteria and destroy them. By destroying them, they can present antigen, which is why they are called antigen-presenting cells [30].
- **Immune network algorithm (INA)**: This type of algorithm focuses on the structure of the network where nodes are represented by the antibody or antibody-producing cells and the training algorithm includes increasing edges between n-odes based on affinity. These algorithms are used in the field of clustering, data visualization, control, and optimization domains [31].

The comparison of AIS approaches is presented in Table 2.

4 Applications of AIS Algorithms

In [32], to diagnosis the hydraulic brake fault in automobile, the authors have used clonal selection classification algorithm inspired from the clonal selection theory. Nine fault conditions were simulated and tested for each fault condition in a brake system. The simulation result shows that the proposed approach gives better classification accuracy as compared to other machine learning approaches. In [33], the concept of artificial immune system is used to identify the faults in a large wireless sensor network. According to the authors their proposed approach is better and efficient because of faster diagnosis. For the fault detection and analysis occurred in machines, Gan et al. proposed a fault detection system based on clonal selection programming [34]. By performing different tests at different conditions the proposed method is found to be suitable for practical industrial applications. In [35], faulty sensor nodes are detected using AIS's clonal selection principle and categorized by probabilistic neural network strategy into corresponding types. The faulty sensor nodes are also isolated in the isolation phase. Performance of the proposed algorithm is compared and simulated with existing algorithms and the simulation result shows that the suggested algorithm provides a better result.

In [36], a fault diagnosis algorithm is proposed by combining both the clonal selection principle and negative selection algorithm, which determines the fault types properly. By optimizing the mutation operator, the convergence rate of antibody generation in the detector set was also improved. The fault diagnosis model was tested by experiments. The vibrating signals were collected and transferred by a WSN. The data were analyzed and diagnosed based on the fault diagnosis model. In [37] using NSA, the authors have proposed a motor fault diagnosis scheme. The motor faults can be encountered using a hierarchical structure. This structure efficiently detects the incipient motor faults as well as the fault types. In the simulation, the authors have examined the fault diagnosis method using two real-world problems.

The authors in [38] proposed a multi-operational algorithm using the negative selection principle. For comparison with other algorithms, they have used the fault model of DC motor as a benchmark. In [39], the authors have proposed two novel negative selection algorithm. Usually, the detectors are generated randomly, but in this work, the detectors are generated in a nonrandom ways which eliminate the training time of detectors. To examine the performances of both the experiments performed on the iris data set, ball bearing fault data set and two-dimensional synthetic data sets. The result shows that in most of the cases they give better results than the others. In [40], the authors have proposed one type of data flow attack called a Sybil attack. They have implemented an improved version of NSA with learning capability, and they have also used r-contiguous bit matching rule. They have compared their work with other works by taking three performance parameters such as false positive, false negative, and detection rate and their work show better results than the others.

In [24], the authors examined the use of AIS to recognize the fault trends in supermarket freezer cabinet temperature information. Their primary objective is to identify the early signs of icing up in supermarket refrigerated cabinets. In addition to information encoding, the r-bit matching rules provide a precise classification rate of faulty data. In [41] inspired by an immune system, fault detection isolation of a wind turbine system have been proposed. To detect and isolate both individually and simultaneously occurring faults, the authors have designed an NSA which is hierarchical in nature. To evaluate the proposed work, a nonparametric statistical comparison test has put under various fault circumstances. The simulation result shows that the NSA and SVM give same performances while in certain fault circumstances NSA gives a better result than SVM.

In [42], the authors have proposed a method for classification of real noise in speech sentences based on NSA. To validate the proposed method they have taken six types of real noise. This method shows better results than the classical classifiers in terms of accuracy. Aydin et al. [43] proposed a chaotic-based NSA for anomaly detection. Both clonal and negative selection approach is used to generate detectors. The detectors generated in the training stage are used to check the performances in the testing stage. The authors have also used the KNN method to generate detectors. They have analyzed their work in the broken rotor bar fault detection and Fisher Iris datasets.

A fault detection and isolation (FDI) method is developed using the dendritic cell algorithm in [44]. The authors have applied this method to a wind turbine test model. The proposed method can detect as well as isolate sensor faults. A statistical comparison test is carried out to compare the performances of the proposed scheme. In [45], a model was proposed which describes the mechanism of biological differentiation of dendritic cells. This model abstracts the information of dendritic cell fusion process, defines the functions of external signals which are applied to WSN and defines the mathematical model of DC. A real-time intrusion detection was performed and the performances were analyzed by scalability, complexity, and robustness which gives better detection with less energy consumption.

In [46], the authors suggested a new method for the diagnosis of faults using artificial immune network. They have combined it with radial basis function (RBF) of neural network and the structure is same. Compared to RBF the suggested method has less hidden layers and the diagnosis rate is better. Wang et al. [47] proposed an artificial immune network coupled with the fuzzy c-means clustering to detect the types of faults in transformer. The experimental results indicate that the suggested algorithm can efficiently classify the power transformer fault types.

In [48], the authors have investigated the fault diagnosis of plant systems using an immune network. The origin of failure can be detected by calculating the failure of each unit locally. The authors have carried out a simulation to validate their proposed method. To improve the capability of interpreting the result of dissolved gas analysis, the authors have proposed an artificial immune network classification (AINC) algorithm [49]. By mimicking the learning and defensive mechanism of immune system AINC responds to the fault samples of power transformer. The proposed algorithm gives better diagnosis accuracy and effectively classifies the faults by testing many real fault samples. Various applications of artificial immune systems are given in Table 3.

Table 3 Typical applications of AIS

Author	Approach	Application
Jagadeeshwaran et al. [32]	Clonal selection algorithm	Brake fault diagnosis
Mohapatra et al. [35]	Clonal selection principle	Fault diagnosis in WSN
Gan et al. [34]	Clonal selection algorithm	Induction machine fault detection
Chen et al. [36]	Negative selection principle	Fault diagnosis of vibrating screen
Gao et al. [37]	Negative selection principle	Motor fault diagnosis
Taylor et al. [24]	Negative selection principle	Fault detection in refrigerator system
Alizadeh et al. [41]	Negative selection principle	Fault diagnosis of wind turbine
Abreu et al. [42]	Negative selection principle	Noise classification in speech signal
Aydin et al. [43]	Negative selection principle	Anomaly detection
Alizadeh et al. [44]	Dendritic cell	Fault detection and isolation of wind turbines
Xiao et al. [45]	Dendritic cell algorithm	Intrusion detection
Wang et al. [47]	Immune network algorithm	Fault diagnosis in transformer
Ishiguro et al. [48]	Immune network algorithm	Fault diagnosis in plant system
Hao et al. [49]	Immune network	Fault diagnosis in power transformer

5 Conclusion

In conclusion, it can be said that different biological inspired approaches such as AIS, PSO, ACO, GA, and ABC have been studied for computational problems. Artificial immune system is one of the active research areas which is used by researchers in fault detection and optimization problems. In this chapter, a detailed explanation of a biological immune system is given. Different approaches of an artificial immune system have been discussed that can be used for fault diagnosis in wireless sensor network. Various applications of artificial immune system algorithms are also discussed.

References

1. Yick J, Mukherjee B, Ghosal D (2008) Wireless sensor network survey. Comput Netw 52(12):2292–2330
2. Akyildiz IF, Su W, Sankarasubramaniam Y, Cayirci E (2002) Wireless sensor networks: a survey. Comput Netw 38:393–422
3. Mohapatra S, Khilar PM (2016) Forest fire monitoring and detection of faulty nodes using wireless sensor network. In: Region 10 Conference (TENCON), 2016 IEEE

4. Mukherjee A et al (2019) A disaster management specific mobility model for flying ad-hoc network. In: Emergency and disaster management: concepts, methodologies, tools, and applications. IGI Global, pp 279–311
5. Das SK, Tripathi S (2018) Adaptive and intelligent energy efficient routing for transparent heterogeneous ad-hoc network by fusion of game theory and linear programming. Appl Intell 48(7):1825–1845
6. Das, SK, Tripathi S (2018) Intelligent energy-aware efficient routing for MANET. Wireless Netw 24(4):1139–1159
7. Fong S et al (2018) Predicting unusual energy consumption events from smart home sensor network by data stream mining with misclassified recall. J Ambient Intell Humanized Comput 9(4):1197–1221
8. Das SK, Tripathi S (2017) Energy efficient routing formation technique for hybrid ad hoc network using fusion of artificial intelligence techniques. Int J Commun Syst 30(16):e3340
9. Roy S et al (2016) Symmetric key encryption technique: a cellular automata based approach in wireless sensor networks. Proc Comput Sci 78:408–414
10. Design frameworks for wireless networks. Springer, Lecture Notes in Networks and Systems, pp 1–439. ISBN: 978-981-13-9573-4
11. Swain RR, Khilar PM (2017) Composite fault diagnosis in wireless sensor networks using neural networks. Wireless Pers Commun 95(3):2507–2548
12. Panda M, Khilar PM (2015) Distributed self fault diagnosis algorithm for large scale wireless sensor networks using modified three sigma edit test. Ad Hoc Netw 25:170–184
13. Sahoo MN, Khilar PM (2014) Diagnosis of wireless sensor networks in presence of permanent and intermittent faults. Wireless Pers Commun 78(2):1571–1591
14. Mourad E, Nayak A (2012) Comparison-based system-level fault diagnosis: a neural network approach. IEEE Trans Parallel Distrib Syst 23(6):1047–1059
15. Preparata FP, Metze G, Chien RT (1967) On the connection assignment problem of diagnosable systems. IEEE Trans Electron Comput 6:848–854
16. Malek M (1980) A comparison connection assignment for diagnosis of multiprocessor systems. In: Proceedings of the 7th annual symposium on computer architecture. ACM
17. Maeng J, Malek M (1981) A comparison connection assignment for self-diagnosis of multiprocessor systems. In: Proceedings of the 11th international symposium on fault-tolerant computing. ACM Press, New York
18. De Castro LN, Timmis J (2002) Artificial immune systems: a new computational intelligence approach. Springer Science & Business Media
19. Janeway CA et al (2001) The immune system in health and disease. Immunobiology. Current Biology Limited (2001)
20. Rizwan R et al (2015) Anomaly detection in wireless sensor networks using immune-based bioinspired mechanism. Int J Distrib Sens Netw 11(10):684952
21. de Castro LN, Timmis J (2002) Artificial immune systems: a novel paradigm to pattern recognition. Artif Neural Netw Pattern Recogn 1:67–84
22. Dasgupta D, Gonzlez F (2002) An immunity-based technique to characterize intrusions in computer networks. IEEE Trans Evol Comput 6(3):281–291
23. Dasgupta D et al (2004) Negative selection algorithm for aircraft fault detection. Artif Immune Syst :1–13
24. Taylor DW, Corne DW (2003) An investigation of the negative selection algorithm for fault detection in refrigeration systems. In: International conference on artificial immune systems. Springer, Heidelberg
25. De Castro LN, Von Zuben FJ (1999) Artificial immune systems: Part Ibasic theory and applications. Universidade Estadual de Campinas, Dezembro de, Tech. Rep, vol 210, issue 1
26. Pinto JCL, Von Zuben FJ (2005) Fault detection algorithm for telephone systems based on the danger theory. In: International conference on artificial immune systems. Springer, Heidelberg
27. Kiang CC, Srinivasan R (2012) An artificial immune system for adaptive fault detection, diagnosis and recovery. In: Int J Adv Eng Sci Appl Math 4(1–2):22–31

28. De Castro LN, Von Zuben FJ (2002) Learning and optimization using the clonal selection principle. IEEE Trans Evol Comput 6(3):239–251
29. Forrest S et al (1994) Self-nonself discrimination in a computer. In: 1994 IEEE computer society symposium on research in security and privacy, Proceedings, IEEE
30. Greensmith J, Aickelin U (2009) Artificial dendritic cells: multi-faceted perspectives. Human-centric information processing through granular modelling. Springer, Heidelberg, pp 375–395
31. Timmis J, Neal M, Hunt J (2000) An artificial immune system for data analysis. Biosystems 55(1–3):143–150
32. Jegadeeshwaran R, Sugumaran V (2015) Brake fault diagnosis using Clonal Selection Classification Algorithm (CSCA)—a statistical learning approach. Eng Sci Technol Int J 18(1):14–23
33. Mohapatra S, Khilar PM (2017) Artificial immune system based fault diagnosis in large wireless sensor network topology. In: Region 10 Conference (TENCON), 2017 IEEE
34. Gan Z, Zhao M-B, Chow TWS (2009) Induction machine fault detection using clone selection programming. Expert Syst Appl 36(4):8000–8012
35. Mohapatra S, Khilar PM, Swain RR (2019) Fault diagnosis in wireless sensor network using clonal selection principle and probabilistic neural network approach. Int J Commun Syst :e4138
36. Chen G, Zhang L, Bao J (2013) An improved negative selection algorithm and its application in the fault diagnosis of vibrating screen by wireless sensor networks. J Comput Theor Nanosci 10(10):2418–2426
37. Gao XZ, Wang X, Zenger K (2014) Motor fault diagnosis using negative selection algorithm. Neural Comput Appl 25(1):55–65
38. Laurentys CA et al (2010) Design of an artificial immune system for fault detection: a negative selection approach. Expert Syst Appl 37(7):5507–5513
39. Li D, Liu S, Zhang H (2015) Negative selection algorithm with constant detectors for anomaly detection. Appl Soft Comput 36:618–632
40. Zeeshan M et al (2015) An immunology inspired flow control attack detection using negative selection with R-contiguous bit matching for wireless sensor networks. Int J Distrib Sens Netw 11(11):169654
41. Alizadeh E, Meskin N, Khorasani K (2017) A negative selection immune system inspired methodology for fault diagnosis of wind turbines. IEEE Trans Cybern 47(11):3799–3813
42. de Abreu CCE, Duarte MAQ, Villarreal F (2017) An immunological approach based on the negative selection algorithm for real noise classification in speech signals. AEU-Int J Electron Commun 72:125–133
43. Aydin I, Karakose M, Akin E (2010) Chaotic-based hybrid negative selection algorithm and its applications in fault and anomaly detection. Expert Syst Appl 37(7):5285–5294
44. Alizadeh E, Meskin N, Khorasani K (2017) A dendritic cell immune system inspired scheme for sensor fault detection and isolation of wind turbines. IEEE Trans Ind Inf 14(2):545–555
45. Xiao X, Zhang R (2017) Study of immune-based intrusion detection technology in wireless sensor networks. Arab J Sci Eng 42(8):3159–3174
46. Jiang WK, Chen YJ, Zhang J (2013) A fault diagnosis method based on artificial immune network. In: Applied mechanics and materials, vol 385. Trans Tech Publications
47. Wang FZ, Shao SM, Dong PF (2014) Research on transformer fault diagnosis method based on artificial immune network and fuzzy c-means clustering algorithm. In: Applied mechanics and materials, vol 574. Trans Tech Publications
48. Ishiguro A, Watanabe Y, Uchikawa Y (1994) Fault diagnosis of plant systems using immune networks. In: Proceedings of IEEE international conference on MFI'94. Multisensor fusion and integration for intelligent systems, IEEE
49. Hao X, Cai-Xin S (2007) Artificial immune network classification algorithm for fault diagnosis of power transformer. IEEE Trans Power Deliv 22(2):930–935

Part II
Swarm Optimization

Chapter 7
Intelligent Routing in Wireless Sensor Network Based on African Buffalo Optimization

Samiran Bera, Santosh Kumar Das and Arijit Karati

1 Introduction

Nowadays, the applications of wireless sensor network (WSN) increase rapidly [1–3]. It is a collection of more than one sensor nodes (SNs) with single or multiple base stations (BSs) depending on the requirement. The topology of this network is dynamic. Basically, it is used to sense and detect several information from its environment. Each SN is directly or indirectly connected with BS which is a main coordinator system having high energy and high processing capacities [4]. One of the modern facilities of WSN is Internet of things (IoT)-based wireless body area network (WBAN) in healthcare system. It paves the gap of traditional system like to visit hospital. IoT allows several facilities like communicating, sensing, processing with biomedical and physical parameters [5, 6]. Besides, the cloud computing also provides significant advantages to the WBAN due to its large processing and storage infrastructure. It helps to process the data and information as offline as well as online by body sensor streams [7–10]. Apart from these, there are several applications of WSN as follows:

S. Bera
Department of Management Studies, Indian Institute of Technology (Indian School of Mines), Dhanbad, Jharkhand 826004, India
e-mail: samironbera@gmail.com

S. K. Das (✉)
School of Computer Science and Engineering, National Institute of Science and Technology (Autonomous), Institute Park, Pallur Hills, Berhampur, Odisha 761008, India
e-mail: sunsantosh2014@gmail.com

A. Karati
Department of Computer Science and Engineering, National Sun Yat-Sen University, Kaohsiung 80424, Taiwan
e-mail: arijit.karati@gmail.com

D. De et al. (eds.), *Nature Inspired Computing for Wireless Sensor Networks*, Springer Tracts in Nature-Inspired Computing,
https://doi.org/10.1007/978-981-15-2125-6_7

(a) Animal movement tracking
(b) Civil structure monitoring system
(c) Commercial application
(d) Consumer application
(e) Entertainment application
(f) Environment monitoring system
(g) Industrial application
(h) Military application
(i) Precision agriculture system
(j) Security and surveillance system
(k) Smart building application
(l) Smart grids and energy control system
(m) Transportation and logistics
(n) Urban terrain tracking system.

Although WSN has several applications in the context of real life, it has also some limitations such as:

(a) **Battery issue**: The sensor nodes consist of limited capacity of batteries that is insufficient during any mission or operation because it needed recharge after some time interval. So, its capacity is low than wired network.
(b) **Complicated configuration**: The configuration is more complicated compared to wired network. Therefore, for any trouble or issue then it is difficult to provide solution.
(c) **Costly**: It is more costly than wired network. So, it is not easily available everywhere based on the requirement.
(d) **Distraction**: In modern situation, this network keeps on distracting by several wireless devices like Bluetooth.
(e) **Low communication speed**: The communication speed of WSN is low compared to wired network. So, it takes more time to survive or collect information.
(f) **Security**: The WSN is not secure compared to other network or wired network. It is easily troubled by its environment. So, hackers can easily hack the required information.

The above-mentioned limitations cause different types of uncertainties in terms of privacy and security. Security provides the safeguard of a different kind of data or data set or a different related information, and privacy assists safeguarding of user information or identity [11]. To solve different types of problems such as outlier/anomaly detection, fault diagnosis, intrusion detection and mobility prediction, several works have been proposed [12, 13]. Most of the outlier detection and prediction algorithms are outperformed with the help of machine learning algorithms [14, 15]. It may be noted that machine learning algorithms help to enhance the critical situation in an optimized way [16–18]. Some of the basic steps that follow by machine learning algorithms are (i) feature selection and output labeling, (ii) sample collection, (iii) offline training and (iv) online classification.

Hence, there is a need to design an efficient routing algorithm for WSN that qualitatively optimizes the network and its metrics. So, in this paper an intelligent routing protocol is designed for WSN using African buffalo optimization (ABO). This ABO algorithm plays the role of intelligent application. This algorithm is based on nature-inspired optimization which helps to optimize each network metric efficiently and enhance the network lifetime.

2 Related Work

In the last decade, several routing protocols have been designed for WSN. Curry and Smith [19] proposed a detailed survey of several optimizations for WSN. In this survey, each literature describes an intelligent algorithm that directly or indirectly uses one or more techniques for enhancing the network lifetime or network metrics, efficiently. It also illustrates impractical situations and their optimal solution with respect to network lifetime and network metrics. Yan et al. [20] proposed a low energy-based node positioning system in WSN. The proposed work is based on PSO technique. The basic aim of this proposal is to optimize the dynamic position of the sensor nodes efficiently in terms of minimum energy consumption based on the particle of optical sensor nodes. Yu et al. [21] proposed a hybrid localization method for WSN. The basic keyword of this method is chicken swarm optimization (CSO). The nature optimization technique CSO is used for deep mining based on the wheel graph. This method helps in transformation of the clusters and improves the precision of the location for the sensor nodes. Phoemphon et al. [22] proposed a hybrid method for localization system in WSN. This method is the fusion of fuzzy logic, machine learning and vector particle swarm optimization. It helps to improve the traditional localization system, i.e., centroid, by using the hybrid mechanism. Finally, it overcomes the limitation of estimation precision. Ravi and Kashwan [23] designed an algorithm named as EASRP for ad hoc network. It uses two exising methods such as AFECA and Span where first method is a fidelity method and second is distance measuring method. In this system, a design hardware circuit is used for optimizing energy of the nodes. Sun et al. [24] proposed an attack localization system in WSN. The basic aim of this method allows task allocation in the network using binary POS mechanism. The complete process is based on the three objective functions: (i) maximization of load balancing system, (ii) minimization of the energy consumption of the network and (iii) minimization of the cost of execution time. The combination of the stated optimization helps to construct the constraints of the received signal strength and enhance the network lifetime. Cao et al. [25] proposed a technique for deployment in WSN using PSO in distributed environment. In this method, there are two types of nodes used such as sensor and relay nodes to prolong the lifetime and maximize the coverage. The stated method is basically used for 3D industrial WSN. Finally, it helps to reduce computational cost and enhance the network lifetime. Das and Tripathi [26] proposed a routing technique for transparent heterogeneous ad hoc network (THANET). This technique basically used non-cooperating game theory optimization for managing several conflicting strategies of the network. This noncooperative game theory optimization used fuzzy logic for making rigid goal to fuzzy goal. So that, dynamic environment of THANET is managed efficiently. Das et al. [27] proposed a routing protocol for multicast ad hoc network in order to create an energy effective path from origin to every multicast set built on two vague parameters such as distance and energy, whereas other parameters were not been considered by this proposed work which leads to its limitation. Hence, Yadav et al. [28] stretched the effort placed on multi-constraint method. They have considered three parameters that are delay, bandwidth and energy. It supports to elect the best route with the help of fuzzy cost. Here too, limitations occurred as it is a point-based membership function and fails to hold the fuzziness information. Henceforth, Das and Tripathi [29] proposed an energy conscious-based routing protocol by considering five parameters such as distance, energy, delay, packets and hop count. The objective of this routing is to search for an optimal route by considering multi-criteria decision making and intuitionistic fuzzy soft set. As it never uses any optimization method leads to its limitation. So, by using above techniques several contradictory objectives may not be optimized. Das and Tripathi [30] brought up a routing technique based on nonlinear optimization technique. This nonlinear optimization technique is based on geometric programming that works with posynomial environment instead of polynomial environment. It helps to determine nonlinear parameters efficiently and enhance the network lifetime. Gu and Zhu [31] designed an energy-efficient protocol named as RECI that handles two issues of the routing such as energy index and minimum hop count. These two basic parameters of the network help to design efficient route of the network. It also helps to increase the network lifetime by consuming less energy during data transmission. Das and Tripathi [32] design a fusion algorithm for managing dynamic and conflicting environment of the hybrid ad hoc network (HANET). This fusion is based on several artificial intelligence (AI) technique such as nonlinear geometric programming, fuzzy logic, multi-objective optimization, aspiration level and optimset. The basic aim of this fusion is to manage multiple nonlinear conflicting objectives of the network efficiently. So that, network lifetime and several network metrics are increased simultaneously in several scenarios of passes. Zahedi et al. [33] proposed an intelligent routing protocol for clustered WSN. This is fuzzy-based routing protocol that used swarm intelligence to manage all cluster head nodes of WSN. In this routing protocol, Mamdani inference system is used for decision making of the fuzzy logic. Finally, it helps to prolong the network lifetime and balance the cluster using firefly swarm algorithm. Shankar et al. [34] proposed a hybrid algorithm for energy efficiency in WSN. The aim of this algorithm is selection of cluster head and is to reduce energy consumption of the network. Both stated operations are performed with the fusion of harmony search algorithm (HAS) and PSO. The dynamic capability of the algorithm is more with respect to topology of the network. It helps to judge alive and dead nodes, increase throughput and reduce residual energy of the network. Azharuddin and Jana [35] proposed an enhanced algorithm for WSN with the help of PSO. The PSO is a machine intelligence technique that is used here to select cluster head and cluster member. The proposed technique helps to manage distributed traffic load. It possesses network lifetime enhancement as well as network metrics enhancement. Sridhar et al. [36] design an efficient routing system named as EN-AODV. In this protocol, the main energy is measured with the help of receiving and sending data packet of the nodes because during sending and receiving data packet energy is consumed. In this routing protocol, network lifetime is increased by maintaining energy capacity of the nodes. Ouchitachen et al. [37] proposed a multi-objective optimization technique in WSN for weighted clustering algorithm. In this algorithm, a genetic algorithm (GA)-based base station (BS) is used to manage consumed energy of the sensor nodes. The proposed algorithm helps to satisfy and fulfill the requirement of the sensor nodes. It also helps to improve the communication between sender and received nodes by crossing neighbor nodes. Gholipour et al. [38] proposed a technique for congestion control in WSN using fusion of genetic algorithm and support vector machine. In this technique, SVM parameters are tuned using GA and match actual data with current in different phases. The purpose of this algorithm is to increase energy efficiency and throughput and decrease the packet loss. Bhatia et al. [39] designed a distance-aware routing protocol for WSN using GA. The proposed routing protocol is based on the LEACH protocol. In this protocol, random probability is added into the GA for improving the CH selection and establishes efficient communication between CH and BS. Finally, it enhances the network lifetime of the network based on different metrics. Ray and De [40] designed an algorithm for WSN using swarm optimization. The main purpose of this algorithm is to solve two issues like energy conservation and coverage. The main keyword of this algorithm is glowworm swarm optimization (GSO) which is a bio-inspired algorithm. It helps to reduce redundant coverage by sensor traveling from one place to another place. Finally, it helps to optimize distance traversal and reduce energy consumption of the sensor nodes. Taherian et al. [41] proposed a secure and optimal routing protocol for WSN using PSO nature-inspired algorithm. In this algorithm, an efficient technique is used to divide each sensor as clustering method and apply there PSO for efficient and safe routing system in WSN. Barekatain et al. [42] designed a routing protocol for WSN using fusion of k-means and GA. In this routing protocol, the CH collects data from all the cluster members, i.e., simple sensor nodes, and sends into BS time to time. This system helps to aggregate the necessary information into single place. The main aim of this routing protocol is to reduce energy consumption and extend the network lifetime. Dhivya and Sundarambal [43] proposed a tabu swarm optimization (TSO) technique for network lifetime maximization in WSN. This is a QoS-based routing optimization which is designed by the fusion of PSO and tabu search. This fusion technique helps to enhance the network lifetime and reduce the energy consumption of the WSN. Das et al. [44] provide a detailed book for wireless network. It consists of several frameworks of wireless network along with WSN and ad hoc network. Some of the works [45, 46] described intelligent routing technique using fuzzy petrinet and strategy management using nonlinear formulation technique. Both works help to estimate uncertainty parameters of the network.

3 Preliminary: African Buffalo Optimization

African buffalo optimization (ABO) is a swarm intelligent (SI) technique, which falls under the domain of meta-heuristics. This algorithm is a part of nature-inspired algorithm which is rapidly used in the modern era [47]. It is rapidly used in WSN as well as ad hoc and wireless body network [48–52]. ABO is inspired by the movement of African buffalos. African buffalos are large herbivorous animals which move across forests, deserts and savannah in search of lush green grass, i.e., food [53]. However, it is quite a challenge to survive on landscape mostly comprising of arid deserts. African buffalo, thus, have adapted certain characteristics allowing them not only to survive but to thrive, even in such harsh environment [54]. These characteristics are (i) communication abilities, (ii) extensive memory capacity and (iii) effective herd management structure [55]. These characteristics are briefly stated as follows.

(a) **Communication abilities**: African buffalos communicate with the herd through "vocalization." Vocalization is of mainly two types: namely "maa" and "waa [56]." African buffalos use "maa" vocalization when present location is either seemingly promising or is safe. This implies that the herd should stay and graze, i.e., to "exploit" present location. Similarly, the African buffalos use "waa" vocalization when present location is unfavorable or dangerous. This informs the herd about lack of pasture or alerts the herd about the presence of a predator, respectively. This causes herd to be alert and employs to seek safer/better grazing areas, i.e., to "explore" other locations. Therefore, through these communicative abilities, African buffalos cooperate to search source of food and safer pastures and defend the herd from predators.
(b) **Extensive memory capacity**: African buffalo can keep track of their routes over thousands of miles across forests, deserts and savannah. It is possible as African buffalo have extensive memory capacity. This allows African buffalos to the following aspects.
    (i) Compare safety and favorability of present location in comparison with previous/former locations, and
    (ii) Follow location of the buffalo grazing in most favorable/safe area.

Further, it enhances the ability to make decision based on personal experiences (from own's previous position) and herd's experience (from present position of the herd). Extensive memory capacity, thus, is highly essential as African buffalo migrates extensively from one place to the other for survival, i.e., in search of food and safety.

(c) **Effective herd management structure**: The democratic nature of the African buffalo is another crucial aspect. It allows them to balance between exploration and exploitation. African buffalo(s) can either elect to follow decision of the majority, i.e., herd, or commit to a choice different from other members of the herd. Herd management structure allows African buffalos to draw benefit from collective intelligence of the herd. This implies that search by African buffalos is not bound, but rather enhanced by the herd management structure. African buffalos thus are considered as one of the most organized and successful herbivores.

In a nutshell, the characteristics mentioned above allow African buffalos to:

(a) escape starving region and move to better grazing pastures (through "vocalization" operators),
(b) undertake informed decision to stay or move away from present location by comparing present and former location, i.e., on the basis of safety of the herd and quality of grazing area (through utilization of extensive memory capacity), and
(c) maintain democratic selection between exploration and exploitation (through effective herd management structure).

3.1 The Component View of the ABO

The ABO mimics the characteristics of African buffalos. These characteristics are captured through components outlined as follows.

(a) **Initialization**: ABO initializes herd or population (group of buffalos) by assigning random location to each individual (i.e., buffalo) that lies within the search space. The location of each buffalo represents a solution in the search space. It is intuitive that a well-diversified population can explore solution space better and reach the optimal solution faster. To this end, a different probability distribution is considered based on the problem characteristics to initialize position of buffalos in the herd.
(b) **Update position**: ABO updates the location of each buffalo regularly based on its personal best location and location of the best buffalo. The updating operation constitutes of two "vocalization" methods, which are "maa" (for exploration) and "waa" (for exploitation).

  (I) **Using "maa" vocalization**: The "maa" operator promotes further exploration of the search space. The "maa" operator constitutes of three key components:
    (i) **Memory**: Individual's awareness of relocation, i.e., movement from one location to another location,
    (ii) **Cooperation**: to track present location of the best buffalo in the herd, and
    (iii) **Intelligence**: to compare present location with the former location(s).

These three components allow buffalos to take informed decisions while updating position. Further, based on the parameter settings, i.e., hyper-parameters, ABO achieves balance between exploration and exploitation in search space.

(II) **Using "waa" vocalization**: The "waa" operator promotes exploitation of present region of search space. It is through actual adjustment of an individual based on two calls: "maa" and "waa," which results in a new location of the individual. The "waa" operator constitutes of two components:

(i) **Adjustment**: It indicates the amount of change of location, i.e., movement of buffalo based on "maa" and "waa" vocalization operators.
(ii) **Degree of adjustment**: It indicates the required degree of adjustment which is determined using a random number.

The "waa" vocalization technique, thus, makes necessary changes post "waa" vocalization to update location of each buffalo.

(c) **Update global and personal best locations**: African buffalo keep tracks of territory visited which aids it to compare each and every succeeding move. Based on the fitness, i.e., quality of solution which depends on the goodness of present location, individual buffalo updates its experience. The information on quality of solution exists in two forms as:

(I) **Personal best**: It is the location of best solution obtained by each buffalo.
(II) **Global best**: It is the location of best solution obtained of all buffalo.

It is to be noted that personal best solution may/may not represent present locations of herd individuals. It can exist in the memory of each buffalo as experiences of visited areas. The global best solution, on the other hand, represents a buffalo currently in the herd with the overall best position so far. The location of global best solution is reported as the optimal solution if ABO meets the termination criteria.

(d) **Re-initialize positions**: To maintain adequate exploration, the entire herd is re-initialized if fitness of the best buffalo (i.e., global best) fails to improve over certain duration. This mechanism allows ABO to escape local optima, thus preventing stagnation.
(e) **Termination**: The ABO terminates when one or a combination of the following conditions is met:

i. **Maximum execution time**: pre-defined CPU/wall clock time,
ii. **Maximum stall limit**: number of iterations allowed without improvement (i.e., global best solution),
iii. **Minimal criteria**: Solution obtained is within acceptable limits, and
iv. **Exhaust solution space**: Entire solution space is visited.

However, besides above-mentioned criteria, others can be used too as suitable for the problem.

3.2 African Buffalo Optimization: The Algorithm

The African buffalo optimization (ABO) algorithm proposed by [57] is provided in this section. This algorithm consists of six steps for solving the global optima. The first step is initialization of the group of buffalos, i.e., their position. The second step computes fitness of individual buffalo, followed by finding the personal and global best in the third step. Based on values obtained from third step, ABO updates positions of individual buffalo in the fourth step. The fifth step evaluates for any improvement in global best. Based on the fifth step, ABO determines whether to re-initialize herd or terminate the algorithm in the sixth step. The detailed algorithm is given as Algorithm 1.

Algorithm 1: African Buffalo Optimization

1. Initialize position of each buffalo randomly in search space.
2. Compute fitness of each buffalo in the herd.
3. Update personal best solution of each buffalo and global best solution of the herd.
4. Update position of each buffalo using Eqs. (1) and (2).

$$m_{n(t+1)} = m_{nt} + C_1\left(x^g - x_{nt}\right) + C_2\left(x_n^p - x_{nt}\right) \tag{1}$$

$$w_{n(t+1)} = \frac{(w_{nt} + m_{nt})}{\lambda} \tag{2}$$

   where w_{nt} and m_{nt} denote exploration and exploitation of search space, C_1 is social learning parameters, C_2 is cognitive learning parameter, λ is a random number in range [0,1], and x^g and x_n^p denote location of buffalo with global and personal best solution, respectively.
5. If global best solution does not improve over say t iterations, and current iteration is less than maximum iteration (say termination criteria), go to step 1. Otherwise, proceed with step 6.
6. Check termination criteria. If termination condition is not met, return to step 2. Otherwise, report global best as optimal and stop.

3.3 Merits of ABO Algorithm

The African buffalo optimization (ABO) was developed in an attempt to provide solution to issues of low speed, premature convergence, usage of several parameters and complicated fitness functions of some already existing algorithms like the genetic algorithm [58], simulated annealing [59], ant colony optimization [60] and particle swarm optimizations [61] to mention a few. The merits of ABO algorithm stand out as follows:

(a) **Simplicity, flexibility and robustness**: ABO algorithm efficiently solves complex optimization problems using simple operators such as simulating communicative and cooperative methods. It can easily be adapted based on problem characteristics, making it flexible over different problems and sizes. Also, in comparison with other meta-heuristic techniques, ABO can escape local optima by re-initializing the herd over time making it highly robust.
(b) **Faster convergence**: ABO algorithm ensures faster convergence in comparison with other meta-heuristics techniques by adjusting the social and cognitive learning parameters, which is modulated by "maa" and "waa" operators. Through "vocalization" technique, ABO search for the optimal solution by balancing between exploration and exploitation.
(c) **Less parameter**: Parameter tuning is very crucial for a faster convergence. However, with increase in number of parameters, time required for tuning increases. Further, it might lead to slower convergence if it is not tuned properly, which again increases with the increase in number of parameters. Therefore, an algorithm with less number of parameters (to tune) is preferable. ABO algorithm, in contrast to other meta-heuristics, employs fewer parameters (which are learning parameters), for herd to explore or exploit depending on progress of the algorithm. Thus, ABO can be considered as a suitable candidate for a large set of optimization problems.

3.4 Application of ABO Algorithm

The ABO algorithm is based on meta-heuristic algorithm which is based on stochastic algorithm. The meta-heuristic algorithm overcomes the drawback of the traditional technique, i.e., heuristic. Heuristic technique is a deterministic technique, whereas meta-heuristic technique is the combination of heuristic and randomization techniques. The ABO algorithm works with global search system. It broadly used any of the domain where parameters are imprecise and complex. Some of the application areas are given below where this algorithm is rapidly used.

(a) **Multi-objective Optimization**: This optimization is used to deal more than one objective functions where nature of each objective is conflicting one to another. ABO plays a vital role in this optimization to maintain objective as well as constraint. This optimization is defined based on the constraints in two ways either minimization or maximization. Finally, it produces a set of solutions that helps to define best trade-off between conflicting objectives.
(b) **Constrained Optimization**: This optimization is used to optimize one specific objective based on some decision variables that are present in the constraints. All constraints indicate conditions of the objective function.

(c) **Discrete Optimization**: This optimization is also based on either minimization or maximization. It is also known as combinatorial optimization. It produces the optimal solution in countable finite number of iterations.
(d) **Binary Integer Programming Optimization**: Binary integer programming is also known as 0–1 integer programming or BIP. In this optimization, all the variables that take a place of decision variable are either 0 or 1.

Hence, the applications of ABO are same as applications of meta-heuristic like: (i) It is used for search problems in many applications such as machine learning, feature selection, automatic clustering, and (ii) it is also used in NP-hard optimization problems such as flow shop scheduling, maximum clique problem, p-median problem, traveling salesman problem. So, ABO algorithm is used to make search more controllable and systematic and to make the performance more scalable. This algorithm is easily implemented in C, C++, Java, Python, MATLAB, Octave, Scilab, etc.

4 Proposed Method

In this section, details of the proposed method are illustrated. ABO is implemented which is nature-inspired meta-heuristic-based algorithm. This algorithm here is applied in the WSN routing problem to enhance the network performance. This section is divided into two subsections in the context of design and formulation.

A typical network model is shown in Fig. 1. This figure contains several types of sensor nodes. Description of different types of nodes is as follows:

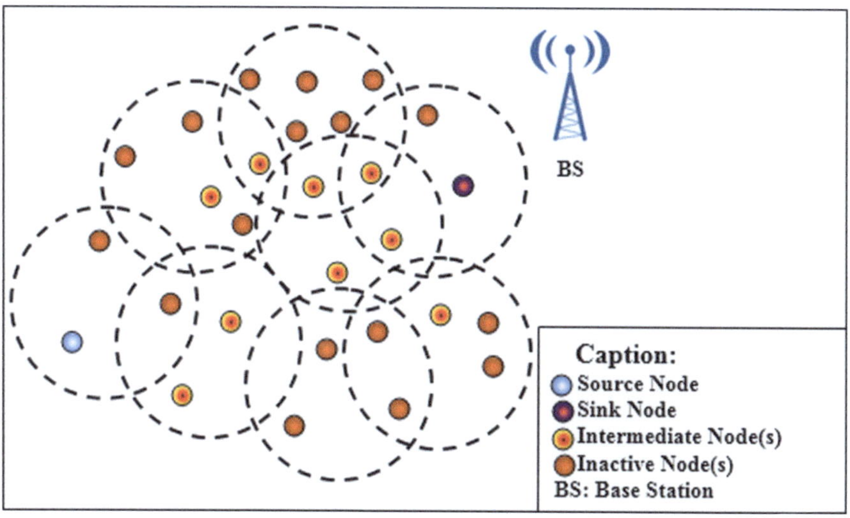

Fig. 1 Network model

Definition 4.1 (Source Node) Source node is the main initiator that sends information data packet.

Definition 4.2 (Sink Node) Sink node is the main receiver that receives information data packet.

Definition 4.3 (Intermediate Node) Intermediate nodes indicate all the nodes and connect the source and sink nodes.

Definition 4.4 (Inactive Node) Inactive nodes do not participate in routing unlike source node, sink node and intermediate node.

Definition 4.5 (Base Station) BS is the main controller of WSN which is used to manage and control overall network. It acts as a gateway of the network which is used to collect data and information from all sensor nodes and transmit to the server node.

The transmission of information from source to destination is relayed through intermediate nodes (shown in orange color). This forms a sequence of nodes which begins at the source node (shown in blue color) and terminates at the sink node (shown in purple color).

The sequence of intermediate nodes is represented as attributes (using continuous or binary variables) for each buffalo, as shown in Table 1. For a herd of buffalo, different paths, i.e., sequence of intermediate nodes, exist. Each buffalo comprises a sequence of node which is different from the others. This is the characteristics of swarm intelligence (SI) techniques. It is to be noted that either all, or a subset, or none of the intermediate nodes can be considered as attributes of individual buffalo. Based on the sequence of nodes, i.e., source–intermediate–sink, the fitness value, i.e., loss of energy, is evaluated.

Figure 2 represents a unidirectional network with 7 nodes and 11 edges. Node 1 represents the source node, node 7 represents the sink node, and remaining nodes act as intermediates. Inactive nodes are absent in this example. The network diagram is

Table 1 List of alternative path from source node to sink node

Sl. no.	Possible path	Loss of energy
1	$1 \rightarrow 2 \rightarrow 3 \rightarrow 4 \rightarrow 5 \rightarrow 7$	$5 + 6 + 7 + 4 + 3 = 25$
2	$1 \rightarrow 2 \rightarrow 3 \rightarrow 4 \rightarrow 6 \rightarrow 7$	$5 + 6 + 7 + 8 + 9 = 35$
3	$1 \rightarrow 2 \rightarrow 3 \rightarrow 6 \rightarrow 7$	$5 + 6 + 11 + 9 = 31$
4	$1 \rightarrow 2 \rightarrow 4 \rightarrow 5 \rightarrow 7$	$5 + 15 + 4 + 3 = 27$
5	$1 \rightarrow 2 \rightarrow 4 \rightarrow 6 \rightarrow 7$	$5 + 15 + 8 + 9 = 37$
6	$1 \rightarrow 2 \rightarrow 5 \rightarrow 7$	$5 + 18 + 3 = 26$
7	$1 \rightarrow 3 \rightarrow 4 \rightarrow 5 \rightarrow 7$	$12 + 7 + 4 + 3 = 26$
8	$1 \rightarrow 3 \rightarrow 4 \rightarrow 6 \rightarrow 7$	$12 + 7 + 8 + 9 = 36$
9	$1 \rightarrow 3 \rightarrow 6 \rightarrow 7$	$12 + 11 + 9 = 32$

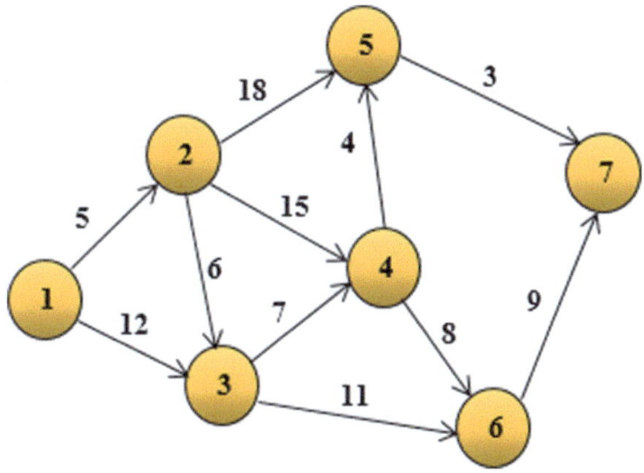

Fig. 2 Network diagram

fairly complex as each node has multiple incoming edges (i.e., receives information from one or more than one nodes) and multiple outgoing edges (i.e., transmits information to one or more than one nodes). Thus, multiple possible paths from source to sink exist, which can be represented by a combinatorial tree as provided next.

All possible combinations of nodes which transmits data from source to sink can be observed in Fig. 3, which is given in Table 1 as shown. The root node represents the source nodes (node 1), followed by intermediate nodes (node 2 to node 6) on the branches, and sink node (node 7) as leaf nodes. It can be easily deduced that with increase in network size, the combinatorial tree will grow exponentially. This implies that a problem with large number of nodes cannot be solved using classical methods. Thus, meta-heuristics such as ABO is implemented to efficiently solve large-scale problems.

ABO represents the sequence of nodes (as possible path shown in Table 1) as variables and evaluates its fitness. Even with the increase in problem size, ABO achieves optimal solution by evaluating only a fraction of all possible paths (or solution space) within reasonable amount of time.

4.1 Problem Formulation

Objective function: Information is transferring from source to sink node through intermediate nodes, which incurs energy at the edges. This is captured by the sum of products of c_{ij} (loss of energy from node i to j while data transfer) and x_{ij} (binary variables connecting two nodes), as shown in Eq. 3. The objective is to minimize loss of energy during transmission, denoted by Z_1.

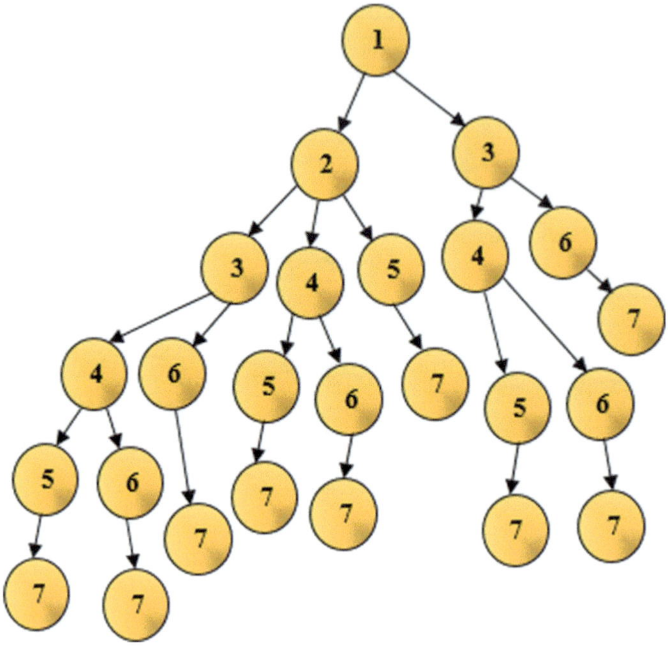

Fig. 3 Combinatorial tree diagram

$$\text{Minimize} : Z_1 = \sum_{i=1}^{N}\sum_{j=1}^{N}(c_{ij}x_{ij}) \tag{3}$$

Subject to constraints:

1. **Source constraint**: Equation 4 ensures information from source (denoted by i = 1) is passed to a node j. Similarly, Eq. 5 captures information transferred to the sink node (denoted by $j = N$), where N is the destination.

$$\sum_{j=1}^{N} x_{ij} = 1; \quad \forall\, i = 1 \tag{4}$$

$$\sum_{i=1}^{N} x_{ij} = 1; \quad \forall\, j = N \tag{5}$$

2. **Flow constraints**: Equation 6 ensures that the number of incoming nodes equals to the number of outgoing nodes. This implies any information received from node (say i) at node j is relayed to another node (say h). Inversely, it also implies that node j relays no information, in the absence of any incoming information.

$$\sum_{i=1}^{N} x_{ij} = \sum_{h=1}^{N} x_{jh}; \quad \forall\, j = 2, 3, 4, \ldots, N - 1 \tag{6}$$

3. **Sub-tour elimination constraints**: Constraint set in Eqs. 7 and 8 ensures discontinuity in sub-tours, i.e., prevention of disjoint graphs. This is captured using a variable T (denoting loss of energy) which increases with increase in edges transferred.

$$T_j = 0; \quad \forall\, j = 1 \tag{7}$$

$$T_j = \sum_{i=1}^{N} \{x_{ij}(T_i + c_{ij})\}; \quad \forall\, j = 1, 2, \ldots, N \tag{8}$$

The above constraint, i.e., Eq. 8, is nonlinear and can be linearized by simply introducing McCormick envelopes.

4. **Allocation constraints**: If node i has no possible path from node j, x_{ij} is restricted to zero. Thus, $x_{ij} = 1$, only when information is transmitted over edge i–j, which is given in Eq. 9.

$$x_{ij} \le c_{ij} - C'; \quad \forall\, i = 1, 2, 3, \ldots, N \tag{9}$$

Here, C' is assumed to be a large number.

5. **Binary and nonnegativity constraints**: The conditions provided below ensure variable x_{ij} to be binary and T_j to be nonnegative.

$$\begin{aligned} x_{ij} &= \{0, 1\} \\ T_j &= R^+ \quad \forall\, j = 1, 2, 3, \ldots, N \end{aligned}$$

5 Performance Evaluation

The proposed method is designed and simulated in Octave 5.1.0. The simulation parameters of the proposed method are shown in Table 2.

The proposed routing protocol is compared with three existing routing methods such as EASRP [23], RECI [31] and EN-AODV [36] and shown in Table 3. EASRP is based on zone routing protocol and uses remote activation switch system. The RECI routing method handles two issues like energy index and hop count of the network. It uses routing metrics for both route discovery and maintenance. The routing protocol EN-AODV is managing energy of each node during two times: First is sending data packet time, and second is receiving data packet time. The proposed method is based on meta-heuristic ABO technique. It is more efficient than three existing methods

Table 2 Simulation parameters

Parameter	Value
Operating system	Windows 7 Professional 32 bit
Operating system type	32 bits
Processor	Intel(R) Pentium(R) CPU G4400 @ 3.30 GHz
RAM	4 GB
Hard disk	300 GB
Monitor	Lenovo
Mouse	Lenovo
Keyboard	Lenovo
Programming environment	Octave 5.1.0
MS Word	2013
MS Excel	2013

Table 3 Feature comparison of the proposed method with some existing methods

Features	Proposed method	EN-AODV	RECI	EASRP
Routing loop avoidance	Yes	Yes	Yes	Yes
Source initiated	Yes	Yes	Yes	Yes
Receiver initiated	No	No	No	No
QoS support	Very good	Less moderate	Medium	Good
Residual energy	High	Very low	Low	Medium
Network lifetime	Very high	Low	Medium	High
Delay	Very low	High	Medium	Low
Packet delivery ratio	Very high	Low	Medium	High
Communication overhead	Very low	High	Medium	Low
Throughput	Very high	Low	Medium	High
Packet loss	Very low	High	Medium	Low
Scalability	Extreme	Less	Moderate	More
Bandwidth	Very high	Low	Medium	High
Robustness	Much more	Less	Moderate	More
Connectivity status	Very good	Bad	Medium	Good
Handling high mobility	Very good	Bad	Medium	Good
Handling high traffic load	Very good	Bad	Medium	Good
Handling mutual interference	Very good	Bad	Medium	Good
Handling imprecise information	Very good	Bad	Medium	Good

because it optimizes several conflicting parameters efficiently. The proposed routing method along with three existing methods source initiated. So, this routing method along with proposed method is routing loop avoidance. Source initiated indicates during communication when source node wants to transfer data packet, then only communication will start; otherwise, no transmission will be held. Routing loop issue indicates route request packet, i.e., RREQ, continuously routed only few of the routes. The reason is duplicate RREQ packet generation.

Although, several feasible or one optimal path are available in the network. Although the existing methods are low efficient than the proposed method, all are based on loop avoidance. Due to ABO meta-heuristic method, the proposed method easily handles imprecise information. So, it is able to manage mutual interference of the network and efficiently handle high mobility and high traffic load of the WSN. So, handling high mobility, high traffic load, mutual interference and imprecise information of the proposed method is very good. EASRP is based on zone-based routing protocol which has mix features such as proactive and reactive routing protocols. So, its several features are lesser than the proposed routing method but higher than two routing protocols, i.e., RECI and EN-AODV. The RECI maintained both route discovery and route maintenance, and it uses some extra parameters of the network, so it is efficient than EN-AODV only.

5.1 Variation of Iterations

To illustrate the nature of convergence curve, a large-scale problem consisting of 20 nodes is considered as shown by Fig. 4. The problem with 20 nodes has $20!(20 \times 19 \times 18 \times \cdots \times 3 \times 2 \times 1)$ possible routes. With increase of number of nodes and edges between nodes, the network problem becomes increasingly difficult to solve. Thus, optimization using ABO is justified.

To simplify representation, Fig. 4 shows flow of information in one direction on edges having least loss of energy. Remaining edges exist (for bidirectional flow of information), but are not shown. Next, follows a brief discussion on convergence curve obtained through optimization.

In this routing protocol, the residual energy is minimized in five rounds such as Round 1 to Round 5. Each round terminates when solution does not improve over 10 iterations. The variation of data rate can be observed from Figs. 5, 6, 7, 8 and 9 where x-axis represents iterations and y-axis shows any improvements, i.e., change in loss of energy. Here, data rate indicates residual energy points of the network lifetime of WSN.

Actually, any data rate shows actual services delivered by a path. In this proposal, the residual energy indicates energy used during this service.

Figure 5 shows first round of the simulation, where x-axis (indicating number of iteration) comprises from 1 to 31 and y-axis (indicating value of objective function) corresponds from 74 to 60, respectively. Thus, saving of 14 units is achieved in the

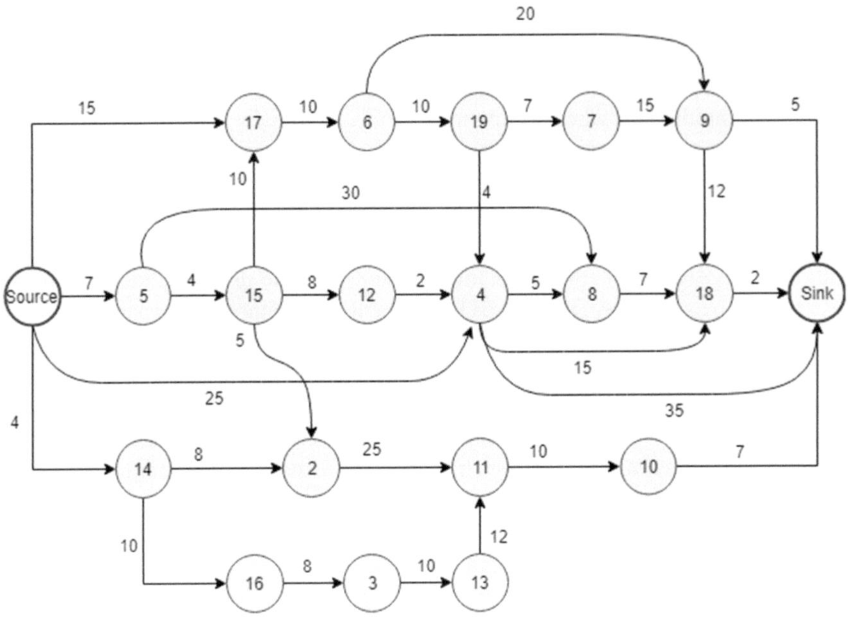

Fig. 4 Network diagram with 20 nodes (source + intermediate + inactive + sink)

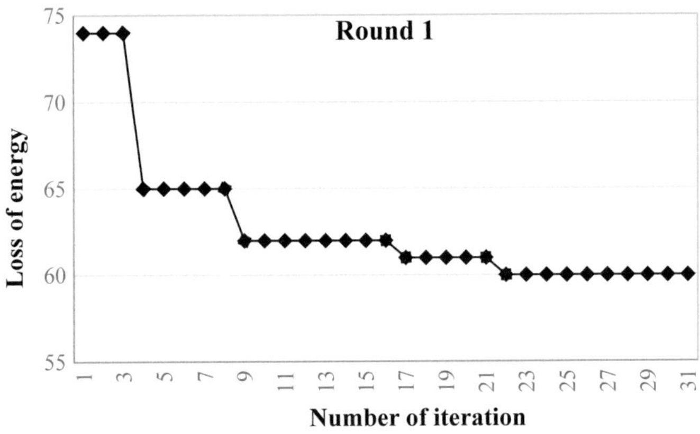

Fig. 5 Graph depicting minimization of loss of energy with increase in iterations (Round 1)

first round. As the solution fails to improve over 10 iterations (i.e., from 22 to 31), ABO re-initializes the herd for Round 2.

Figure 6 shows second round of the simulation. It starts with an objective value of 58, thereby improving over 60 (which is optimized value from Round 1). With increase in iteration, the loss of energy is optimized to 53, which is an improvement

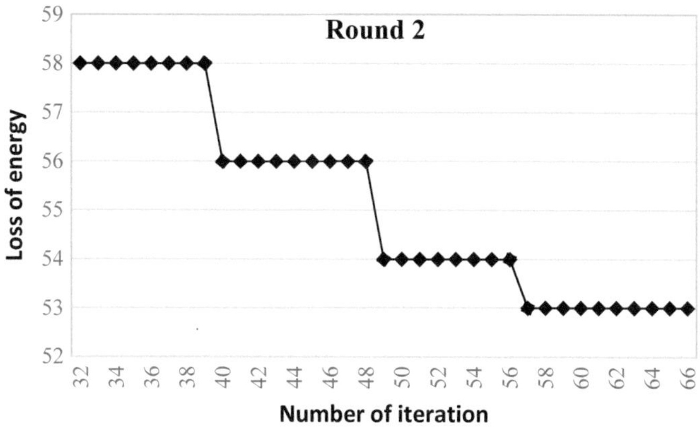

Fig. 6 Graph depicting minimization of loss of energy with increase in iterations (Round 2)

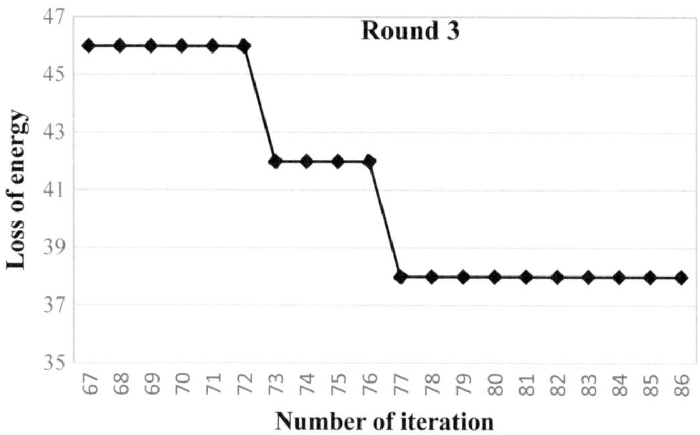

Fig. 7 Graph depicting minimization of loss of energy with increase in iterations (Round 3)

of 7 units over 35 iterations. Similar to Round 1, Round 2 terminates when objective value fails to improve over 10 iterations, i.e., from 57 to 66.

Figure 7 shows third round of the simulation. It starts with an objective value of 46, thereby improving over 53 (which is optimized value from Round 2). With increase in iteration, the loss of energy is optimized to 38, which is an improvement of 15 units over 20 iterations. Similar to previous rounds, Round 3 terminates when objective value fails to improve over 10 iterations, i.e., from 77 to 86.

Figure 8 shows fourth round of the simulation. It starts with an objective value of 35, thereby improving over 38 (which is optimized value from Round 3). With increase in iteration, the loss of energy remains at 35, which implies no improvement over 10 iterations. Similar to previous rounds, Round 4 terminates when objective value fails to improve over 10 iterations, i.e., from 87 to 96.

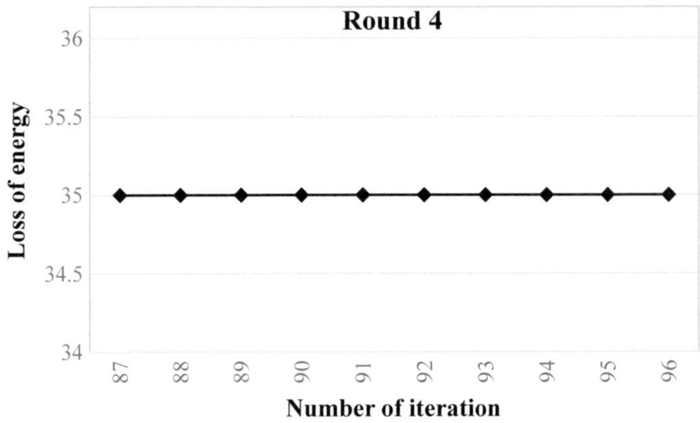

Fig. 8 Graph depicting minimization of loss of energy with increase in iterations (Round 4)

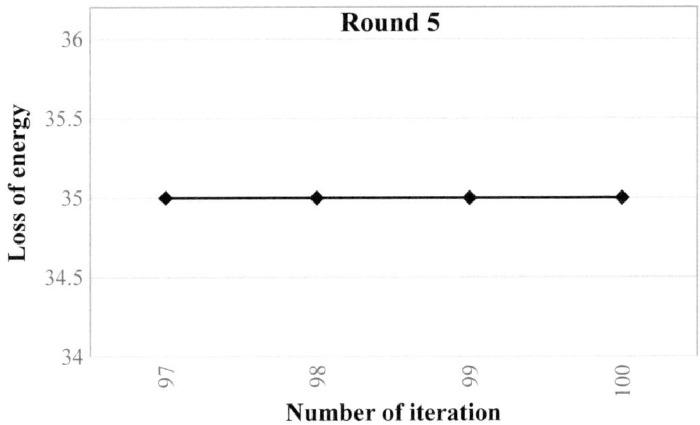

Fig. 9 Graph depicting minimization of loss of energy with increase in iterations (Round 5)

Figure 9 shows fifth round of the simulation. It starts with an objective value of 35, thereby without any improvement over previous round. With increase in iteration (which is limited to maximum of 100 iterations), the optimal value of loss of energy remains at 35 during remaining 4 iterations, i.e., 97–100.

It is to be noted that the number of iterations is different in each round, i.e., 31, 35, 20, 10 and 4 for Round 1, 2, 3, 4 and 5, respectively. It is because the terminating condition for each round is based on improvement, i.e., change in objective function value over pre-defined number of iterations. If quality of solution improves within pre-defined number of iterations, the number of rounds will increase.

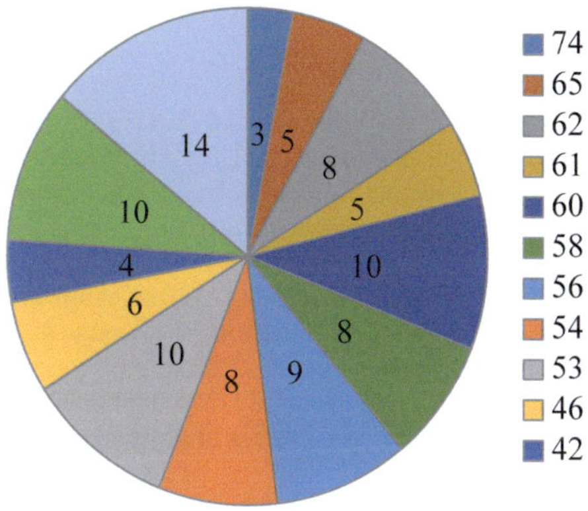

Fig. 10 Type of energy level in unique variation

5.2 *Unique Variation of Iterations*

Figure 10 shows unique variation of the proposed routing protocol. As in Figs. 5, 6, 7, 8 and 9, the iteration round is different that distributed in four parts. In each round, solution varies except in Round 3 and Round 4 because these two rounds are stuck mode rounds. So, Fig. 10 represents each different iteration of the data rate uniquely. It clearly shows the actual unique variations of the solution. Finally, it also represents the global optima, i.e., global minimization value, i.e., 35. This can be found in Fig. 10, as shown.

6 Conclusion

In this proposed routing protocol, meta-heuristic-based optimization technique is used, i.e., ABO. It is most powerful optimization technique than other meta-heuristic approaches. This optimization technique consists of multiple groups of iterations where each group again consists of multiple sub-iterations. This is a unique feature of the meta-heuristic approach which is not available in other meta-heuristic approaches. This optimization technique efficiently optimizes different network parameters of the WSN to pave the gap of traditional approach. It helps to reduce the network traffic globally and reduce the energy consumption of the nodes as well as link of the WSN. Hence, during data transmission, network estimates its energy consumption by selecting an optimal path. Future work is to analyze this algorithm and compares with other meta-heuristic methods in terms with mathematical and simulation.

References

1. Watt AJ, Phillips MR, Campbell CA, Wells I, Hole S (2019, June 1) Wireless sensor networks for monitoring underwater sediment transport. Sci Tot Env 667:160–165
2. Skiadopoulos K, Tsipis A, Giannakis K, Koufoudakis G, Christopoulou E, Oikonomou K, Stavrakakis I (2019, June 1) Synchronization of data measurements in wireless sensor networks for IoT applications. Ad Hoc Netw 89:47–57
3. Fong S, Li J, Song W, Tian Y, Wong RK, Dey N (2018, Aug) Predicting unusual energy consumption events from smart home sensor network by data stream mining with misclassified recall. J Ambient Intell Hum Comput 9(4):1197–1221
4. Jain PK, Quamer W, Pamula R (2018) Electricity consumption forecasting using time series analysis. In: International conference on advances in computing and data sciences. Springer, Singapore, pp 327–335
5. Karati A, Biswas GP (2019) Provably secure and authenticated data sharing protocol for IoT-based crowdsensing network. Trans Emerg Telecommun Technol 30(4):e3315, 1–22
6. Karati A, Islam SH, Karuppiah M (2018) Provably secure and lightweight certificateless signature scheme for IIoT environments. IEEE Trans Ind Inf 14(8):3701–3711
7. Panda SK, Jana PK (2019) An energy-efficient task scheduling algorithm for heterogeneous cloud computing systems. Clust Comput 22(2):509–527
8. Panda SK, Jana PK (2018) Normalization-based task scheduling algorithms for heterogeneous multi-cloud environment. Inf Syst Front 20(2):373–399
9. Panda SK, Pande SK, Das S (2018) Task partitioning scheduling algorithms for heterogeneous multi-cloud environment. Arab J Sci Eng 43(2):913–933
10. Karati A, Amin R, Islam SH, Choo KKR (2018, May 8) Provably secure and lightweight identity-based authenticated data sharing protocol for cyber-physical cloud environment. IEEE Trans Cloud Comput, IEEE 1–14. https://doi.org/10.1109/TCC.2018.2834405
11. Karati A, Islam SH, Biswas GP (2018) A pairing-free and provably secure certificateless signature scheme. Inf Sci 450:378–391
12. Jain PK, Pamula R (2019) Two-step anomaly detection approach using clustering algorithm. International conference on advanced computing networking and informatics. Springer, Singapore, pp 513–520
13. Mishra G, Agarwal S, Jain PK, Pamula R (2019) Outlier detection using subset formation of clustering based method. International conference on advanced computing networking and informatics. Springer, Singapore, pp 521–528
14. Kumari P, Jain PK, Pamula R (2018) An efficient use of ensemble methods to predict students academic performance. In: 4th international conference on recent advances in information technology (RAIT), IEEE, pp 1–6
15. Punam K, Pamula R, Jain PK (2018) A two-level statistical model for big mart sales prediction. In: 2018 international conference on computing, power and communication technologies (GUCON), IEEE, pp 617–620
16. Das SP, Padhy S (2018) A novel hybrid model using teaching–learning-based optimization and a support vector machine for commodity futures index forecasting. Int J Mach Learn Cybernet 9(1):97–111
17. Das SP, Padhy S (2017) Unsupervised extreme learning machine and support vector regression hybrid model for predicting energy commodity futures index. Memet Comput 9(4):333–346
18. Das SP, Padhy S (2017) A new hybrid parametric and machine learning model with homogeneity hint for European-style index option pricing. Neural Comput Appl 28(12):4061–4077
19. Curry RM, Smith JC (2016) A survey of optimization algorithms for wireless sensor network lifetime maximization. Comput Ind Eng 101:145–166
20. Yan Z, Goswami P, Mukherjee A, Yang L, Routray S, Palai G (2019) Low-energy PSO-based node positioning in optical wireless sensor networks. Optik 181:378–382
21. Yu X, Zhou L, Li X (2019) A novel hybrid localization scheme for deep mine based on wheel graph and chicken swarm optimization. Comput Netw 154:73–78

22. Phoemphon S, So-In C, Niyato DT (2018) A hybrid model using fuzzy logic and an extreme learning machine with vector particle swarm optimization for wireless sensor network localization. Appl Soft Comput 65:101–120
23. Ravi G, Kashwan KR (2015) A new routing protocol for energy efficient mobile applications for ad hoc networks. Comput Electr Eng 48:77–85
24. Sun Z, Liu Y, Tao L (2018) Attack localization task allocation in wireless sensor networks based on multi-objective binary particle swarm optimization. J Netw Comput Appl 112:29–40
25. Cao B, Zhao J, Lv Z, Liu X, Kang X, Yang S (2018) Deployment optimization for 3D industrial wireless sensor networks based on particle swarm optimizers with distributed parallelism. J Netw Comput Appl 103:225–238
26. Das SK, Tripathi S (2018) Adaptive and intelligent energy efficient routing for transparent heterogeneous ad-hoc network by fusion of game theory and linear programming. Appl Intell 48(7):1825–1845
27. Das SK, Yadav AK, Tripathi S (2017) IE2M: design of intellectual energy efficient multicast routing protocol for ad-hoc network. Peer-to-Peer Netw Appl 10(3):670–687
28. Yadav AK, Das SK, Tripathi S (2017) EFMMRP: design of efficient fuzzy based multi-constraint multicast routing protocol for wireless ad-hoc network. Comput Netw 118:15–23
29. Das SK, Tripathi S (2018) Intelligent energy-aware efficient routing for MANET. Wirel Netw 24(4):1–21
30. Das SK, Tripathi S (2019) Energy efficient routing formation algorithm for hybrid ad-hoc network: a geometric programming approach. Peer-to-Peer Netw Appl 12(1):102–128
31. Gu C, Zhu Q (2014) An energy-aware routing protocol for mobile ad hoc networks based on route energy comprehensive index. Wirel Pers Commun 79(2):1557–1570
32. Das SK, Tripathi S (2017) Energy efficient routing formation technique for hybrid ad hoc network using fusion of artificial intelligence techniques. Int J Commun Syst 30(16):e3340
33. Zahedi ZM, Akbari R, Shokouhifar M, Safaei F, Jalali A (2016) Swarm intelligence based fuzzy routing protocol for clustered wireless sensor networks. Expert Syst Appl 55:313–328
34. Shankar T, Shanmugavel S, Rajesh A (2016) Hybrid HSA and PSO algorithm for energy efficient cluster head selection in wireless sensor networks. Swarm Evol Comput 30:1–10
35. Azharuddin M, Jana PK (2016) Particle swarm optimization for maximizing lifetime of wireless sensor networks. Comput Electr Eng 51:26–42
36. Sridhar S, Baskaran R, Chandrasekar P (2013) Energy supported AODV (EN-AODV) for QoS routing in MANET. Proc Soc Behav Sci 73:294–301
37. Ouchitachen H, Hair A, Idrissi N (2017) Improved multi-objective weighted clustering algorithm in wireless sensor network. Egypt Inf J 18(1):45–54
38. Gholipour M, Haghighat AT, Meybodi MR (2017) Hop-by-hop congestion avoidance in wireless sensor networks based on genetic support vector machine. Neurocomputing 223:63–76
39. Bhatia T, Kansal S, Goel S, Verma AK (2016) A genetic algorithm based distance-aware routing protocol for wireless sensor networks. Comput Electr Eng 56:441–455
40. Ray A, De D (2016) An energy efficient sensor movement approach using multi-parameter reverse glowworm swarm optimization algorithm in mobile wireless sensor network. Simul Model Pract Theor 62:117–136
41. Taherian M, Karimi H, Kashkooli AM, Esfahanimehr A, Jafta T, Jafarabad M (2015) The design of an optimal and secure routing model in wireless sensor networks by using PSO algorithm. Proc Comput Sci 73:468–473
42. Barekatain B, Dehghani S, Pourzaferani M (2015) An energy-aware routing protocol for wireless sensor networks based on new combination of genetic algorithm & k-means. Proc Comput Sci 72:552–560
43. Dhivya M, Sundarambal M (2012) Lifetime maximization in wireless sensor networks using Tabu swarm optimization. Proc Eng 38:511–516
44. Das SK, Samanta S, Dey N, Kumar R (2020) Design frameworks for wireless networks. Lecture notes in networks and systems. Springer, Singapore, pp 1–439. ISBN: 978-981-13-9573-4
45. Samantra A, Panda A, Das SK, Debnath S (2020) Fuzzy petri nets-based intelligent routing protocol for Ad Hoc network. In: Design frameworks for wireless networks, Springer, Singapore, pp 417–433

46. Das SK, Sachin T (2020) A nonlinear strategy management approach in software-defined ad hoc network. In: Design frameworks for wireless networks. Springer, Singapore, pp 321–346
47. Dey N, Ashour AS, Bhattacharyya S (2019) Applied nature-inspired computing: algorithms and case studies, pp 1–275
48. Dey N, Ashour AS, Shi F, Fong SJ, Sherratt RS (2017) Developing residential wireless sensor networks for ECG healthcare monitoring. IEEE Trans Consum Electron 63(4):442–449
49. Elhayatmy G, Dey N, Ashour AS (2018) Internet of things based wireless body area network in healthcare. In: Internet of things and big data analytics toward next-generation intelligence. Springer, Cham, pp 3–20
50. Mukherjee A, Dey N, Kausar N, Ashour AS, Taiar R, Hassanien AE (2019) A disaster management specific mobility model for flying ad-hoc network. In: Emergency and disaster management: concepts, methodologies, tools, and applications. IGI Global, pp 279–311
51. Roy S, Karjee J, Rawat US, Dey N (2016) Symmetric key encryption technique: a cellular automata based approach in wireless sensor networks. Proc Comput Sci 78:408–414
52. Das SK, Tripathi S, Burnwal AP (2015) Intelligent energy competency multipath routing in wanet. In: Information systems design and intelligent applications. Springer, New Delhi, pp 535–543
53. Odili JB, Kahar MNM (2016) African buffalo optimization. Int J Soft Eng Comput Syst 2(1):28–50
54. Paul JD, Roberts GG, White N (2014) The African landscape through space and time. Tectonics 33(6):898–935
55. Lorenzen ED, Heller R, Siegismund HR (2012) Comparative phylogeography of African savannah ungulates 1. Mol Ecol 21(15):3656–3670
56. Odili JB, Kahar MNM, Anwar S, Ali M (2017) Tutorials on African buffalo optimization for solving the travelling salesman problem. Int J Softw Eng Comput Syst 3(3):120–128
57. Odili JB, Kahar MNM, Anwar S (2015) African buffalo optimization: a swarm-intelligence technique. Proc Comput Sci 76:443–448
58. Mokshin AV, Mokshin VV, Sharnin LM (2019) Adaptive genetic algorithms used to analyze behavior of complex system. Commun Nonlinear Sci Numer Simul 71:174–186
59. Franzin A, Stützle T (2019) Revisiting simulated annealing: a component-based analysis. Comput Oper Res 104:191–206
60. Jia ZH, Wang Y, Wu C, Yang Y, Zhang XY, Chen HP (2019, May) Multi-objective energy-aware batch scheduling using ant colony optimization algorithm. Comput Ind Eng 131:41–56
61. Latchoumi TP, Balamurugan K, Dinesh K, Ezhilarasi TP (2019) Particle swarm optimization approach for waterjet cavitation peening. Measurement 141:184–189

Chapter 8
On the Development of Energy-Efficient Distributed Source Localization Algorithm in Wireless Sensor Networks Using Modified Swarm Intelligence

Harikrushna Gantayat and Trilochan Panigrahi

1 Introduction

In wireless sensor networks (WSNs), the tiny inexpensive sensor nodes with sensing, processing and wireless communication capability are deployed arbitrarily for specific applications [1]. The framework for designing WSNs is provided in [2]. Accurate direction of arrival (DOA) (also referred as angle-of-arrival) estimation is required for localizing a source in sensor network. The DOA of signals at the sensor array can be determined from the outputs of a set of sensors. In literature, array signal processing methods have been reported, but all algorithms are processed data in centralized processor [3].

The source localization problem has many applications in WSNs, such as localization and tracking of vehicles, surveillance, localization of pollution sources, disaster rescue and so on. It may be an important task in Internet of Things-based health care system [4]. However, in general, it is a difficult problem. Firstly, sensor nodes in a WSN are densely deployed. The topology of a dynamic sensor network changes very frequently. Question is how to deploy and organize these sensors in an efficient way for the source localization. WSNs are also used for disaster management using flying ad hoc network [5]. Definitely, it is a challenging problem in WSN. The sensor nodes in a WSN are limited in power, computational capacities and memory. Hence, resource management for the sensor nodes is another issue. Several researcher proposed efficient methodology for sensor placement, sensor selection, sensor pairing, data aggregation, etc [2].

H. Gantayat (✉)
National Institute of Science and Technology, Berhampur, Odisha, India
e-mail: harifranky2006@gmail.com

T. Panigrahi
National Institute of Technology Goa, Ponda, Goa, India
e-mail: tpanigrahi80@gmail.com

D. De et al. (eds.), *Nature Inspired Computing for Wireless Sensor Networks*, Springer Tracts in Nature-Inspired Computing,
https://doi.org/10.1007/978-981-15-2125-6_8

The distributed sensor nodes in a network measure useful information such as received signal strength, direction of arrival, time difference of arrival of the acoustic, radio-frequency (RF), seismic or thermal signals emitted by a source [6]. Another issue in a WSN is all sensors are battery powered and of limited energy distributed in a unattended area. The fact is that data communications among the sensor nodes consume maximum energy. But much more energy efficient if the estimation algorithm performs distributed information processing than to do central processing which requires extensive communications which makes the network energy inefficient [7].

In this study, DOA-based source localization techniques are discussed. It is reported in literature that the maximum-likelihood (ML) is an efficient statistical method for DOA estimation [8]. In ML methods, the cost function is formulated from the data model, the covariance matrix of the input data vector of the array and the noise distribution. Then, the global maxima of ML function are obtained when the estimated DOAs are equal to actual angles during the optimization process. Standard iterative approaches are used to solve the nonlinear ML optimization problem which have in general huge computational complexity [9]. Therefore, swarm intelligence algorithms are used to optimize the ML function [10].

In fact, a single sensor alone cannot estimate the DOA. It needs an array of sensors (in sensor network, sensor itself with the immediate neighbours forms a subarray), and the algorithm for array signal processing can be used to assess the DOA. An alternative approach is considering the whole network as an arbitrary array, and central processor can estimate the DOA after accumulating the data from each sensor node. But there are two major issues in central processing methods. The first problem is the excessive communication overhead in the network. The second issue is difficult to maintain the coherence between signals received from widely separated sensors in large sensor network [11].

Recent attention has been paid to the distributed estimation of DOA of sources. It is because of adaptability of the distributed method with the change in environment and energy efficient. Here, the diffusion mode of cooperation is used where the sensors communicate with all of their neighbours [12]. A particular issue is regarded here in which each node in the network estimates the node-specific DOA from its locally defined log-likelihood cost functions.

Each sensor node in the sensor network forms an arbitrary array with the neighbours to estimate the DOA. After collecting spatially uncorrelated data, each node shares to its neighbours once in an experiment with their relative positions to form the local maximum-likelihood cost function. Then, the cost function is optimized at each sensor node using the evolutionary algorithm locally. And then, the nodes cooperate themselves with the diffusion mode of operation. The estimated parameters are diffused for further estimation and to achieve the global DOA with local optimization. The performance of the distributed algorithm is evaluated in terms of probability of resolution (PR) and root-mean-square error (RMSE). It is observed that the distributed estimation algorithm gives better performance at each sensor node compared to that obtained without cooperative. But the performance reduces when the network is sparse, that is the connectivity is less. Especially, the edge nodes suffer most as the degree is very less and unable to estimate the DOA properly.

To overcome the issues in distributed DOA estimation, a standard clustering approach is used to achieve efficient and scalable performance. Clustering approach saves energy and reduces networks communication overhead further [13]. It is because sensors communicate their data over shorter distance to their respective cluster heads (CH) either by one-hop or multi-hop communication perspectives. Distributed DOA estimation techniques are formulated here by inheriting the clustering idea. Each sensor shares the observation data once in each experiment to the CH. Each of the clusters cooperates themselves which is similar to the distributed in-network scheme to achieve the global estimates [10].

The organization of the book chapter is as follows. In Sect. 2, related work is given which reflects the need of the study. In Sect. 3, details of the maximum-likelihood DOA estimation of narrow-band far-field signal are provided that describes the distributed DOA estimation methods. The diffusion particle swarm optimization algorithm is described in Sect. 5. How the diffusion PSO algorithm is used for ML-DOA estimation in sensor network is explained in Sect. 4. Section 7 presents clustering-based distributed DOA estimation in WSN. Section 9 provides the future direction of the research work, and the chapter is concluded in Sect. 8.

2 Related Works

In this section, the work done in the area of distributed DOA estimation is described briefly. More emphasis will be given to swarm intelligence-based algorithm-based DOA estimation.

In literature, iterative search methods used to estimate DOA from the maximum-likelihood cost function. For example, alternating projection-approximated maximum-likelihood (AP-AML) [14], expectation-maximization (EM) and space-alternating generalized expectation-maximization (SAGE) algorithms [15] are applied for optimization of the ML function. Since ML function is multimodal in nature, the mathematical techniques fall into the local minima. Whereas, in literature, it has been found that the multiagent-based swarm intelligence algorithm like genetic algorithms (GAs), particle swarm optimization (PSO) and simulated annealing (SA) can easily handle multimodal cost functions. The PSO has been applied antenna array synthesis [16], electromagnetic optimization [17, 18]. This is also fact that the PSO algorithm provides competitive or even better results in a faster way compared with other heuristic methods like GA. Many researchers have used the GA and PSO and its variants [9, 19, 20] to estimate the DOA. The ML-PSO DOA estimation method is around 20 times more efficient than the GA-EM [9].

In WSNs literature, to overcome the communication overhead and computational burden in centralized processing, distributed approaches have been proposed. The distributed methods reduce the communication and adaptive to the environment change. Long back, a decentralized method has been proposed where the large array is divided into subarrays for the estimation of DOA [21]. Local processor uses the subarray information to detect the source and its location and shares to the central processor. Similarly, two decentralized array processing algorithms using MODE are proposed in [22]. The estimates of all local array are combined by using the theoretical bounds in order to achieve the global optimum.

It has been found that the source locations are estimated by measuring DOA with an antenna array at each sensor node [6, 23–25] in the sensor network. All these methods are called as decentralized where the task is distributed among different group of sensors but not distributed. It is because, in distributed estimation, each sensor node estimates the parameters locally and shares among the neighbouring nodes at each iteration to achieve the global DOA.

Distributed DOA estimation is formulated in [11]. Each sensor node forms subarray with the immediate neighbours. Then, each subarray collaborates themselves to locate the source DOA by optimizing a local ML function. Diffusive mode of cooperation is incorporated. Further, clustering-based approach is applied in WSNs [10]. For making the DOA estimation algorithm fully distributed, an alternative search techniques are to be applied for optimization of the ML function. Here, the PSO algorithm in its diffusive distributed version is used for the estimation of the DOA.

3 Maximum-Likelihood DOA Estimation of Narrow-Band Far-Field Signal

Let us consider a sensor network with N number of nodes distributed arbitrarily. There are $M (> N)$ number of narrow-band far-field sources impinge on the sensor network. Let the true location of the unknown sources is at $\boldsymbol{\theta} = [\theta_1, \theta_2, \ldots, \theta_M]$. Let $\mathbf{a}(\theta) \in \mathbb{C}^{N\times 1}$ be the complex response of the sensor array to a source in direction θ. The array response matrix of sensor array formed by the sensor network $\mathbf{A}(\boldsymbol{\theta}) = [\mathbf{a}(\theta_1), \ldots \mathbf{a}(\theta_M)]$ depends on the positions of sensor nodes. Let λ is the wavelength of the source. The complex gain of the lth signal at the kth node (with respect to a reference node), which is the (k, l)th element of matrix $\mathbf{A}(k, l)$, is [26]

$$\mathbf{A}(k,l) = \exp\left\{j\frac{2\pi}{\lambda}[x_k \sin\theta_l + y_k \cos\theta_l]\right\} k = 1, 2, \ldots, N, \quad l = 1, 2, \ldots, M \tag{1}$$

where, (x_k, y_k) is the position of kth sensor. The output $\mathbf{y}(i) \in \mathcal{C}^{N\times 1}$ from sensors array formed by whole sensor network is given as [27]

$$\mathbf{y}(i) = \mathbf{A}(\boldsymbol{\theta})\mathbf{s}(i) + \mathbf{v}(i), \qquad i = 1, 2, \ldots, L \tag{2}$$

where $\mathbf{s}(i) \in \mathbb{C}^{M\times 1}$ source signal. The noise vector $\mathbf{v}(i)$ is complex and normally distributed with zero mean and covariance matrix of $\sigma^2\mathbf{I}_N$, where $\mathbf{I}_N$ is the identity matrix of order N, σ^2 is the noise variance, and L is the number of snapshots. It is further assumed that the signal and noise vectors are assumed to be uncorrelated, and their covariance matrices are provided below as per the standard properties of complex uncorrelated signal vectors.

$$\begin{aligned} E[\mathbf{s}(i)\mathbf{s}(j)^{\mathrm{H}}] &= \mathbf{S}\delta_{ij} \\ E[\mathbf{s}(i)\mathbf{s}(j)^{\mathrm{T}}] &= 0 \\ E[\mathbf{v}(i)\mathbf{v}(j)^{\mathrm{H}}] &= \sigma^2\mathbf{I}\delta_{ij} \\ E[\mathbf{v}(i)\mathbf{v}(j)^{\mathrm{T}}] &= 0 \end{aligned} \tag{3}$$

where $\mathbf{S} = E[\mathbf{s}(i)\mathbf{s}^{\mathrm{H}}(i)]$ is the source signal covariance matrix, $(\cdot)^{\mathrm{H}}$ represents complex conjugate transpose, and $E(\cdot)$ denotes expectation operator. The array covariance matrix is given by

$$\mathbf{R} = E[\mathbf{y}(i)\mathbf{y}^{\mathrm{H}}(i)] = \mathbf{A}(\boldsymbol{\theta})\mathbf{S}\mathbf{A}^{\mathrm{H}}(\boldsymbol{\theta}) + \sigma^2\mathbf{I}_N \tag{4}$$

The symbols used in developing the algorithms are described in Table 1.

3.1 Formulation of ML-DOA Estimation Problem

In literature, authors have presented two ML models: conditional and unconditional model. Signals are assumed to be deterministic in conditional model, whereas in unconditional model, the signals are random in nature. But it has been proved in [28] that the performance of the unconditional model is significantly better. Therefore, the unconditional method is used here.

Following [27], the unconditional estimates of DOA for the arbitrary sensor array are obtained by using the array response matrix $\mathbf{A}$ and the signal covariance matrix

$$\hat{\mathbf{R}} = \frac{1}{L}\sum_{i=1}^{L}\mathbf{y}(i)\mathbf{y}^{\mathrm{H}}(i).$$

The global ML function is given by

$$f(\boldsymbol{\theta}) = \log\left|\mathbf{P}_{\mathbf{A}}\hat{\mathbf{R}}\mathbf{P}_{\mathbf{A}}^{H} + \frac{\mathbf{P}_{\mathbf{A}}^{\perp}}{N-M}\mathrm{tr}\left[\mathbf{P}_{\mathbf{A}}^{\perp}\hat{\mathbf{R}}\right]\right| \tag{5}$$

where $\mathrm{tr}[\bullet]$ denotes the trace of the matrix; $\mathbf{P}_{\mathbf{A}} = \mathbf{A}(\mathbf{A}_g^H\mathbf{A})^{-1}\mathbf{A}^H$ is the projection of matrix $\mathbf{A}$, and $\mathbf{P}_{\mathbf{A}}^{\perp} = \mathbf{I} - \mathbf{P}_{\mathbf{A}}$ is the orthogonal complementary.

Table 1 Symbols and their description

Symbols	Description
N	Number of nodes
M	Number of sources
λ	Wavelength of source signal
$s(i)$	Source signal
$v(i)$	Noise vector
I_N	Identity matrix of order N
σ^2	Noise variance
L	Number of snapshots
$\mathbf{S}$	Source signal covariance matrix
$()^H$	Complex conjugate transpose
$E()$	Expectation operator
$\mathbf{R}$	Array covariance matrix
$\mathbf{A}$	Array response matrix
$\hat{\mathbf{R}}$	Sample covariance matrix
$tr[]$	Trace of the matrix
$\mathbf{P_A}$	The projection of matrix A
N_k	Collection of sensor nodes in the kth node neighborhood
n_k	Degree of kth node
$\mathbf{A}_k()$	Steering matrix
$v_k(i)$	Noise process at the kth subarray
$\hat{\mathbf{R}}_k$	Sample covariance matrix of the measured data at kth subarray
$f_k(\boldsymbol{\theta})$	ML cost function at kth sensor
$\mathbf{A}_k(\boldsymbol{\theta})$	Response matrix of kth subarray
$p_k(\mathbf{y}_k(i)\|\boldsymbol{\theta})$	The probability density function (pdf) for single snapshots at kth subarray
$p_k(\mathbf{y}_k\|\boldsymbol{\theta})$	The joint pdf for L number of independent snapshot
$f(\boldsymbol{\theta})$	Global log-likelihood function
c_{kl}	Combiner coefficients between node k and l
$\hat{\boldsymbol{\theta}}_j$	DOA estimates of neighbouring sensors
$\mathbf{p}^i_{jk}$	The position vectors of the jth particle for the kth sensor
$\mathbf{v}^i_{jk}$	The velocity vectors of the jth particle for the kth sensor
n_k and n_l	The degree for nodes k and l
K	The maximum iteration number

4 Distributed DOA Estimation

The issue of concern is to predict DOA $\boldsymbol{\theta}$ in a distributed way from the observations $\{\mathbf{y}(i)\}_1^L$. In the sensor network, if they can interact immediately with each other, we can tell that two nodes are linked. The data communication in WSN is followed by a routing algorithm. Further, routing is a challenging task when the network is hybrid. A geometrical programming and many more intelligence approach for routing such kind of ad hoc network are given in [29, 30]. A node is linked to itself at all the times. Let $\mathcal{N}_k$ represents a collection of sensor nodes in the kth node neighbourhood. The number of nodes linked to node k is referred to as the degree of kth node and is referred to as n_k.

4.1 Local Cost Function for DOA Estimation

Each sensor in a network can estimate the DOA by forming a local array with the one-hop immediate neighbours. Each node in the network with its instant neighbours creates a random array [11]. Now, the complex n_k-vector $\mathbf{y}_k(i)$ for the output of kth subarray becomes

$$\mathbf{y}_k(i) = \mathbf{A}_k(\boldsymbol{\theta})\mathbf{s}(i) + \mathbf{v}_k(i), \qquad i = 1, 2, \ldots, L \tag{6}$$

where $\mathbf{A}_k(\boldsymbol{\theta}) \in \mathbb{C}^{n_k \times M}$ is steering matrix and the vector $\mathbf{v}_k(i) \in \mathbb{C}^{n_k}$ is the noise process at the kth subarray.

Now, it is possible to estimate the DOA in the kth subarray using subspace-based algorithms like MUSIC, ESPIRIT, the Capon beam former, etc. The covariance matrix of the measured data is to be estimated first for any algorithm as

$$\hat{\mathbf{R}}_k = \frac{1}{L}\sum_{i=1}^{L} \mathbf{y}_k(i)\mathbf{y}_k^{\mathrm{H}}(i) \tag{7}$$

For MUSIC algorithm, the eigenvalues and corresponding eigenvectors of $\hat{\mathbf{R}}_k$. The noise subspace is assumed to orthogonal to the signal space. Further, the noise subspace is determined to give an estimate of the spatial spectrum. Then, the spatial spectrum is used to estimate the DOA.

But these methods may not provide best estimate as the size of subarray is small, and subspace methods need more sample to estimate the DOA efficiently. The size may increase by increasing the connectivity of sensor nodes in sensor network. But, the number of snapshots may not be possible in sensor network with limited time. Thus, we prefer an algorithm which can provide best estimate even though the number of snapshots is less and for smaller array size. For this, the statistically efficient algorithm is chosen here, i.e. maximum-likelihood algorithm. Now, the ML cost function at kth sensor is given as

$$f_k(\boldsymbol{\theta}) = \log\left|\mathbf{P}_{\mathbf{A}_k}\hat{\mathbf{R}}_k\mathbf{P}_{\mathbf{A}_k}^H + \frac{\mathbf{P}_{\mathbf{A}_k}^{\perp}}{N-M}\mathrm{tr}\left[\mathbf{P}_{\mathbf{A}_k}^{\perp}\hat{\mathbf{R}}_k\right]\right| \tag{8}$$

where

- $\mathbf{A}_k$ is the response matrix of kth subarray
- $\hat{\mathbf{R}}_k$ is the sample covariance matrix defined in (7)
- $\mathbf{P}_{\mathbf{A}_k} = \mathbf{A}_k(\mathbf{A}_k^H\mathbf{A}_k)^{-1}\mathbf{A}_k^H$ is the projection of matrix $\mathbf{A}_k$
- $\mathbf{P}_{\mathbf{A}_k}^{\perp} = \mathbf{I} - \mathbf{P}_{\mathbf{A}_k}$ is the orthogonal complementary.

The local cost function $f_k(\boldsymbol{\theta})$ defined in (8) is obtained by replacing $\mathbf{A}$ by $\mathbf{A}_k$ and sample covariance matrix $\hat{\mathbf{R}}$ by $\hat{\mathbf{R}}_k$ in the ML cost function for whole array defined in (5).

4.2 *Distributed DOA Estimation Using Local ML Functions*

In distributed estimation, each sensor should have local cost function with same objective. Here, every sensor node in the network forms subarray with the neighbours. By doing this, the array factor is different which leads to different local ML function $f_k(\boldsymbol{\theta})$. Each node tries to estimate the DOA by following an iterative optimization method. But the difficulty to estimate to achieve the common objective that is determining the source DOA using the multimodal nature of the ML cost function. Mathematical techniques may not be feasible here. Thus, evolutionary algorithm like PSO-based ML solutions [9] is proposed in literature. Distributed version of the PSO known as diffusion PSO is used here. Each node shares its estimated values with its neighbours after estimating the local DOA. After combining all the received estimates by a node, then update own estimate.

Now, the objective is to formulate the distributed ML function. As per the theory of distributed estimation, the global cost function should be formulated as the sum of local cost functions [31–33]. In fact, each sensor node formulates its own ML cost function after accumulating the received data from the source. A sensor node is connected to any other neighbouring nodes. So, there are common nodes among the subarrays. To approximate the global function as the sum of local ML, assumption like the data is temporally and spatially independent to be considered.

4.2.1 Problem Formulation for Distributed ML-DOA Estimation

Let us consider the data accumulated at kth sensor (that is k subarray). For multivariate data, the probability density function (pdf) for single snapshots at kth subarray is given as [27]

$$p_k(\mathbf{y}_k(i)|\boldsymbol{\theta}) = \frac{1}{\det[\pi \mathbf{R}_k]} \exp\left\{-(\mathbf{y}_k^H \mathbf{R}_k^{-1} \mathbf{y}_k)\right\} \tag{9}$$

Since each array receives L number of independent snapshots, the joint pdf $p_k(\mathbf{y}_k(i)|\boldsymbol{\theta})$ is defined as

$$p_k(\mathbf{y}_k|\boldsymbol{\theta}) = \prod_{i=1}^{L} \frac{1}{\det[\pi \mathbf{R}_k]} \exp\left\{-\mathbf{y}_k^H(i)\mathbf{R}_k^{-1}\mathbf{y}_k(i)\right\} \tag{10}$$

Log-likelihood cost function for DOA estimation by the sensors for L independent snapshots is defined as

$$f(\boldsymbol{\theta}) = \ln \prod_{i=1}^{L} p(\mathbf{y}(i)|\boldsymbol{\theta}) = \sum_{i} \ln p(\mathbf{y}(i)|\boldsymbol{\theta}) \tag{11}$$

where $p(\mathbf{y}(i)|\boldsymbol{\theta})$ is the joint pdf for whole sensor array.

Since it is assumed that the snapshots measured are independent, simple product of probabilities is used here. Then, to transform the product to summation, the log-likelihood ML function is considered. Further, the log function inverts the exponential function in Gaussian probability density function. To make the ML-DOA estimation as distributed, the approximation made in (12) is valid.

$$p(\mathbf{y}_k(i)|\boldsymbol{\theta}) \approx \prod_{i=1}^{L} p_k(\mathbf{y}_k(i)|\boldsymbol{\theta}) \qquad k = 1, 2, \ldots, N \tag{12}$$

Then,

$$p(\mathbf{y}(i)|\boldsymbol{\theta}) \approx \prod_{k=1}^{N} p_k(\mathbf{y}_k(i)|\boldsymbol{\theta}) \tag{13}$$

The data at each subarray are not independent because each sensor node shares the observed data to its neighbouring nodes. To make completely independent, clustering-based approach is proposed [10]. But certain assumption able to form the approximation by substituting (13) in (11) yields

$$f(\boldsymbol{\theta}) \approx \sum_{k=1}^{N} \left\{ \sum_{i=1}^{L} \ln p_k(\mathbf{y}(i)|\boldsymbol{\theta}) \right\} \tag{14}$$

Now, this cost function can be used for distributed optimization as the global cost function is approximately equal to the sum of local cost functions at each subarray [32]. In fact, this approximation in (14) will have some loses in its performance which will be realized in the numerical simulation result. The global function has now been given as

$$f(\boldsymbol{\theta}) = \text{minimize} \sum_{k=1}^{N} f_k(\boldsymbol{\theta}) \tag{15}$$

where each $f_k(\theta)$ is representing the local cost function which is known to kth senor node only. The local cost function for kth subarray is given as

$$f_k(\boldsymbol{\theta}) \approx \sum_{i=1}^{L} \ln p_k(\mathbf{y}(i)|\boldsymbol{\theta}) \tag{16}$$

The local ML function given in (8) can be formulated as the right-hand side of (16). Therefore, the global log-likelihood function can be approximated as

$$f(\boldsymbol{\theta}) \approx \sum_{k=1}^{N} \log \left| \mathbf{P}_{\mathbf{A}_k} \hat{\mathbf{R}}_k \mathbf{P}_{\mathbf{A}_k}^H + \frac{\mathbf{P}_{\mathbf{A}_k}^{\perp}}{N-M} \mathrm{tr}\left[\mathbf{P}_{\mathbf{A}_k}^{\perp} \hat{\mathbf{R}}_k \right] \right| \tag{17}$$

In a sensor network, the degree of each of the sensors is different, and the sensors are distributed randomly. Thus, the size and geometry of each of the subarray formed are completely different from others. It shows that each sensor has a different objective function which depends on the topology of the sensor network and the position of sensor nodes. In fact, all are forming the local maximum-likelihood function based on the actual source direction. Therefore the shape of the function may different, but global optima are common.

4.2.2 Need of Iterative Optimization Algorithm

The problem in the log-likelihood function is multimodal in nature which is not convex. But, all sensor node shares a common global maximum in their respective ML cost function. Therefore, one can assume that in all the local ML functions are approximately convex in a small region around this global maximum [7]. If an iterative multiagent search algorithm searches the convex region where the cost function is convex, then the distributed estimation theory [32] can be applied.

The evolutionary optimization algorithm has that beauty, it can search the global maxima in a multimodal cost function. Since the present approach is distributed estimation of the DOA, now diffusion PSO may be used where individual sensors interact with other neighbouring sensors to make the search process faster and accurate [34]. Inherently in diffusion approach, each sensor node interacts with the neighbours to update the estimated parameter with the help of local observation and cost function. The proposed algorithm is to minimize the multimodal cost function (15) in distributed way to estimate the direction of arrival.

Briefly, one can understand the process in the following way. Each sensor node starts with an initial estimate of the unknown variable $\boldsymbol{\theta}_k(0)$ and updates its estimate at discrete times $i = 1, 2, \ldots$ using PSO learning algorithm. Let $\boldsymbol{\theta}_k(i)$ denotes the DOA vector estimated by the kth sensor at $i$the time instant. Each sensor receives the estimated DOA from the neighbours $j \in \mathcal{N}_k$ first. Then, the sensor node combines its current estimate with the received estimates from its neighbours. The combining is done by the following equation.

$$\boldsymbol{\theta}_k(i+1) = \sum_{j \in \mathcal{N}_k} c_{kj}(i)\boldsymbol{\theta}_j(i) \tag{18}$$

where the combiner coefficients $c_{kj}(i)$ are the factor by which the DOA estimates of neighbouring sensors $\boldsymbol{\theta}_j(i), j \in \mathcal{N}_k$ contribute to the kth sensor.

The combiner coefficient depends on the connectivity of the sensor nodes which is assigned as positive nonzero value when the neighbour sensor j connected with k. The coefficient is zero, if neighbour j is not communicated with k. In literature, different methods are proposed to calculate these weights according to the network topology [12]. Coefficients are assumed to be constant for network during the estimation time. The details of the diffusion particle swarm optimization are given in the next section.

5 Diffusion Particle Swarm Optimization (DPSO)

The particle swarm optimization (PSO) is a multiagent heuristic search technique developed in 1995 [35, 36]. In PSO, each particle represents random gausses in the solution space known as particle position. The position of each particle updates with its associated random velocity. The particle has its memory of own best experience. The particle iteratively updates its velocity according to its own and group's experiences known as the *pbest* and *gbest*, respectively.

Let us consider a swarm of P particles at each sensor. Each particle is having M dimensional positions and velocity vectors. For jth particle of the kth sensor nodes, position and velocity are represented by $\mathbf{x}^i_{jk} = [x^i_{jk1}, x^i_{jk2}, \ldots, x^i_{jkM}]^T$ and $\mathbf{v}^i_{jk} = [v^i_{jk1}, v^i_{jk2}, \ldots, v^i_{jkM}]^T$, respectively. These particles change their positions in the hyperspace (i.e. $\mathbb{R}^M$) with two best estimates which are *pbest* and *gbest*. Let $\mathbf{p}^i_{jk} = [p^i_{jk1}, p^i_{jk2}, \ldots, p^i_{jkM}]$ denotes the *pbest* of the jth particle at the kth node and that of *gbest* is denoted as $\mathbf{p}^i_{gk} = [p^i_{gk1}, p^i_{gk2}, \ldots, p^i_{gkM}]$. During the searching process, each particle updates its velocity and position vectors according to the following update equations:

$$\mathbf{v}^{i+1}_{jk} = \omega^i \mathbf{v}^i_{jk} + c_1 \mathbf{r}^i_1 \odot (\mathbf{p}^i_{jk} - \mathbf{x}^i_{jk}) + c_2 \mathbf{r}^i_2 \odot (\mathbf{p}^i_{gk} - \mathbf{x}^i_{jk}) \tag{19}$$

$$\mathbf{x}^{i+1}_{jk} = \mathbf{x}^i_j + \mathbf{v}^{i+1}_{jk} \tag{20}$$

where

- $\odot$ represents the element-wise product,
- $j = 1, 2, \ldots, P$ is the particle index;
- $i = 1, 2, \ldots,$ denotes the iterations and
- $k = 1, 2, \ldots, N$ is used as the sensor number.

In weighted PSO algorithm, the velocity of each particle is updated based on the current velocity scaled by *inertial weight* ω. Every particles are always trying to move towards own best that is *pbest* and global best known as *gbest*. The constants c_1 and c_2 are acceleration factors used to determine the relative pull of *pbest* and *gbest*. These constants are called as *cognitive* and *social* parameters, respectively [17, 36]. Two independent random vectors $\mathbf{r}_1$ and $\mathbf{r}_2$ (M-dimensional) uniformly distributed between 0 and 1. These random vectors determine the relative pull of *pbest* and *gbest* randomly [17]. The performance of the PSO algorithm can be improved with certain modifications in the update equation. Such variant PSO algorithms are used in literature for the ML-DOA estimation such as adaptive PSO [20] and comprehensive learning PSO [19].

While optimizing in a distributed DOA estimation, in each iteration, sensor nodes mutually share their *gbest* with the neighbouring nodes. To do so, diffusion cooperation strategy among the sensor nodes is incorporated in the conventional PSO algorithm which is known as diffusion PSO (DPSO). The steps in DPSO are

- Each particle receives the *gbest* from the neighbours and combines them.
- Each sensor updates its particle's velocities,
- then updates particle positions,
- Then, updates one best that is *pbest* and
- All particle together finds their *gbest*,
- Shares its *gbest* vector with the neighbours.

The parameters *pbest* and *gbest* are measured with local cost function ML. There are two ways a node can utilize the received *gbest*. At every sensor, a consensus mechanism given in (21) is used to combine the received *gbest* estimates from the neighbours,

$$\mathbf{p}_{gk}^{(i-1)} = \sum_{l \in \mathcal{N}_k} c_{kl} \mathbf{p}_{gk}^{(i-1)} \tag{21}$$

In diffusion process, each sensor node receives the estimated direction of arrival from the neighbours. These estimated values are combined with weighted coefficients where the sum of the coefficients is one. Equal weights are the simplest way to chose the coefficients. But in metropolis, more weights are giving to the array having more number of sensors (i.e more degree), and it is the fact that the array having more sensors has better direction of arrival estimation capability. The metropolis weight is used to calculate the coefficients c_{kl} as [12]

$$c_{kl} = \begin{cases} \frac{1}{\max(n_k, n_l)}, & \text{if } k \neq l \text{ are linked} \\ 0, & \text{for } k \text{ and } l \text{ are not linked} \\ 1 - \sum_{l \in \mathcal{N}_k / k} c_{kl}, & \text{for } k = l \end{cases} \tag{22}$$

where n_k and n_l are the degree for nodes k and l.

Each sensor node fuses the received DOA estimates from the immediate neighbours as per the metropolis weigh defined in (22). The fused DOAs $\mathbf{p}_{gk}^{(i-1)}$ help the node to do the local optimization process and to have global DOA. In each iteration, the global estimated DOA shares beach node to their neighbours for further diffusion process.

6 Diffusion PSO Algorithm for ML-DOA Estimation in Sensor Network

In distributed estimation, every sensor node in the network estimates the required information locally by using local cost function. And then share the estimate with neighbours for further processing. Here, each sensor node in the network forms an arbitrary array of size equal to its degree with its immediate neighbours. At each subarray, the sensor node accumulates the data from all the participating nodes of the subarray and then formulates its local ML cost function. By following the diffusion consensus in the network, each sensor node attempts to estimate DOA locally by cooperating with other sensor nodes in the network. For this, each sensor node uses the diffusion PSO algorithm to compute the local ML function. The distributed DPSO algorithm starts by initializing a predefined number of particles at each node in the search range of 0 to π with random positions in each dimensions. Then, the associated velocity of each particle is initialized randomly in the range of 0 to π. The position vector of the kth sensor node and jth particle is defined as $x_{kj} = [\theta_{kj1}, \ldots, \theta_{kjM}]$.

A suitable mapping is used to determine the candidate solution from particle position vector in the problem space. In the very first iteration of the algorithm, the particle's random positions are regarded as the *pbest*. Each particle calculates its fitness at each node by using the local ML function (8). The *gbest* is computed from the fitness of all the particle at each node. For diffusion purpose, now the *gbest* is shared by each node with their neighbours. At every node, a diffusion mechanism given in (21) is used to fuse the received *gbest* from the neighbours in tye network. Now, the fused *gbest* is used for subsequent update of velocity and position vectors by using the Eqs. (19) and (20).

In diffusion PSO algorithm, the key equation in the search process is the velocity update equation. It has been seen that there are three components which contribute to the change of current velocity to new velocity. The updated velocity further used to change the position of a particle. The first part of the velocity update equation is proportional to the present velocity. This guides each particle to proceed in the same direction. The convergence behaviour of DPSO algorithm is defined from the inertial weight w which is multiplied to the past velocity decides [37]. During the initial iterations, a larger w value facilitates global exploration. After certain iteration, a smaller w facilitates local exploitation in the current search area which is required at the end of the optimization process. It is because, during the last iterations, the particle is very close to its global solution. In conventional weighted PSO, w decreases during the optimization process.

Let the maximum and minimum value of w are $w_{\max}$ and $w_{\min}$, respectively. The $w(i)$ is updated with respect to the iterations as follows:

$$w^i = \begin{cases} w_{\max} - \frac{w_{\max}-w_{\min}}{rK}(i-1), & \text{if } 1 \le i \le [rK] \\ w_{\min}, & \text{for } [rK]+1 \le i \le K \end{cases} \tag{23}$$

where

- $[rK]$ is the number of iterations up to which inertial weight decreases,
- r is the ratio $0 < r < 1$,
- K is the maximum iteration number, and
- $[\cdot]$ denotes rounding operator.

If we look to the literature of PSO [38], similar values are selected as $w_{\max} = 0.9$, $w_{\min} = 0.4$, and $r = 0.4 \sim 0.8$.

Now look to the second and third components of the velocity update equation. These terms force the particles to move towards the particle's own best that *pbest* and the group's best positions which is *gbest*. The constants c_1 and c_2 are used to bias the particle's search towards the two best locations pbest and gbest, respectively. Further, the velocity should not go beyond certain limit to avoid the swarm from divergence [39]. The velocity limit is defined at each node as

$$v^i_{jk} = \begin{cases} V_{\mathrm{MAX}}, & \text{if } v^i_{jk} > V_{\mathrm{MAX}} \\ V_{\mathrm{MIN}}, & \text{if } v^i_{jk} < V_{\mathrm{MIN}} \end{cases} \tag{24}$$

The value of V_{MAX} should be limited to half of the dynamic range. The particle's new position is updated by using (20). Just like the velocity is limited in between its maximum and minimum value, the position also bounded to its actual range. If any dimension of the new position vector is going beyond the boundary that is in between 0 and π, then it is adjusted to lie within this range only. The termination criteria are defined here are the maximum number of iterations K. After completion of K iterations, the final gbest p_{kg} is the ML estimates of source DOA. The main steps of ML-DPSO algorithms are outlined in Algorithm 1 .

6.1 Performance Measure

The performances of DOA estimation algorithms are measured in two ways. One is root-mean-square error (RMSE), and other one is probability of resolution (PR). The RMSE provides how close the estimated angles with the true value. Similarly, the PR gives us the ability of the algorithm to resolve the closely spaced sources. These parameters are defined as follows.

Algorithm 1: Main steps of Diffusion PSO algorithm

```
Setup Problem:
• Define sensor network with topology.
• Calculate metropolis weights for diffusion process.
• Set up random array at each node with its immediate neighbours.
• Define local fitness function at each node.
• Select PSO parameter.
Swarm Initialization at each node:
• Random positions.
• Random velocities.
for each iteration do
    for each node do
        for each particle do
            Map particle location to solution vector in solution space;
            Evaluate the objective function of current iteration of the node according to its
            local ML function (8);
            According to fitness value update particles best location p_k^i and group best
            location g_k^i;
            Update particles velocity according to (19);
            if velocity exceeds the limits then
                Limit particles velocity using (24);
            end
            Update particle position using (20);
            if particles position out of solution boundaries then
                Clip or adjust particles position;
            end
        end
        Share its gbest to all neighbour nodes;
    end
    Do the local diffusion according to (21);
    Check termination criterion;
end
```

- Root-mean-square error (RMSE): The RMSE is defined as

$$\text{RMSE} = \sqrt{\frac{1}{MN_{\text{run}}} \sum_{i=1}^{N_{\text{run}}} \sum_{l=1}^{N} \left(\hat{\theta}_l(i) - \theta_l \right)^2} \tag{25}$$

  where M is the number of source signals, $\hat{\theta}_l(i)$ is the estimate of the lth DOA achieved in the ith run, θ_l is the true DOA of the lth source.
- Probability of Resolution (PR): The PR is the ability of DOA estimation algorithm to resolve closely spaced sources. Two sources are said to be resolved in a particular run if both $|\hat{\theta}_1 - \theta_1|$ and $|\hat{\theta}_2 - \theta_2|$ are smaller than $\Delta\theta/2$ where $\Delta\theta = |\theta_1 - \theta_2|$ is the separation between two sources.

6.2 Example

In this example, the performance of the proposed distributed DOA estimation method for the network with less connectivity is validated and compared with decentralized algorithm like ML-PSO and MUSIC. The average degree of connectivity is less. The network topology is given in Fig. 1. The network can be enlarged to any number of sensor nodes, but to show the validity of the proposed distributed ML-DOA estimation algorithm using DPSO, a sensor network with 18 nodes is considered as an illustration. The cognitive parameters c_1 and c_2 are chosen 2. The DPSO algorithm is run for 200 number of iterations.

The simulation results are plotted in Figs. 2 and 3. The index used to describe the algorithms in figures is provided in Table 2. Since the performance of the algorithm depends upon noise and power of additive random noise with respect to the signal power is defined by the signal to noise ratio (SNR). Therefore, all the numerical results are provided here by varying the SNR. The SNR varies from -20 dB to 30 dB with a step size of 1 dB. The simulation results are shown in the figures by using centralized ML-PSO, distributed ML-DPSO and centralized MUSIC algorithms. Average over 500 Monte Carlo simulations results for the MUSIC algorithm is plotted. Hundred numbers of snapshots are used in MUSIC algorithm for which the RMSE asymptotically closes to the theoretical bound known as Cramer-Rao lower bound (CRLB). Whereas, the distributed ML-DPSO and centralized ML-PSO algorithms are repeated for 100 Monte Carlo simulations by choosing only 20 snapshots in order to reduce the communication overhead in distributed estimation.

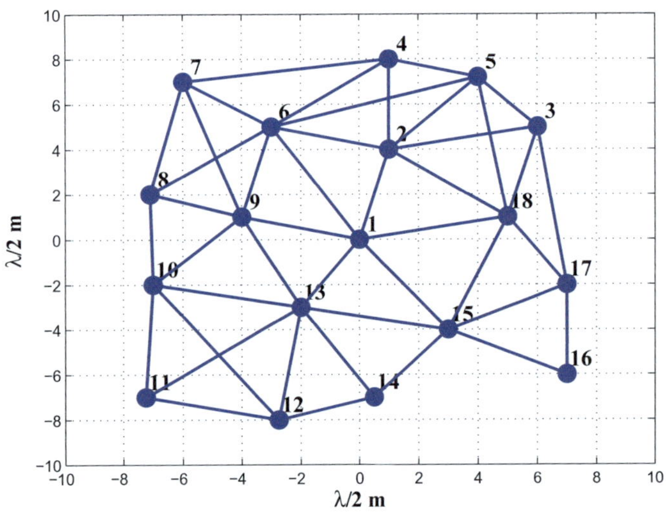

Fig. 1 Network topology used in Example 1

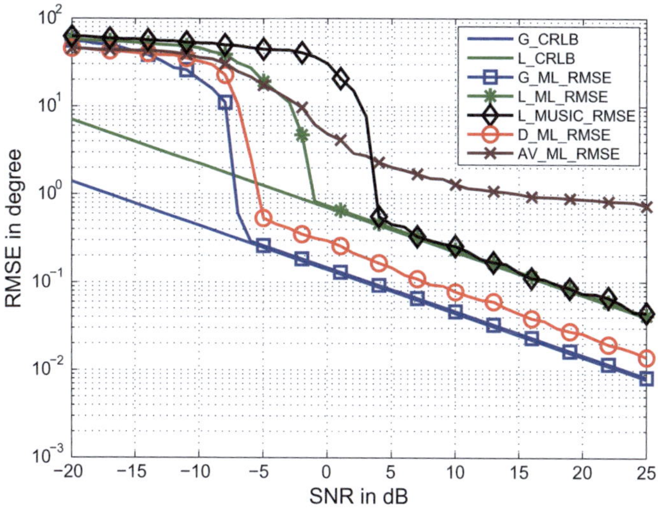

Fig. 2 RMSE versus SNR plots in DOA estimation by global ML-PSO, Diffusion ML-PSO and MUSIC methods at Subarray 1 with and without cooperation

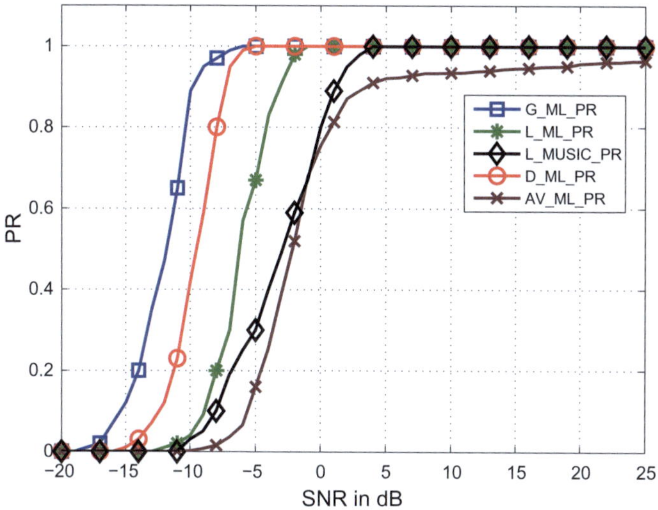

Fig. 3 PR versus SNR plots in DOA estimation by global ML-PSO, Diffusion ML-PSO and MUSIC methods at Subarray 1 with and without cooperation

Table 2 Index and their description used in the figures

Index	Description
G_CRLB	CRLB for global array
L_CRLB	CRLB for local subarray
G_ML_RMSE	RMSE performance of ML-PSO for global array
L_ML_RMSE	RMSE performance of ML-PSO for local subarray without diffusion
L_MUSIC_RMSE	RMSE performance of MUSIC at local subarray
D_ML_RMSE	RMSE performance of ML-DPSO at subarray 1
AV_ML_RMSE	Average RMSE performance of ML-DPSO over all the subarrays
G_ML_PR	PR performance of ML-PSO for global array
L_ML_PR	PR performance of ML-PSO for local subarray without diffusion
L_MUSIC_PR	PR performance of MUSIC at local subarray
D_ML_PR	PR performance of ML-DPSO at subarray 1
AV_ML_PR	Average PR performance of ML-DPSO over all the subarrays

It is observed from Fig. 3 that the performance of the distributed ML-DPSO at Sensor 1 in terms of PR is at par with that of the centralized ML-PSO. In centralized ML-PSO, the estimation is done by using all sensor nodes in the network by forming a global array. Now if we compare the performance of Subarray 1 at Sensor 1, the performance with cooperation is much better than without cooperation. In fact, the global MUSIC with higher number of snapshots is not having better performance.

Similar analysis can be done using the RMSE performance in Fig. 2. The RMSE performance of ML-DPSO at Sensor 1 is in between the centralized ML-PSO and Subarray 1 without cooperation among other subarrays. At lower SNR (in between -20 dB to -7dB), at Sensor 1, the overall performance of ML-DPSO and centralized ML-PSO is nearly the same.

When the Subarray 1 estimates the DOA without any cooperation, then the best achievable performance would be the CRLB. But the same subarray is providing better performance when the estimation is carried out by distributed manner. It is because, in distributed estimation, every subarray tries to attain global estimation. Therefore, at Subarray 1, DPSO is lower than CRLB of Subarray 1.

The convergence performance of the PSO algorithms at 10 dB SNR in a particular experiment is shown in Figs. 6 and 7. It is observed from the figure that global PSO (i.e. centralized method) is giving better performance than distributed PSO. But in centralized approach, the amount of communication overhead is more. The distributed approach needs less number of iterations to achieve PR is 1 which is evident from the Fig. 6. Similarly, the RMSE also asymptotically approaches to the CRLB faster than non-distributed approach which is shown in Fig. 7. The steady-state value of the RMSE in distributed approach is in between the CRLB of global array and the Subarray 1.

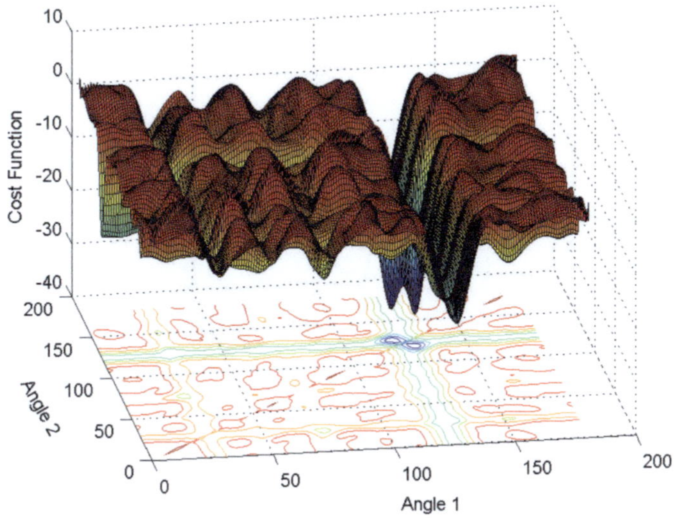

Fig. 4 ML functions at Subarray 1 of Example at 10 dB SNR

In general, the DOA estimation ability of an sensor array increases with the number of elements presents in the array. If the degree of a sensor node is two and trying to estimate the DOA of two sources, then whatever may be the algorithm used the performance will be very poor. It can be seen from the ML function plotted in and Fig. 5 at Sensor 16. The local minima are dominating the global one. But, in the ML cost function at Sensor node 1 plotted in Fig. 4, the global minimum is dominating over all local minima. If we see the global ML function at Node 1, the global minima look like a quantum well (especially at high SNR when signal completely dominates the noise). Sometimes, the PSO algorithm also unable to search the solution. Variants of PSO like adaptive PSO and CLPSO are used to handle that problem.

Practically, the sensors present in the interior of the network have degree more than two compared to the sensors lies at the edges of the network. In a network, if the average degree of the nodes is less, then the overall performance is poor. But at the same time, the overall communication is also less. So, we have to compromise between communication overhead and performance. For any connectivity, the performance of the distributed algorithm is better compared to individual subarray produced at each sensor. In addition to the independent issue, while deriving the distributed cost function, the connectivity problem also can be overcome by using the clustering technique.

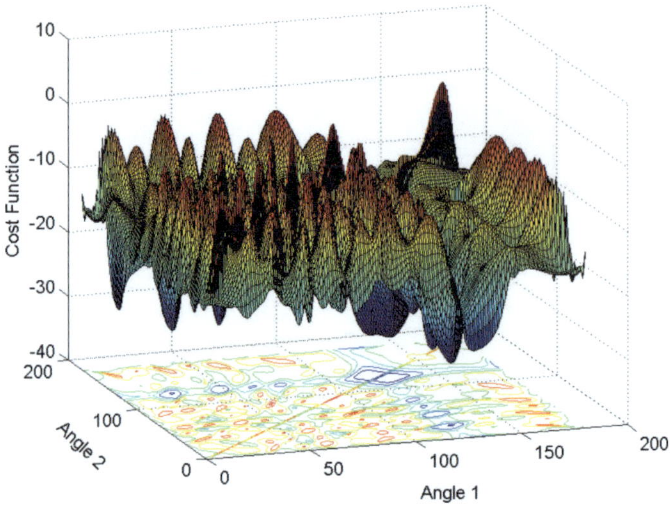

Fig. 5 ML functions at Subarray 16 of Example at 10 dB SNR

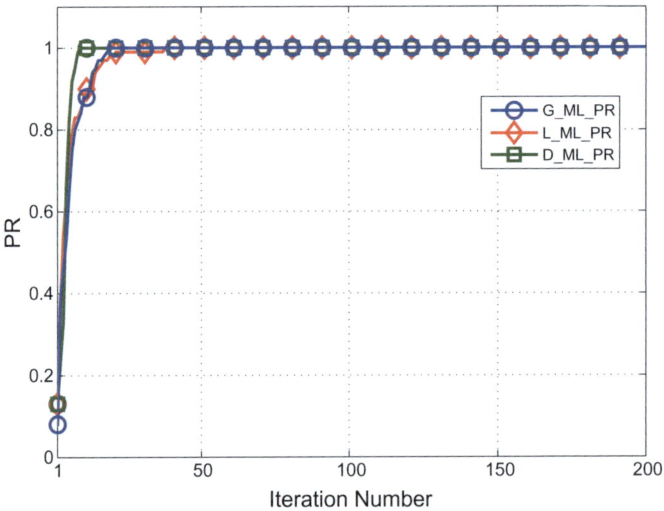

Fig. 6 PR versus Iterations plots in DOA estimation by global ML-PSO and Diffusion ML-PSO methods at Subarray 1 with and without cooperation

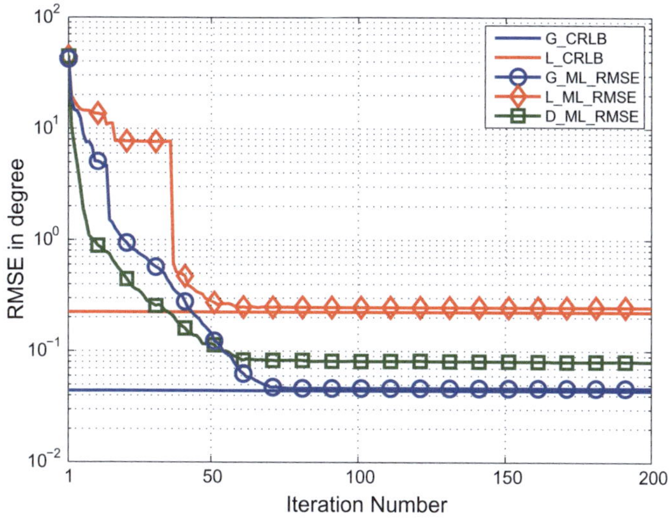

Fig. 7 RMSE versus Iterations plots in DOA estimation by global ML-PSO and Diffusion ML-PSO methods at Subarray 1 with and without cooperation

7 Clustering-Based Distributed DOA Estimation in Wireless Sensor Networks

The performance obtained by individual sensors with cooperation increases (later will see from the simulation results) compared to that of non-cooperative approach but less than the global estimates where the whole sensor network is considered as an arbitrary sensor array. It is due to the assumption (16) taken for the distributed ML formulation. In fact, the formulation of centralized cost function as the sum of cost functions at each subarray is a problem. This is required for consensus algorithm in a intelligent sensor network. Further, the performance of the distributed algorithm is also inclined by the connectivity of the sensor in the network. It is because the size of the array is a crucial factor for the ability of an array to find DOA. For lower size subarray, the cost function cannot provide better performance. In WSNs, the edges have less connectivity. Therefore, clustering-based diffusion cooperation is the best solution to overcome above-said problems.

The sensors organize themselves into groups known as clusters and collaborate to locate and multiple sources. Each cluster is considered as an arbitrary array. Individual cluster can attempt to estimate the DOA by optimizing its local ML function using. Then, diffusive mode of cooperation among the clusters can be incorporated to get the global estimates. It is the fact that the centralized algorithm in WSN always gives more accurate estimation but not energy efficient because of more communication overhead. But the distributed in-network algorithm may not provide accurate estimate for lesser connectivity sensor network. Whereas, the advantage of distributed in-cluster algorithm is that the best suitable when the application demands energy efficiency with nearly equal performance to that of centralized algorithm.

Clustering-based distributed approach reduces the computational burden on individual sensor nodes compared to the distributed in-network algorithm discussed before. In distributed in-network algorithm, each sensor optimizes the local ML function using PSO. Whereas, in distributed in-cluster algorithm, the PSO algorithm runs at the cluster head only. Further, the number of communications among the cluster can be minimized by using block concepts in [40]. Initially, the PSO algorithm needs more information from the global network to search the global optimum in their local cost functions. After certain number of iterations, clusters may share their estimated information after fixed interval in order to reduce the communication overheads. As a result, the communication overhead is reduced which makes the algorithm energy efficient. This clustering-based source localization is one of the good solutions for home application [41].

7.1 *Clustering-Based Distributed DOA Estimation*

In this section, the distributed DOA estimation algorithm among the clusters is developed. Let us divide the whole sensor network into N_c number of clusters. Let

- $\mathcal{N}_j^c$ is the set of sensors belonging to jth cluster,
- n_j^c is the number of sensors in jth cluster, and
- n_c is the average cluster size

After clustering of the sensor nodes using any conventional clustering algorithm, each sensor node shares the received data to the cluster head by using either single-hop or multi-hop communication mode which is shown in Fig. 8. Let $\mathbf{y}_j^c(i)$ is the data vector of the sensor array formed at jth cluster is given as

$$\mathbf{y}_j^c(i) = \mathbf{A}_j^c(\boldsymbol{\theta})\mathbf{s}(i) + \mathbf{v}_j^c(i), \qquad i = 1, 2, \ldots, L \tag{26}$$

where $\mathbf{A}_j^c(\boldsymbol{\theta})$ denotes the array response matrix of that arbitrary array of the jth cluster and vector $\mathbf{v}_j^c(i)$ represents the random noise. This data model is similar to that of (6) for subarray. The difference here is that each of the sensor shares to one of the cluster head only unlike to all the neighbour in the previous case.

The joint pdf $p_c(\mathbf{y}_c(i)|\boldsymbol{\theta})$ for L snapshots is defined as

$$p_j^c(\mathbf{y}_j^c|\boldsymbol{\theta}) = \prod_{i=1}^{L} \frac{1}{\det[\pi R_{\mathbf{y}_j^c}(\theta)]} \exp\left\{-(\mathbf{y}_j^c(i)^H R_{\mathbf{y}_j^c}^{-1} \mathbf{y}_j^c(i))\right\} \tag{27}$$

, and the log-likelihood cost function of the jth cluster is given by

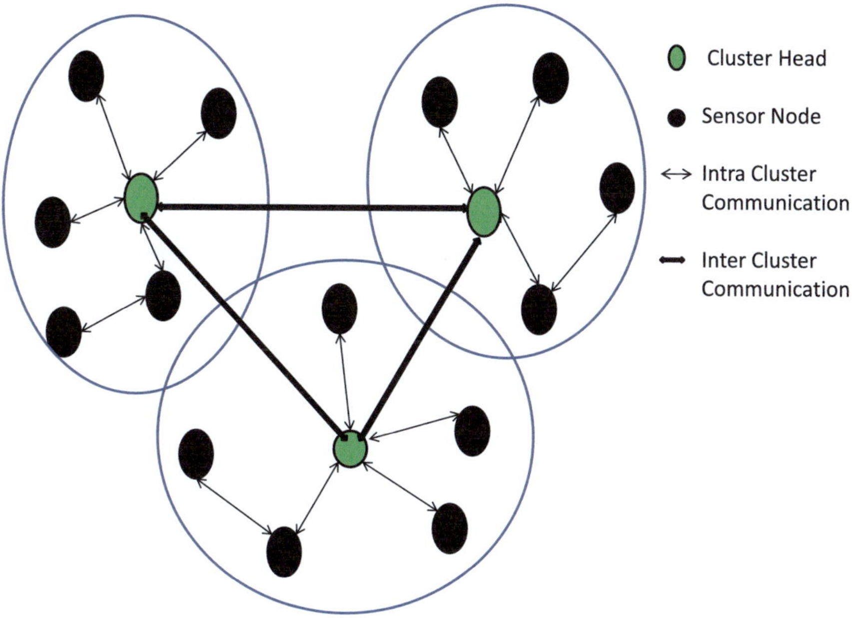

Fig. 8 Random array formation at each cluster. Cluster head estimates the DOA [26]

$$f_j^c(\boldsymbol{\theta}) = \ln \prod_{i=1}^{L} p_j^c(\mathbf{y}_c(i)|\boldsymbol{\theta}) = \sum_i \ln p_j^c(\mathbf{y}_j^c(i)|\boldsymbol{\theta}) \tag{28}$$

Now for the distributed DOA estimation, the global joint pdf is decomposed as

$$p_g(\mathbf{y}_g(i)|\boldsymbol{\theta}) = \prod_{j=1}^{N_c} p_j^c(\mathbf{y}_j^c(i)|\boldsymbol{\theta}) \tag{29}$$

Here, no approximation is made as each sensor shares its observation to the cluster head. Since the observations in sensor network are spatially and temporarily independent, thus the data are clearly uncorrelated among clusters. Taking log-likelihood of the function as

$$f_g(\boldsymbol{\theta}) = \sum_{j=1}^{N_c} \left\{ \sum_{i=1}^{L} \ln p_j^c(\mathbf{y}(i)|\boldsymbol{\theta}) \right\} \tag{30}$$

Finally, by using (28) the global cost function is decomposed into the sum of local individual clusters cost functions as

$$f_g(\boldsymbol{\theta}) = \sum_{j=1}^{N_c} f_j^c(\mathbf{y}_j^c(i)|\boldsymbol{\theta}) \tag{31}$$

where $f_j^c(\mathbf{y}_j^c(i)|\boldsymbol{\theta})$ is the ML estimates of DOA at jth cluster is

$$f_j^c(\boldsymbol{\theta}) = \log\left|\mathbf{P}_{\mathbf{A}_j^c}\hat{\mathbf{R}}_j^c\mathbf{P}_{\mathbf{A}_j^c}^H + \frac{\mathbf{P}_{\mathbf{A}_j^c}^{\perp}}{N-M}\text{tr}\left[\mathbf{P}_{\mathbf{A}_j^c}^{\perp}\hat{\mathbf{R}}_j^c\right]\right| \tag{32}$$

where $\mathbf{P}_{\mathbf{A}_j^c} = \mathbf{A}_j^c(\mathbf{A}_j^{cH}\mathbf{A}_j^c)^{-1}\mathbf{A}_j^{cH}$ is the projection of matrix $\mathbf{A}_j^c$ and $\mathbf{P}_{\mathbf{A}_j^c}^{\perp} = \mathbf{I} - \mathbf{P}_{\mathbf{A}_j^c}$ is the orthogonal complementary. The sample covariance matrix for $j$the cluster $\hat{\mathbf{R}}_j^c$ is given as

$$\hat{\mathbf{R}}_j^c = \frac{1}{L}\sum_{i=1}^{L}\mathbf{y}_j^c(i)\mathbf{y}_j^c(i)^{\mathrm{H}} \tag{33}$$

Just like the distributed in networking case, each subarray at every sensor has a local cost function, and here, each cluster has own set of sensor nodes. This makes the cost function at each cluster $f_j^c(\boldsymbol{\theta})$ is different for different clusters. Since ML function is multimodal nature, PSO-based ML solution [9, 42] is used to overcome this. Each cluster head shares their estimated DOA to other clusters for diffusion purpose.

7.1.1 Distributed DOA Estimation Using Clusters

It is a common assumption that the cluster heads are having high energy, and they are communicating with each other by inter-cluster path [43]. The problem is to minimize the global cost function $f_g(\boldsymbol{\theta})$.

$$f_g(\boldsymbol{\theta}) = \sum_{j=1}^{N_c} f_j^c(\boldsymbol{\theta}) \tag{34}$$

subject to $\theta \in \mathrm{R}^M$. Here also the distributed version of the diffusion PSO algorithm is used for solving (34). Here, each cluster runs the PSO algorithm individually to estimate the DOA locally. Simultaneously, in each iteration, each cluster head shares the best solution to other cluster heads. Let the DOA estimation vector by jth cluster at time i is $\boldsymbol{\theta}_j(i)$. Each cluster combines its current estimate and the estimates received from other clusters. The diffusion equation is

$$\boldsymbol{\theta}_j^c(i+1) = \sum_{l=1}^{N_c}\frac{1}{N_c}(i)\boldsymbol{\theta}_l^c(i) \tag{35}$$

The factor by which the estimates of others clusters contribute to the present cluster is $\frac{1}{N_c}$. It is because, all clusters are linked to each other.

7.1.2 Example

In this example, the advantage of clustering-based DOA estimation over without clustering is presented. The same diffusion PSO algorithm is used here for different network. But the parameters of the PSO remain same as before for distributed in-network DOA estimation algorithm.

Consider a sensor network of $N = 24$ homogenous sensor nodes distributed randomly in an area of 20×20 units. The sensor nodes position co-ordinates are assumed to be known. The connectivity of sensor nodes in the network is shown in Fig. 9. Distributed in-clustering algorithm is independent of connectivity, but clustering may depend. The simplest K-means clustering algorithm is applied to divide the network into three numbers of non-overlapping clusters which is shown in Fig. 10. The sensor nodes 8, 18 and 16 are the cluster heads (CH). The choice of CH is such that the cluster head distance should be minimized. This depends upon the clustering algorithm used. But in the present case, the central node of cluster chose as the CH. It may require more number of multi-hop communications to interact with other CHs if proper CH is not chosen.

Two uncorrelated and equal signals power sources impinge on the network. The true positions of the sources are 130° and 138°. The number of snapshots taken: $L = 20$ for ML-PSO and ML-DPSO, $L = 100$ for MUSIC algorithm. The performance parameters are measured by varying the SNR from −20 to 20 dB with step size of 1 dB. The average values for 100 Monte Carlo simulation are plotted in figures. Estimated RMSE algorithms distributed in-clustering, centralized ML-PSO

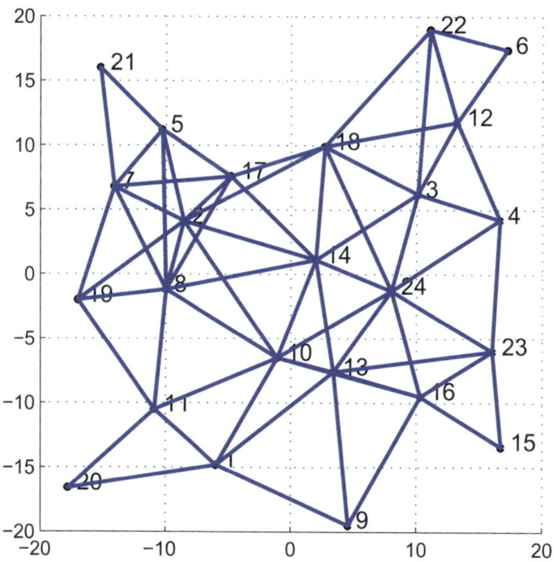

Fig. 9 Network topology used in example for clustering-based DOA estimation [10]

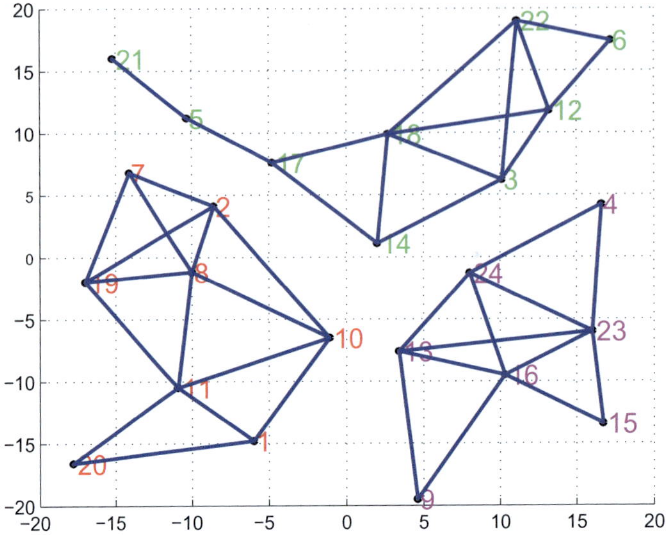

Fig. 10 Clusters

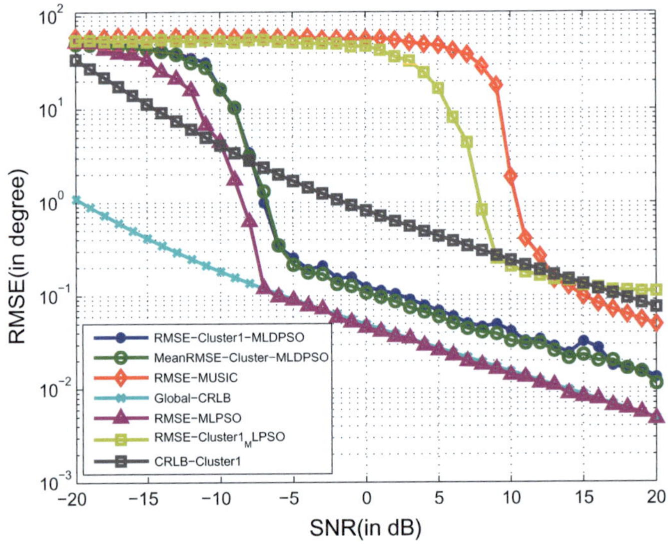

Fig. 11 RMSE versus SNR plots in DOA estimation by centralized ML-PSO, centralized MUSIC, distributed in-clustering, distributed in-network algorithms and CRLB

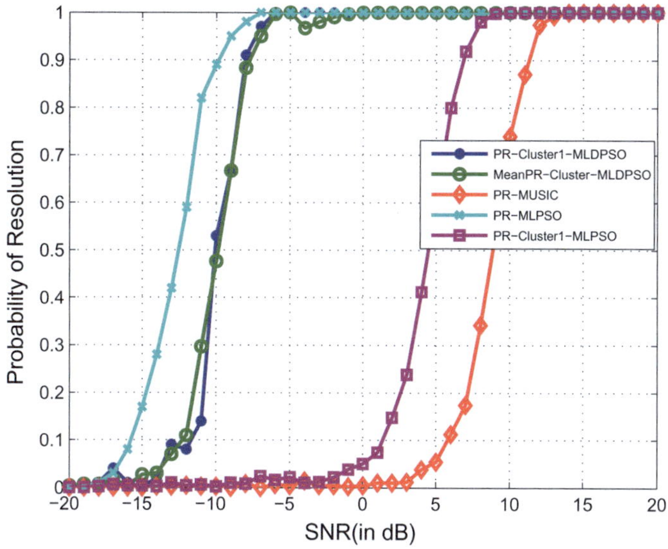

Fig. 12 PR versus SNR plots in DOA estimation by centralized ML-PSO, centralized MUSIC, distributed in-clustering distributed in-network

and MUSIC and distributed in-network are plotted in Fig. 11. The RMSE is compared with the Cramer-Rao lower bound (CRLB) [27]. The optimum performance by any estimation algorithm of a sensor array is the CRLB. Similarly, the PR performance by all the algorithm versus SNR is depicted in Fig. 12.

It has been observed from Figs. 11 and 12, when we measure the DOA by using the whole network as a sensor array (centralized ML-PSO) always gives better RMSE and PR performance over other algorithms. The performance of the clustering-based distributed DOA estimation algorithm is closer to global performance. The performances are much better than the distributed in-network algorithm.

8 Conclusion

The source localization algorithm using DOA estimation is discussed in this chapter. The distributed DOA estimation problem in wireless sensor network is formulated. Every sensor node in the sensor network forms a subarray array with its immediate neighbours. Each sensor node formulates its local maximum-likelihood function and then tries to estimate DOA individually by cooperating with other sensors in the network after coveting the data from the neighbours. A diffusion particle swarm optimization (DPSO) algorithm is used at each sensor to estimate the DOA. In each iteration of the DPSO algorithm, each particle at individual node computes its fitness using local ML function and updates their positions and velocities with the help of fused information from the neighbours. In this way, each node runs the PSO locally by using its own local fitness function, but finally, they attempt to achieve the global estimate due to the fusion mechanism. Further, the distributed in-network algorithm is extended to clustering-based distributed DOA estimation algorithm. The distributed in-clustering algorithm is independent of node connectivity in the network and reduces the computational and communication overhead.

The RMSE and PR performance are compared among the distributed ML-DPSO algorithms with centralized ML-PSO and MUSIC algorithm by doing the Monte Carlo simulation. It is observed from the results that distributed in-clustering algorithm provides the best performance compared to that of distributed in-network and non-cooperative algorithms. The performance of the distributed approach improves but less than the centralized algorithms. The communication overheads for all the algorithms are compared and found that, in worst case, the distributed algorithm needs less number of messages to be communicated in order to attain the steady state in the network. This makes the algorithm energy efficient.

9 Future Direction

Present discussion is on the distributed DOA estimation which is used for source localization and tracking of uncorrelated and far-field sources. A diffusion cooperative strategy is used among the sensor nodes in the network to estimate the global DOA locally. The diffusion PSO algorithm is used to optimize the log-likelihood function with cooperation among the neighbours. Further, the distributed in-network algorithm is extended to distributed in-clustering to make the cost function is independent among the clusters. But the same diffusion PSO algorithm has been used at each of the cluster.

If the signals are correlated, then the algorithm has to be developed based on correlation data model [44]. Special attention also required how to deal the mixture of near and far-field signal [25, 45]. This situation is common in cellular communication to localize the mobile user position. Most of the present algorithm in literature is centralized-based. But in WSN, a distributed algorithm is preferred to make the network energy efficient and also to minimize the phase error problem.

Further, it has been seen that the number of message communication depends on the maximum number of iterations required to get the steady-state performance in the network. The aim is always to reduce the communication overhead in the network because communication consumes most of the power. It is better to search for variant of PSO algorithms or other evolutionary algorithm like harmonic search, grey-wolf optimizer which can further reduce the number of iteration required to converge the algorithm [46, 47].

In recent past, different diffusion algorithms are proposed by researches to enhance the estimation performance and reduce the communication overhead [48]. The diffusion algorithms also should work in adverse network environment like in the presence of impulsive noise or in dynamic network. Multi-hop diffusion may be used in case of sparse sensor network to overcome the connectivity problem [49]. Efficient clustering methods may also be used to optimize the size of cluster and to minimize the inter-cluster multi-hop communication between CH and sensor node.

Acknowledgements This work was supported in part by the Science and Engineering Research Board (SERB), Govt. of India (Ref. No.SB/S3/EECE/210/2016 Dated 28/11/2016).

References

1. Akyildiz IF, Su W, Sankarasubramaniam Y, Cayirci E (2002) A survey on sensor networks. IEEE Commun Mag 40(8):102–114
2. Das SK, Samanta S, Dey N, Kumar R (eds) (2020) Design frameworks for wireless networks. Springer, Lecture Notes in Networks and Systems
3. Krim H, Viberg M (1996) Two decades of array signal processing research: the parametric approach. IEEE Signal Process Mag 13(4):67–94
4. Elhayatmy G, Dey N, Ashour AS (2018) Internet of things based wireless body area network in healthcare. In: Dey N, Hassanien AE, Bhatt C, Ashour AS, Satapathy SC (eds) Internet of things and big data analytics toward next-generation intelligence. Springer International Publishing, Cham, pp 3–20
5. Dey N, Mukherjee A, Kausar N, Ashour AS, Taiar R, Hassanien AE (2016) A disaster management specific mobility model for flying ad-hoc network. Int J Rough Sets Data Anal 3(3):72–103
6. Chen JC, Yao K, Hudson RE (2002a) Source localization and beamforming. IEEE Signal Process Mag 30–39
7. Prasad MS, Panigrahi T (2019) Distributed maximum likelihood doa estimation algorithm for correlated signals in wireless sensor network. Wirel Pers Commun 105(4):1527–1544
8. Stoica P, Sharman KC (1990) Maximum likelihood methods for direction-of-arrival estimation. IEEE Trans Acoust, Speech Signal Process 38(7):1132–1143
9. Minghui L, Lu Y (2008) Maximum likelihood DOA estimation in unknown colored noise fields. IEEE Trans Aerosp Electron Syst 44(3):1079–1090
10. Trilochan P, Ganapati P, Bernard M, Babita M (2013) Distributed DOA estimation using clustering of sensor nodes and diffusion pso algorithm. Swarm Evol Comput 9:47–57
11. Panigrahi T, Panda G, Mulgrew B (2012a) Distributed bearing estimation technique using diffusion particle swarm optimisation algorithm. IET Wirel Sens Syst 2(4):385–393
12. Xiao L, Boyd S, Lall S (2005) A scheme for robust distributed sensor fusion based on average consensus. In: Proceedings of 4th international symposium on information processing in sensor networks. Loss Angles, CA, pp 63–70
13. Abbasi A, Younis M (2007) A survey on clustering algorithms for wireless sensor network. Comput Commun 30:2826–2841
14. Ziskind I, Wax M (1988) Maximum likelihood localization of multiple sources by alternating projection. IEEE Trans Acoust, Speech Signal Process 36(10):1553–1560
15. Chung PJ, Böhme JF (2002) DOA estimation using fast EM and SAGE algorithms. Signal Process 82(11):1753–1762. ISSN 0165-1684
16. Boeringer DW, Werner DH (2004) Particle swarm optimization versus genetic algorithms for phased array synthesis. IEEE Trans Antennas Propag 52(3):771–779
17. Robinson J, Rahmat-Samii Y (2004) Particle swarm optimization in electromagnetics. IEEE Trans Antennas Propag 52(2):397–407
18. Panigrahi BK, Pandi VR, Das S (2008) Adaptive particle swarm optimization approach for static and dynamic economic load dispatch. Energy Convers Manag 49(6):1407–1415
19. Trilochan P, Srinivas R, Harikrishna G (2016) Application of comprehensive learning particle swarm optimization algorithm for maximum likelihood doa estimation in wireless sensor networks. Int J Swarm Intell 2(2–4):

20. Panigrahi T, Rao DH, Panda G, Mulgrew B, Majhi B (2011a) Maximum likelihood DOA estimation in distributed wireless sensor network using adaptive particle swarm optimization. In: The proceeding of ACM international conference on communication, computing and security (ICCCS2011), pp 134–136
21. Wax M, Kailath T (1985) Decentralized processing in sensor array. IEEE Trans Acoust, Speech Signal Process 33:1123–1129
22. Stoica P, Nehorai A, Söderström T (1995) Decentralized array processing using the MODE algorithm. Circuits Syst Signal Process 14(1):17–38. ISSN 0278-081X. https://doi.org/10.1007/BF01183746
23. Moses RL, Moses OL, Krishnamurthy D, Patterson R (2002) A self-localization method for wireless sensor networks. EURASIP J Appl Signal Process 4:348–358
24. Chen JC, Hudson RE, Yao K (2002b) Maximum-likelihood source localization and unknown sensor location estimation for wideband signals in the near-field. IEEE Trans Signal Process 50(8):1843–1854
25. Abedin MJ, Sanagavarapu AM (2008) Localization of near-field radiating sources with an arbitrary antenna array. In: Antennas and propagation society international symposium, ed NA. Marrakech, Morocco (2008) IEEE Computer Soc San Diego, USA, pp 572–577
26. Panigrahi T (2012) On the development of distributed estimation estimation techniques in wireless sensor networks. PhD thesis, National Institute of Technology Rourkela, India
27. Van Trees HL (2002) Optimum array processing. Wiley-Interscience Publication
28. Stoica P, Nehorai A (1990) Performance study of conditional and unconditional direction-of-arrival estimation. IEEE Trans Acoust, Speech Signal Process 38(10):1783–1795 Oct
29. Dash SK, Tripathi S (2019) Energy efficient routing formation algorithm for hybrid ad-hoc network: a geometric programming approach. Peer-To-Peer Netw Appl 12(1):102–128
30. Das SK, Tripathi S (2018) Adaptive and intelligent energy efficient routing for transparent heterogeneous ad-hoc network by fusion of game theory and linear programming. Appl Intell 48(7):1825–1845
31. Marano S, Matta V, Willett P (2008) Distributed estimation in large wireless sensor networks via a locally optimum approach. IEEE Trans Signal Process 56(2):748–756
32. Nedic A, Ozdaglar A (2008) Cooperative distributed multi-agent optimization. In: Eldar Y, Palomar D (eds) Convex optimization in signal processing and communications. Cambridge University Press, pp 340–386
33. Nedic A, Ozdaglar A (2009) Distributed subgradient methods for multi-agent optimization. IEEE Trans Autom Control 54(1):48–61
34. Majhi B, Panda G, Mulgrew B (2009) Distributed identification of nonlinear processes using incremental and diffusion type pso algorithms. In: CEC'09: proceedings of the eleventh conference on congress on evolutionary computation. Piscataway, NJ, USA. IEEE Press, pp 2076–2082. ISBN 978-1-4244-2958-5
35. Kennedy J, Eberhart R (1995) Particle swarm optimization. IEEE Int Conf Neural Netw 4:1942–1948
36. Eberhart, Shi Y (2001) Particle swarm optimization: developments, applications and resources. In: Proceedings of the 2001 congress on evolutionary computation, vol 1, pp 81–86
37. Parsopoulos KE, Vrahatis MN (2002) Recent approaches to global optimization problems through particle swarm optimization. Nat Comput 1:235–306
38. Shi Y, Eberhart RC (1998) Parameter selection in particle swarm optimization. In: EP '98: Proceedings of the 7th international conference on evolutionary programming VII. Springer-Verlag, London, UK, pp 591–600. ISBN 3-540-64891-7
39. Clerc M, Kennedy J (2002) The particle swarm - explosion, stability, and convergence in a multidimensional complex space. IEEE Trans Evol Comput 6(1):58–73
40. Panigrahi T, Pradhan PM, Panda G, Mulgrew B (2012b) Block least-mean square algorithm over distributed wireless sensor network. J Comput Netw Commun, 2012
41. Simon F, Jiaxue L, Wei S, Yifei T, Wong Raymond K, Nilanjan D (2018) Predicting unusual energy consumption events from smart home sensor network by data stream mining with misclassified recall. J Ambient Intell HumIzed Comput 9:1197–1221

42. Panigrahi T, Panda G, Mulgrew B, Majhi B (2011b) Maximum likelihood source localization in wireless sensor network using particle swarm optimization. In: The proceeding of international conference on electronics systems (ICES-11), pp 111–115
43. Younis O, Krunz M, Ramasubramanian S (2006) Node clustering in wireless sensor networks: recent developments and deployment challenges. IEEE Netw 20(3):20–25
44. Shree Prasad M, Panigrahi T (2017) Computationally efficient near-field source localization algorithm using single snapshot. In: Accepted 14th edition of the IEEE India council international conference INDICON-2017
45. Shree Prasad M, Panigrahi T, Dubey A (2017) Computationally efficient near-field source localization algorithm using single snapshot. In: Accepted 14th edition of the IEEE India Council International Conference INDICON-2017
46. Panigrahi BK, Ravikumar Pandi V, Das S, Cui Z, Sharma R (2011c) Economic load dispatch using population-variance harmony search algorithm. Trans Inst Meas Control
47. Faris H, Aljarah I, Al-Betar MA, Mirjalili S (2017) Grey wolf optimizer: a review of recent variants and applications. Neural Comput Appl. https://doi.org/10.1007/s00521-017-3272-5
48. Kong J-T, Lee J-W, Kim S-E, Shin S, Song W-J (2017) Diffusion lms algorithms with multi combination for distributed estimation: Formulation and performance analysis. Digit Signal Process, 71(Supplement C):117 – 130, 2017
49. Nayak M, Panigrahi T, Sharma R (2015) Distributed estimation using multi-hop adaptive diffusion in sparse wireless sensor networks. In: International conference on microwave, optical and communication engineering (ICMOCE), pp 318–321 https://doi.org/10.1109/ICMOCE.2015.7489756

Chapter 9
Quasi-oppositional Harmony Search Algorithm Approach for Ad Hoc and Sensor Networks

Chandan Kumar Shiva and Ritesh Kumar

1 Introduction

In the present time, wireless sensor network (WSN) is going to be commercialized due to getting more success in the technical advancement and low power consumption of the embedded computing devices. WSN is composed of the number of distributed, small and low-power application devices as sensor node [1]. In the real-time application, the involvement of sensor node is to measure the change in environmental conditions like temperature, humidity, vibration and so on. It can also be used in terms of data storage and signal conditioning devices. In the manufacturing process, most of the currently adopted technologies for WSNs are based on low-cost processor. In many applications, it is expected that the sensor node lasts for a long time. It is because of its uses in remote areas as recharging and/or replacing power supply units are quite difficult [2]. WSN has a problem of network lifetime when sensors are used in large geographical areas as there is difficult to exchange or recharge batteries. WSN is composed of several small battery-based sensors, and these sensors are connected by each other through nodes. Sensor nodes can communicate with each other and with the base station. The basic goals of WSN are to (a) monitor the specific selected area (b) record the occurrence of events and (c) measure the required parameters [2]. The diagram of WSN is shown in Fig. 1. An intelligent energy-efficient multicast routing protocol has been proposed in [3]. The work has been shown in this paper to show the enhanced performance over on-demand multicast routing protocol. To control the uncertainties issues in wireless mobile ad hoc network, fuzzy logic tool has been studied in [4] in order to conserve the network resources. In this process,

C. K. Shiva (✉) · R. Kumar
Department of Electrical and Electronics Engineering, S R Engineering College Warangal, Ananthsagar, Hasanparthy, Warangal, Telangana 506371, India
e-mail: chandankumarshiva@gmail.com

R. Kumar
e-mail: riteshraj3336@gmail.com

D. De et al. (eds.), *Nature Inspired Computing for Wireless Sensor Networks*, Springer Tracts in Nature-Inspired Computing,
https://doi.org/10.1007/978-981-15-2125-6_9

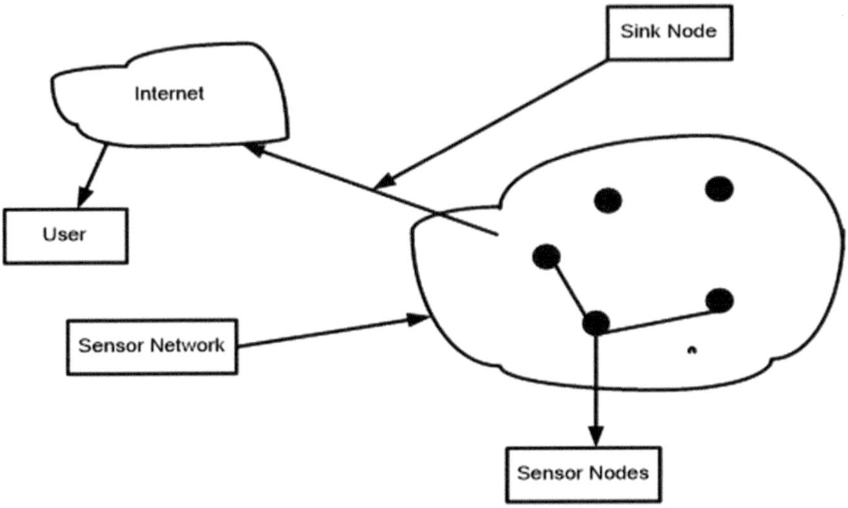

Fig. 1 Wireless sensor network

all the available network metrics of the routes are converted into fuzzy cost matrix or communication cost. To design an energy-efficient routing protocol in mobile ad hoc networks is a challenging task with some limitations. Regarding this, an intelligent energy-aware efficient routing protocol has been studied in [5]. The application of geometric programming-based energy-efficient routing protocol has been proposed for hybrid ad hoc network in [6]. It optimizes two sets of objectives in terms of maximizing network lifetime and minimizing packet loss and routing overhead.

The above study supports that optimization techniques are quite essential for the enhanced performance of ad hoc and sensor network. The application of some powerful hybrid optimization techniques may further show the substantial improvement in ad hoc and sensor network. Therefore, it may be a case of study to explore new optimization techniques in this field.

This chapter outlines issues in WSNs, introduces quasi-oppositional harmony search (QOHS) algorithm and discusses its suitability for WSN applications. Regarding this, the rest of chapter is organized as follows. A brief discussion on need of optimization has been shown in Sect. 2. The detailed description of QOHS algorithm is presented in Sect. 3. Pertaining to engineering optimization task, WSN optimization task is focused in Sect. 4. Finally, Sect. 5 concludes the overall work of this chapter.

2 Need of Optimization

In the present prospect, the concept of optimization has been used in every field of engineering whether the field may be applied engineering, medicine, economics or basic science. The reason is it gives better solution or optimal solution as required by the designer. The term optimal may be understood as maximum or minimum as decided by the nature and size of the program. For the optimization, a mathematical posture of the problem needed to be formulated, and it is an essential part of the optimization task to solve the given problem. However, in most of the cases, the physical realization of the model is too difficult to model due to various uncertainty, system constraint and environmental conditions [7]. The benefits of the optimization techniques are that it does not depend upon the structure of the model and size of the problem. In the line of the above interactions, it is quite interesting for the researchers to develop such models and use optimization task for the problem. Network optimization is one of the essential needs to improve the performance of the peripheral devices and the network. In the present work of WSN, it is needed to improve the life cycle of the network and minimize energy consumption as far as possible [8]. This may be possible with the utilization of powerful optimization technique. The application of artificial intelligence-based power optimization techniques has been shown in [9–12].

Description of QOHS Algorithm

The QOHS algorithm is the modified form of harmony search algorithm (HSA). The HSA has been improvised by the implementation of opposition-based HSA and termed as OHSA. The OHSA is further improved by adding the concept of quasi-opposition-based HSA algorithm and termed as QOHS algorithm. The study of HSA, OHSA and QOHS algorithm has been explained as followed.

2.1 *Basic HSA*

HSA is finding of the music improvisation process as the musician always tries to find the best tuning process for the better harmony of the music. It is a derivative-free algorithm, little mathematics, less computation time as compared to genetic algorithm, finds high-performance region in less time and easy to apply in most of the problem are the reason to implement in most of the optimization problem of the engineering subjects [13, 14]. As similar to other evolutionary optimization technique, HSA starts with randomly generated population vector called as harmony vectors and stored in HM. Following to this, the generated population vector is improvised by memory consideration rule, pitch adjustment rule and random re-initialization of the population vector. The harmony in music is termed as solution vector, and the improvement in solution vectors is termed as local and global solutions. The obtained solution of HSA is termed as harmony and may be presented by the n-dimension solution vector. The application of QOHS algorithm in different fields of applications has been shown in [15–17].

Time to time, HSA has been modified in order to improve its optimizing capability, convergence profile and algorithmic speed. It has been done to improve the tuning efficiency of the algorithm as most of the problems are too complex. In [14], an improved HSA has been proposed by modifying the control variables of the algorithm. The improved HSA using the swarm intelligence concept has been shown in [18]. A self-adaptive global best HSA for solving optimization problems has been shown in [19]. The steps of HSA may be explained as follows.

Phase (a) Problem initialization

In general, a global optimization problem may be stated as Eq. (1):

$$\min f(x) \text{ in the interval of } x_j \in [\text{para}_j^{\min},\ \text{para}_j^{\max}] \tag{1}$$

where $j = 1, 2, \ldots, n$.

In (1), $f(x)$ is the objective function that needs to be minimized or maximized, $X = [x_1, x_2, \ldots, x_n]$ is the set of control variables; n is the number of control variables, $\text{para}_j^{\min}$, $\text{para}_j^{\max}$ are the minimum and maximum limit of the design variable, respectively. Set the parameters of the HSA and decide the stopping criterion.

A termination condition of a QOHS algorithm is important in determining the harmony vector subjected to QOHS algorithm end. The termination condition is usually needed when such that our solution is close to the optimal at the end of the run. Usually, the following termination conditions may be adopted as follows.

(a) When no improvement in the population for the given iterations is observed.
(b) When specified number of generations is reached.
(c) When the fitness value has reached a certain pre-defined value.

Phase (b) generation of initial solution

An initially randomly generated population vector is created as an initial solution for the optimization task. This is essential for the improvisation of the solution vector. Equation (1) shows the randomly generated harmony vector in the range of [$\text{para}_j^{\min}$,$\text{para}_j^{\max}$]. The position of HM matrix is shown in Eq. (2).

$$X_j = [x_1^j, x_2^j, \ldots, x_n^j] \tag{2}$$

where $j = 1, 2, \ldots, \text{HMS}$

$$\text{HM} = \begin{bmatrix} x_1^1 & x_2^1 & \ldots x_n^1 \\ x_1^2 & x_2^2 & \ldots x_n^2 \\ \ldots & & \\ x_1^{\text{HMS}} & x_2^{\text{HMS}} & \ldots x_n^{\text{HMS}} \end{bmatrix} \tag{3}$$

Phase (c) Improvisation of the solution

An improved harmony vector is generated by improvisation process with the implementation of a memory consideration, pitch adjustment and random selection. Initially, a uniform random number r_1 is generated in the range of [0, 1]. If $r_1 < \text{HMCR}$, the designed variables x_j^{new} are generated, otherwise, obtained by a random selection. In the memory consideration, x_j^{new} is selected from harmony vector. Secondly, each decision variable x_j^{new} undergoes a pitch adjustment with a probability of PAR. The pitch adjustment rule is given by Eq. (4).

$$X^{\text{new}} = \left(x_1^{\text{new}}, x_2^{\text{new}} \ldots x_n^{\text{new}}\right) \tag{4}$$

$$x_j^{\text{new}} = x_j^{\text{new}} \pm r_3 \times \text{BW} \tag{5}$$

where r_3 is a uniform random number between [0, 1].

Phase (d) Upgradation of solution

The HM may be updated by the survival of the fitter vector between X^{new} and the worst harmony vector X^{worst}. X^{new} is updated by X^{worst} if the fitness value of X^{new} is better than X^{worst}.

Phase (e) Process of computation

Based upon the above computational procedure, Algorithm 1 shows the pseudocode of HSA.

Algorithm 1: HSA

1. Set the parameters of the HSA.
2. Generate HM and find the fitness function value for each harmony vector.
3. Improvise HM with new harmony X^{new} vectors as follows:

```
for (j = 0; j < n; j++))
    if (r1 < HMCR) then
        x_j^new = x_j^a                      // a ∈ (1,2,......,HMS)
        if (r2 < PAR) then
            x_j^new = x_j^new ± r3 × BW      // r1, r2, r3 ∈ [0,1]
        end if
    else
        x_j^new = para_j^min + r×(para_j^max − para_j^min)      // r ∈ [0,1]
    end if
end for
```

4. Update the HM as $X^{worst} = X^{new}$ if $f(X^{new}) < f(X^{worst})$.
5. If *NI* is completed, return the best harmony vector X^{best}; otherwise go back to **Step 3.**

2.2 Improved HSA

The improved HSA applies the memory consideration, pitch adjustment and random selection but updates the values of PAR and BW as shown in Eqs. (6) and (7), respectively [14].

$$\mathrm{PAR(gn)} = \mathrm{PAR}^{\min} + \frac{\mathrm{PAR}^{\max} - \mathrm{PAR}^{\min}}{\mathrm{NT}} \times \mathrm{gn} \tag{6}$$

$$\mathrm{BW(gn)} = \mathrm{BW}^{\max} \times \mathrm{e}^{\left(\frac{\ln\left(\frac{\mathrm{BW}^{\min}}{\mathrm{BW}^{\max}}\right)}{\mathrm{NT}} \times \mathrm{gn}\right)} \tag{7}$$

In (6), PAR(gn) is the pitch adjustment rate in the current generation (gn), $\mathrm{PAR}^{\min}$ and $\mathrm{PAR}^{\max}$ are the minimum and the maximum adjustment rate, respectively. In (7), BW(gn) is the distance bandwidth at generation (gn), $\mathrm{BW}^{\min}$ and $\mathrm{BW}^{\max}$ are the minimum and the maximum bandwidths, respectively.

2.3 Opposition-Based Learning

Most of the optimization techniques execute by the initialization of randomly generated solution and step by step move toward optimal solution(s). The initially generated solution may be improved by introducing the concept of opposition-based learning (OBL). It is based on the social revolution of human beings and one of the powerful optimization tool for enhancing the convergence rate and speed of response of different optimization techniques. OBL concept states that randomly generated solution may be improved by sequentially testing the opposite solution [20, 21]. In line with this, the better of the two cases may be chosen as the starting solution. The similar methodology may be applied to each solution in the current population. The ideas of OBL have been shown in [20]. The method of OBL has been applied in different field of engineering as shown in [21–24]. The idea of opposite number may be incorporated during the harmony memory (HM) initialization and also for generating the new improved solution. The definition of opposite number and opposite point is defined as follows.

2.3.1 Definition of Opposite Number and Opposite Point

Let $x \in [ub, lb]$ be a real number. The opposite number of x may be defined by Eq. (8).

$$\breve{x} = ub + lb - x \tag{8}$$

whereas opposite point may be defined by (9).

$$\breve{x}_i = ub_i + lb_i - x_i \tag{9}$$

lb and ub are the lower and upper bound of the control variable.

2.3.2 Opposition-Based Optimization

In the present chapter, the concept of OBL is introduced in two steps. HSA is chosen as the parent algorithm, and opposition-based ideas are incorporated to improve initial solution. Algorithm 2 shows the implementation of OHSA.

Algorithm 2: OHS

---

1. Set the parameters HMS, $HMCR$, $PAR^{\min}$, $PAR^{\max}$, $BW^{\min}$, $BW^{\max}$, and NI.
2. Initialize the HM with $X_{0_{i,j}}$.
3. **Opposition based HM initialization.**

for $(i = 0;\ i < HMS;\ i{+}{+})$
  for $(j = 0;\ j < n;\ j{+}{+})$
    $OX_{0_{i,j}} = para_j^{\min} + para_j^{\max} - X_{0_{i,j}}$ // OX_0: Opposite of initial X_0
  end for
end for
// **End of opposition-based HM initialization.**
Select HMS fittest individuals from set of $\{X_{0_{i,j}}, OX_{0_{i,j}}\}$ as initial HM; HM being the matrix of fittest X vectors

4. Improvise a new harmony X^{new} as follows:

Update $PAR(gn)$ by (6) and $BW(gn)$ by (7).
for $(i = 0;\ i < HMS;\ i{+}{+})$
  for $(j = 0;\ j < n;\ j{+}{+})$
    if $(r_1 < HMCR)$ then
      $X_{i,j}^{new} = X_{i,j}^{a}$ // $a \in (1,2,\ldots,HMS)$
      if $(r_2 < PAR(gn))$ then
        $X_{i,j}^{new} = X_{i,j}^{new} \pm r_3 \times BW(gn)$ // $r_1, r_2, r_3 \in [0,1]$
      end if
    else
      $X_{i,j}^{new} = para_{i,j}^{\min} + r \times (para_{i,j}^{\max} - para_{i,j}^{\min})$ // $r \in [0,1]$
    end if
  end for
end for

Algorithm 3: Pseudo code for quasi-opposite number

$M = (para^{\min} + para^{\max})/2$
if $(OX_o < M)$ % OX_0: Opposite of initial X_0
 $QOX_o = M + (OX_0 - M) \times r_1$ % $r_1 \in [0,1]$
else
 $QOX_o = OX_0 + (M - OX_0) \times r_1$
end if-else

2.4 *Quasi-Opposition-Based Learning: A Concept*

Population initialization is a key concept for the performance of an optimization algorithm. A good initialization method may help a lot in finding better solutions and improving convergence rate. Earlier study reveals that utilization of opposite numbers in the population initialization and generation jumping enhances the optimization performance of an optimization algorithm [25]. Instead of opposite numbers, quasi-opposite points are used in the population initialization of the basic HSA. The concept of quasi-oppositional generation jumping is utilized in the HSA. To generate quasi-opposite population from the current solution, quasi-opposite jumping rate is utilized. This novel HSA is termed as quasi-oppositional HS (QOHS) algorithm.

Regarding QOBL, the description of two important definitions (such as quasi-opposite number and quasi-opposite point) is given in the next two sub-sections.

2.4.1 Quasi-opposite Number

It is defined as the number between the center of the search space and the opposite number. Mathematically, it is expressed by Eq. (10) [26].

$$\mathrm{QOX}_0 = \mathrm{rand}\left(\frac{\mathrm{para}^{\min} + \mathrm{para}^{\max}}{2},\ \mathrm{para}^{\min} + \mathrm{para}^{\max} - X_0\right) \tag{10}$$

Algorithm 4 shows the implementation of quasi-opposite number.

Algorithm 4: Pseudo code for quasi-opposite point

$M_j = (para_j^{\min} + para_j^{\max})/2$

if $(OX_{0j} < M_j)$ % OX_0: Opposite of initial X_0

$QOX_{0j} = M_j + (OX_{0j} - M_j) \times r_1$ % $r_1 \in [0,1]$

else

$QOX_{0j} = OX_{0j} + (M_j - OX_{0j}) \times r_1$

end if-else

2.4.2 Quasi-opposite Point

It is the point between the center of the search space and the opposite number. It may be stated by Eq. (11).

$$\mathrm{QOP}_{0_i} = \mathrm{rand}\left(\frac{\mathrm{para}_i^{\min} + \mathrm{para}_i^{\max}}{2}, \mathrm{para}_i^{\min} + \mathrm{para}_i^{\max} - X_i\right), i = 1, 2, \ldots, d \quad (11)$$

where d is the dimension of the problem.

Algorithm 4 shows the implementation of quasi-opposite point

Algorithm 5: Pseudo code for the QOHS algorithm

1. Set the parameters: HMS, $HMCR$, $PAR^{\min}$, $PAR^{\max}$, $BW^{\min}$, $BW^{\max}$, and NI.
2. Initialize the HM with $X_{0_{i,j}}$.
3. **% Quasi-oppositional HM initialization**

```
for (i = 0; i < HMS; i++)
    for (j = 0; j < d; j++)
        OX0_{i,j} = para_{i,j}^min + para_{i,j}^max - X0_{i,j};      % OX0: Opposite of initial X0
        M_{i,j} = (para_{i,j}^min + para_{i,j}^max) / 2;
        if (OX0_{i,j} < M_{i,j})
            QOX0_{i,j} = M_{i,j} + (OX0_{i,j} - M_{i,j}) × r1;      % r1 ∈ [0,1]
        end if
        else
            QOX0_{i,j} = OX0_{i,j} + (M_{i,j} - OX0_{i,j}) × r1;
        end else
    end for
end for
```

% End of quasi-oppositional HM initialization.
Select HMS fittest individuals from set of $\{X_{0_{i,j}}, QOX_{0_{i,j}}\}$ as initial HM; HM being the matrix of fittest X vectors

4. Improvise a new harmony X^{new} as follows:
Update $PAR(gn)$ and $BW(gn)$.

```
for (i = 0; i < HMS; i++)
    for (j = 0; j < d; j++))
        if (r2 < HMCR) then
            X_{i,j}^new = X_{i,j}^a;                                  % a ∈ (1,2,.....,HMS)
            if (r3 < PAR(gn)) then
                X_{i,j}^new = X_{i,j}^new ± r4 × BW(gn);             % r2, r3, r4 ∈ [0,1]
            end if
            else
                X_{i,j}^new = para_{i,j}^min + r5 × (para_{i,j}^max - para_{i,j}^min);   % r5 ∈ [0,1]
            end else
        end if
    end for
end for
```

5. Update the HM as $X^{worst} = X^{new}$ if $f(X^{new}) < f(X^{worst})$
6. **Quasi-opposition based generation jumping**

```
if (r6 < Jr)                                                          % r6 ∈ [0,1], Jr: Jumping rate
    for (i = 0; i < HMS; i++)
        for (j = 0; j < n; j++)
            OX_{i,j} = para_{i,j}^min(gn) + para_{i,j}^max(gn) - X_{i,j};
```

% $para_{i,j}^{\min}(gn)$: minimum value of jth variable of ith parameter in the current generation (gn)

% $para_{i,j}^{\max}(gn)$: maximum value of jth variable of ith parameter in the current generation (gn)

2.4.3 Quasi-oppositional Population Initialization

A randomly generated population vector may be improvised by utilizing the quasi-oppositional concept in the HSA. Initially, a set of initial population is generated within the solution space. Corresponding to this, fitness function value is evaluated for the individual solution. Afterward, the opposite population vector is generated by utilizing the oppositional learning concept. After this, quasi number is calculated and implemented in generating quasi-opposite population by using jumping rate.

2.4.4 Quasi-oppositional Generation Jumping

The process of optimization gained its effectiveness to a new candidate solution with some mechanism. In the QOHS algorithm, based on the concept of jumping rate (J_r), a new population vector is generated [27]. The pseudocode for QOHS algorithm may be shown in Algorithm 5. The flowchart for the same is shown in Fig. 2.

$M_{i,j} = \{para_{i,j}^{\min}(gn) + para_{i,j}^{\max}(gn)\}/2;$

if $(X_{i,j} < M_{i,j})$

$QOX0_{i,j} = M_{i,j} + (OX0_{i,j} - M_{i,j}) \times r_7;$ % $r_7 \in [0,1]$

end if

else

$QOX0_{i,j} = OX0_{i,j} + (M_{i,j} - OX0_{i,j}) \times r_7;$

end else

end for

end for

end if

Select *HMS* fittest HM from the set of $\{X_{i,j}, QOX_{i,j}\}$ as current HM.

% End of quasi-oppositional generation jumping.

If *NI* is completed, return the best harmony vector X^{best} in the *HM*; otherwise go back to **Step 4.**

---

5. Update the HM as $X^{worst} = X^{new}$ if $f(X^{new}) < f(X^{worst})$
6. **Opposition based generation jumping**

if $(rand_2 < J_r)$ // $rand_2 \in [0,1]$, J_r: Jumping rate

for $(i = 0;\ i < HMS;\ i++)$

for $(j = 0;\ j < n;\ j++)$

$OX_{i,j} = \min_j^{gn} + \max_j^{gn} - X_{i,j}$

// $\min_j^{gn}$:minimum value of the *j*th variable in the current generation (gn)

// $\max_j^{gn}$:maximum value of the *j*th variable in the current generation (gn)

end for

end for

end if

Select *HMS* fittest HM from the set of $\{X_{i,j}, OX_{i,j}\}$ as current HM.

// End of opposition-based generation jumping.

7. If *NI* is completed, return the best harmony vector X^{best}; otherwise go back to **Step 4.**

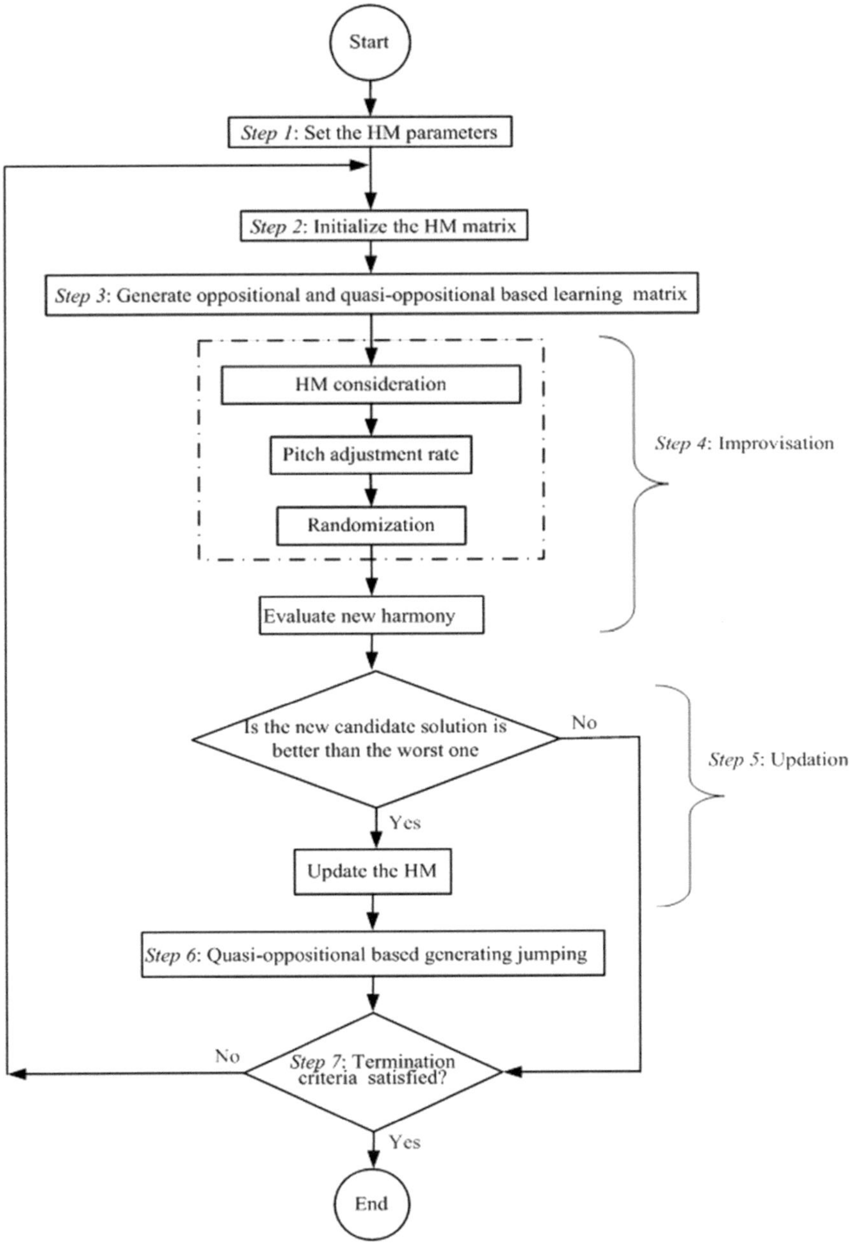

Fig. 2 Flowchart of the QOHS algorithm

2.4.5 QOHS for Benchmark Functions

The performance of QOHS algorithm is tested in this section with five benchmark functions. The details of these benchmark functions are tabulated in Table 1 [28, 29]. QOHS algorithm is implemented to these functions, and correspondingly, mean and standard deviations are shown in Table 1. The parameters of QOHS algorithm are shown in the Appendix section. The QOHS-based numerical results have been tabulated with the obtained results using GA, PSO, group search optimizer [29], continuous quick GSO (CQGSO) and real coded GA approach with random transfer vectors-based mutation (RCGA-RTVM). Table 2 showed that QOHS algorithm converges to better results.

The steps of optimizing QOHS algorithm are summarized as follows:

Step (a): Initialization: Similar to other population-based optimization techniques, initially, randomly generated population of harmony vectors is created and stored in HM. After this, oppositional and then quasi-oppositional-based HM are initialized. Each parameter in the problem is called as a harmony. A HM consists of harmony and represents a solution to the problem.

Step (b): Improvisation: A new candidate harmony is generated from all of the solutions in the HM by adopting a memory consideration rule, a pitch adjustment rule and random re-initialization.

Step (c): Update HM: The degree of fitness of a solution is qualified by assigning a value to it. This is done by defining a proper fitness function to the problem. The HM is updated by comparing the new candidate harmony vector and the worst harmony vector. The worst harmony vector is replaced by the new candidate vector if it is better than the worst harmony vector in the HM. After this, the concept of quasi-oppositional-based generation jumping is implemented in this algorithm. The above process is repeated until a certain termination criterion is met.

ACO: basic concept

Ant colony optimization technique came in early of 1990; it is a powerful technique of optimization which inspired from foraging behavior of ants. This technique helps to find an approximate solution which optimizing problem from discrete to continuous. It also helps to optimize problem occurring in telecommunication such as routing and balancing of load.

Table 1 Benchmark functions

Benchmark functions	n	Search space	Global minimum
$f_1(x) = \sum_{i=1}^{n-1} \left(100\left(x_{i+1} - x_i^2\right)\right)^2 + (x_i - 1)^2$	30	$[-30, 30]^n$	0
$f_2(x) = \sum_{i=1}^{n} \left(\sum_{j=1}^{i} x_j\right)^2$	30	$[-100, 100]^n$	0
$f_3(x) = \sum_{i=1}^{n} ([x_i + 0.5])^2$	30	$[-100, 100]^n$	0
$f_4(x) = 4x_1^2 - 2.1x_1^4 + \frac{1}{3}x_1^6 + x_1x_2 - 4x_2^2 - 4x_2^4$	2	$[-5, 5]^n$	−1.0316285
$f_5(x) = \max_j\{\lvert x_i\rvert,\quad 1 \le i \le n\}$	30	$[-100, 100]^n$	0

Table 2 Comparison of different algorithms mean and standard deviation for benchmark functions

Algorithm	Measure	$f_1(x)$	$f_2(x)$	$f_3(x)$	$f_4(x)$	$f_5(x)$
GA [28]	Mean	338.5516	9749.9145	3.697	−1.0298	7.9610
	Std.	361.497	2594.9593	1.9517	3.1314Std.10^{-3}	1.5063
PSO [23]	Mean	37.3582	1.1979×10^{-3}	0.146	−1.0160	0.4123
	Std.	32.1436	2.1109×10^{-3}	0.4182	1.2786×10^{-2}	0.2500
GSO [28]	Mean	49.8359	5.7829	1.6000×10^{-2}	−1.0316280	0.1078
	Std.	30.1771	3.6813	0.1333	0	3.9981×10^{-2}
CQGSO [27]	Mean	34.4281	0.0404	0.0040	NaN	NaN
	SD	24.5366	0.0291	0.0015	NaN	NaN
RCGA-RTVM [30]	Mean	28.988454719	7.5456×10^{-242}	0.0002	−1.0316284535	7.4950×10^{-24}
	Std.	0.6739399580	0	0.0141421356	2.8796×10^{-11}	1.0434×10^{-23}
QOHS [proposed]	Mean	26.1428	6.1714×10^{-245}	0.0001	−1.0112	6.1211×10^{-23}
	Std.	0.5114	0	0.0112	2.1458×10^{-12}	1.0112×10^{-21}

PSO: basic concept
Particle swarm optimization (PSO) has been discovered by Kennedy and Eberhart. This optimization technique is based on flocking of bird or fish. It has been used to find the optimized solution of problem by iterative method which is trying to improve candidate solution (called as particle) from population (named as swarm) with regards of given measured quantity. It helps to get the solution of problem using two key words, i.e., position and velocity. Movement of each particle is influenced by local best called as position. And, the velocity is nothing but movement of position of particle searching for an optimal solution.

Fuzzy
Fuzzy logic was first termed in 1965 during proposal of fuzzy set theory by Lotfi Zadeh. It is based on decision making such as true (1) and false (0). This technique has an ability to perceive complex nonlinear output-input relation into synthesis of multiple simple input-output relations. Many researchers are using fuzzy theory, fuzzy sets, fuzzy logic and fuzzy measure to recognize, manipulate, interpret and utilize data and information to get appropriate solution. This technique can apply to many fields such as control theory and artificial intelligence.

3 Optimization Techniques Applied in WSN

WSNs are the fretwork of sovereign nodes which used to supervise an environment. An innovator of WSNs has a deal with many difficulties that emanated from ruptures of communication channel, memory and computational constraints and restricted energy. Numerous issues in WSNs are articulated as problems of multi-dimensional optimization and accessed through various metamorphosis optimization techniques. Some of the techniques are as fuzzy logic, PSO, ant colony optimization (ACO), etc. Mazinani et al. [31] proposed a multi-cluster routing method for WSN. The basic tool used in this proposal is fuzzy logic. The proposed name of this method is Fuzzy Multi-cluster-based Routing Protocol (FMRP). In this method, the author is used a threshold value for utilizing and clustering multi-hop system using base station (BS) of the WSN. Finally, this proposal helps to decrease dead nodes of the networks and increase network lifetime also in term of conflicting parameters of the WSN. Taherian et al. [32] proposed a Secure and Optimal Routing Method (SORM) for WSN using PSO. This method is used to enhance the self-directed nodes of the network by using aggregation as well as discovering task of the WSN. The proposed method of this article helps to manage battery power of the nodes by optimizing it using PSO. Finally, it also helps to manage network parameters using intelligently because PSO is one of the artificial intelligence techniques. Arya and Sharma [33] proposed an Optimized Efficient Analysis Method (OEAM) for WSN. This method is used ant colony optimization (ACO) for optimizing metrics of the WSN. The purpose of the sensor nodes is broadcasting, processing and sensing environmental data and information. This method balances the energy utilization of the network by optimizing different resource of the WSN. Finally, the proposed method helps to produce feasible routing solution among source node, sink node and different hop nodes. A QOHS algorithm is an effective, simple, less calculative and efficient computational optimization algorithm. Ease of execution and implementation, the caliber of qualities solution, computationally efficient and speed of convergences are the potencies of QOHS. Proposed algorithm may apply to communicate WSN issues such as lifetime optimization of network, coverage and lifetime optimization of WSNs and dynamic deployment optimization in WSN, communication protocols for large-scale WSN. For the optimization, the following mentioned scenarios have been discussed as followed.

Scenario (a) Network lifetime optimization in WSN
Scenario (b) Optimization of communication protocols for large-scale WSN
Scenario (c) Optimization of WSNs with coverage and lifetime
Scenario (d) Dynamic deployment optimization in WSN.

Scenario (a): Network lifetime optimization in WSN

Network lifetime (NL) is playing a crucial role while designing of critical metric design of energy-constrained WSN [34]. In this method of optimization, medium access control (MAC), joints of optimal design of physical and addressing layers to maximizing NL of multiple sources of the single sink (MSSS) WSN with the spirit of energy constraints may be explored. The problem of NL maximization (NLM) could be enunciated using problem in the optimization of energy-convex with adaption time division multiple access (TDMA) technique. When the constraints of the integer are slackened to catch substantial values, the problem could be carried over into a convex problem, and the outcome achieves into the greater-limits. To provide a methodological frame of reference for an unperturbed NLM problem of WSN in general planar topology, first inhibit the topologies to the planer networks on small-scale, particularly regular quadrangle and triangular topology. Incorporation of QOHS algorithm with D&C approach can handle large-scale planar network as QOHS algorithm has the capability of handling extended large-scale planar network with optimum performance.

Scenario (b): Optimization of communication protocols for large-scale WSN

The sustainable fault, trustworthy and vigorous design amenities in WSNs are challenging as well as a trendy research area. Under these circumstances, an optimization technique may envisage and used to tune-up the parameters of Middleton services to provide optimized results [35]. The method of optimization is based on simulation and subjected to estimate the noisy error surface. QOHS is an optimization algorithm technique, may exemplify using the spanning-tree algorithm, can efficiently operate even if there are any links present in between asymmetrical nodes. In forthcoming, sensors network in large scale will play the vital role of the embedded system used in challenges associated with aviation and space such as smart surface, smart dust, monitoring and controlling of the safety-critical system and can be used to make daily life more comfortable. Such sensor networks frequently used in the distributed operating system, in the manner of services (termed as middleware) as compared to the wireless communication protocols including adaptive sustainable fault owing to the design of dynamic network topology, an architect of such middleware services are not impartial. Since the sensors have restricted resource, thus used protocols are basics and simple as compared to one which has used in wired communication. An un-predictive nature of the environment also plays a crucial role and making the design more complex.

Scenario (c): Optimization of WSNs with coverage and lifetime

The functionality of the WSN is to use sensing mechanism to sense date at a regular interval of time as per the requirement form the long period. The two main constraints for the WSN are the coverage and lifetime which are an essential phenomenon for longer life of the battery. The previous work done is based on the randomly generated distribution of sensors which is not so exact for the better performance of a WSN. In this chapter, an analytical work done has been proposed for the coverage and lifetime of a WSN. The QOHS algorithm may be used effectually to increase the life-span of WSN.

Scenario (d): Dynamic deployment optimization in WSN

In the study of WSN, sensor deployment is one of the keynotes to be concerned. This mechanism is quite effective for improving the capability of WSN in terms of

Table 3 Different features of the proposed method QOHS with respect to three existing methods

QOHS	ACO	PSO	Fuzzy
Randomly generated population initialization may be improved	Need large computation time to reach the solution	Too difficult to initialize the control variables	A number of logic are involved that decrease the chances of better initialization
Few mathematical calculation and better convergences may be obtained	Have uncertain time to convergence	Due to scattering problem, it does not work well	Tins technique is quite different from stochastic and probability techniques
Chances of getting local solutions may be avoided	Probability distribution can change for each iteration	Converge prematurely with large problems	

unsteady and steady nodes [15]. The unsteady node adjusts itself in the path of best deployment under the condition of various uncertainties to cover the largest area. Here, the work is to find the optimum location of the unsteady node. This is to be done here by the QOHS algorithm which is a very authoritative technique of optimization and may be applied to solve multi-objective multi-dimensional optimization problems. Here, QOHS algorithm is still useful with numerous nodes which can be self-adjusted according to its requirement.

4 Performance Evaluation

QOHS algorithm uses the concept of HSA that can be used extensively for finding global search or true approximate method. The characteristics of QOHS algorithm in reference to ACO, PSO and fuzzy methods have been shown in Table 3. In this chapter, the method QOHS algorithm has been compared with the studied existing methods such as OEAM [33], SORM [32] and FMRP [31]. The method OEAM is based on ACO, the method SORM is based on PSO, and the method FMRP is based on fuzzy logic. The feature comparisons of QOHS algorithm with other features have shown in Table 4.

5 Conclusion

In this chapter, the applications of QOHS algorithm toward solving the problem of WSN have been shown with the different case study. It has been observed that WSN has some restrictions that need to resolve such as energy and communications. For the avoidance of such limitations, optimal devices are needed that consume less

Table 4 Feature comparison of the proposed method with some existing methods

Features	QOHS	OEAM [25]	SORM [24]	FMKP [23]
1	√	√	√	√
2	√	√	√	√
3	×	×	×	×
4	4	3	2	1
5	4	3	2	1
6	4	3	2	1
7	1	2	3	4
8	4	3	2	1
9	1	2	3	4
10	4	3	2	1
11	1	2	3	4
12	4	3	2	1
13	4	3	2	1
14	4	3	2	1
15	4	3	2	1
16	4	3	2	1
17	4	3	2	1
18	4	3	2	1

Caption

1: Routing loop avoidance 4: QoS support 7: Delay 10: Throughput 13: Bandwidth 16: Handling high mobility Yes: √ No: ×	2: Source initiated 5: Residual energy 8: Packet delivery ratio 11: Packet loss 14: Robustness 17: Handling traffic load Very high: 4 High: 3	3: Receiver initiated 6: Network lifetime 9: Communication overhead 12: Scalability 5: Connectivity status 18: Handling mutual interference Medium: 2 Low: 1

energy. In response to this, an effective technique is needed that may overcome the problems of optimization of WSN. For this case study, QOHS algorithm, ACO, PSO and fuzzy method have been implemented for this. The comparison study shows that QOHS algorithm is quite effective for this WSN issues. It may be said based on the comparison of different algorithms for shown benchmark functions. The computational efficiency and optimizing proficiency of the suggested QOHS method with the studied three existing methods also verify this. The feature comparison of the QOHS method with studied methods also showed the better performance of QOHS algorithm. Hybrid optimization techniques may be implemented for optimization in the wireless sensor for better performance. In this chapter, an overview of QOHS algorithm has been shown to different aspects of WSN. This shows that QOHS-based approaches may be an effective approach for the issues of WSNs.

Appendix

Parameters of QOHS

Number of parameters depends on problem variables, population size = 50, total number of iteration = 100, HMCR = 0.9, PAR_{min} = 0.45, PAR_{max} = 0.98, BW_{min} = 0.0005, BW_{max} = 50, J_r = 0.8,

References

1. Kulkarni RV, Venayagamoorthy GK (2010) Particle swarm optimization in wireless-sensor networks: a brief survey. IEEE Trans Syst Man Cybern Part C Appl Rev 41(2):262–267
2. Aggarwal R, Mittal A, Kaur R (2016) Various optimization techniques used in wireless sensor networks. Int Res J Eng Technol 3(6):2085–2090
3. Das SK, Yadav AK, Tripathi S (2017) IE2M: design of intellectual energy efficient multicast routing protocol for ad-hoc network. Peer-to-Peer Netw Appl 10(3):670–687
4. Yadav AK, Das SK, Tripathi S (2017) EFMMRP: design of efficient fuzzy based multi-constraint multicast routing protocol for wireless ad-hoc network. Comput Netw 118:15–23
5. Das SK, Tripathi S (2018) Intelligent energy-aware efficient routing for MANET. Wirel Netw 24(4):1139–1159
6. Das SK, Tripathi S (2019) Energy efficient routing formation algorithm for hybrid ad-hoc network: a geometric programming approach. Peer-to-Peer Netw Appl 12(1):102–128
7. Seifi H, Sepasian MS (2011) Electric power system planning: issues, algorithms and solutions. Springer, Berlin
8. Engelbrecht AP (2007) Computational intelligence: an introduction. Wiley, London
9. Chatterjee S, Hore S, Dey N, Chakraborty S, Ashour AS (2017) Dengue fever classification using gene expression data: a PSO based artificial neural network approach. In: Proceedings of the 5th international conference on frontiers in intelligent computing: theory and applications. Springer, Singapore, pp 331–341
10. Jagatheesan K, Anand B, Samanta S, Dey N, Ashour AS, Balas VE (2017) Particle swarm optimisation-based parameters optimisation of PID controller for load frequency control of multi-area reheat thermal power systems. Int J Adv Intell Paradigms 9(5–6):464–489
11. Kaliannan J, Baskaran A, Dey N, Ashour AS (2016) Ant colony optimization algorithm based PID controller for LFC of single area power system with non-linearity and boiler dynamics. World J Model Simul 12(1):3–14
12. Jagatheesan K, Anand B, Dey KN, Ashour AS, Satapathy SC (2018) Performance evaluation of objective functions in automatic generation control of thermal power system using ant colony optimization technique-designed proportional–integral–derivative controller. Electr Eng 100(2):895–911
13. Geem ZW, Kim JH, Loganathan GV (2001) A new heuristic optimization algorithm: harmony search. Simulations 76:60–68
14. Mahdavi M, Fesanghary M, Damangir E (2007) An improved harmony search algorithm for solving optimization problems. Appl Math Comput 188:1567–1579

15. Lee KS, Geem ZW (2004) A new structural optimization method based on the harmony search algorithm. Comput Struct 82:781–798
16. Kim JH, Geem ZW, Kim ES (2001) Parameter estimation of the nonlinear Muskingum model using harmony search. J Am Water Resour Assoc 37:1131–1138
17. Geem ZW, Kim JH, Loganathan GV (2002) Harmony search optimization: application to pipe network design. Int J Model Simul 22:125–133
18. Omran MGH, Mahdavi M (2008) Global-best harmony search. Appl Math Comput 198:643–656
19. Pan QK, Suganthan PN, Tasgetiren MF, Liang JJ (2010) A self-adaptive global best harmony search algorithm for continuous optimization problems. Appl Math Comput 216:830–848
20. Tizhoosh HR (2005) Opposition-based learning: a new scheme for machine intelligence. In: Proceedings of international conference on computational intelligence for modelling, control and automation, vol 1, pp 695–701
21. Tizhoosh HR (2005) Reinforcement learning based on actions and opposite actions. In: Proceedings of ICGST international conference on artificial intelligence and machine learning, Egypt
22. Tizhoosh HR (2006) Opposition-based reinforcement learning. J Adv Comput Intell Inf 10:578–585
23. Ventresca M, Tizhoosh HR (2006) Improving the convergence of backpropagation by opposite transfer functions. In: Proceedings of IEEE world congress on computational intelligence, Vancouver, BC, Canada, pp 9527–9534
24. Rahnamayan S, Tizhoosh HR, Salama MMA (2008) Opposition-based differential evolution. IEEE Trans Evol Comput 12:64–79
25. Nandi M, Shiva CK, Mukherjee V (2017) TCSC based automatic generation control of deregulated power system using quasi-oppositional harmony search algorithm. Eng Sci Technol 20(4):1380–1395
26. Shiva CK, Mukherjee V (2016) Automatic generation control of hydropower systems using a novel quasi-oppositional harmony search algorithm. Electr Power Compon Syst 44(13):1478–1491
27. Shiva CK, Mukherjee V (2015) A novel quasi-oppositional harmony search algorithm for automatic generation control of power system. Appl Soft Comput 35:749–765
28. Moradi-Dalvand M, Mohammadi-Ivatloo B, Najafi A, Rabiee A (2012) Continuous quick group search optimizer for solving non-convex economic dispatch problems. Electr Power Syst Res 93:93–105
29. He S, Wu QH, Saunders J (2009) Group search optimizer: an optimization algorithm inspired by animal searching behavior. IEEE Trans Evol Comput 13(5):973–990
30. Haghrah A, Mohammadi-ivatloo B, Seyedmonir S (2015) Real coded genetic algorithm approach with random transfer vectors-based mutation for short-term hydro-thermal scheduling. IET Gener Trans Distrib 9:75–89
31. Mazinani A, Mazinani SM, Mirzaie M (2019) FMCR-CT: an energy-efficient fuzzy multi cluster-based routing with a constant threshold in wireless sensor network. Alex Eng J 58(1):127–141
32. Taherian M, Karimi H, Kashkooli AM, Esfahanimehr A, Jafta T, Jafarabad M (2015) The design of an optimal and secure routing model in wireless sensor networks by using PSO algorithm. Proc Comput Sci 73:468–473
33. Arya R, Sharma SC (2015) Analysis and optimization of energy of sensor node using ACO in wireless sensor network. Proc Comput Sci 45:681–686
34. Wang H, Agoulmine N, Ma M, Jin Y (2010) Network lifetime optimization in wireless sensor networks. IEEE J Sel Areas Commun 28(7):1127–1137
35. Simon G, Volgyesi P, Maróti M, Ledeczi A (2003) Simulation-based optimization of communication protocols for large-scale wireless sensor networks. In: IEEE aerospace conference, vol 3, pp 31339–31346

Part III
Multi-objective Optimization

Chapter 10
A Comprehensive Survey of Intelligent-Based Hierarchical Routing Protocols for Wireless Sensor Networks

Nabil Sabor and Mohammed Abo-Zahhad

1 Introduction

Recently, WSNs become an important tool and are used in a wide range of exciting applications, such as real-time multimedia communication, medical application, surveillance application, and home monitoring applications. A WSN typically contains a large number of sensor nodes that are cheaper and have limit power capability [1–3]. These nodes are used for sensing, processing, and transmitting data. The architecture of each node consists of four main units as shown in Fig. 1. The first unit is the processor that processes all data and controls the operations of other units. The sensing unit, the second part, is used for sensing an event using a sensor and converts it into digital form using analog-to-digital converter (ADC). The third unit is the transceiver that is used for transmitting and receiving data within a limited transmission range. Finally, the power unit supplies power to all units [4].

The small size of the sensor node limits its resources such as storage, processing, power, and transmission range. Therefore, effective mechanisms are required to utilize the limited resources of WSNs in an efficient way. Routing is one of these mechanisms that control routes for transmitting data based on reducing the energy consumption in communication which enhances the network lifetime. In general, the developed routing protocols can be classified into hierarchical-based, flat-based, and location-based routing protocols [1]. According to the literature [2, 3], the hierarchical-based routing protocols are better than the other routing types in

N. Sabor (✉)
Electrical and Electronics Engineering Department, Faculty of Engineering, Assiut University, Assiut 71516, Egypt
e-mail: nabil_sabor@aun.edu.eg

M. Abo-Zahhad
Department of Electronics and Communications Engineering, E-JUST-Alexandria, New Borg El Arab, Egypt
e-mail: mohammed.zahhad@ejust.edu.eg

D. De et al. (eds.), *Nature Inspired Computing for Wireless Sensor Networks*, Springer Tracts in Nature-Inspired Computing,
https://doi.org/10.1007/978-981-15-2125-6_10

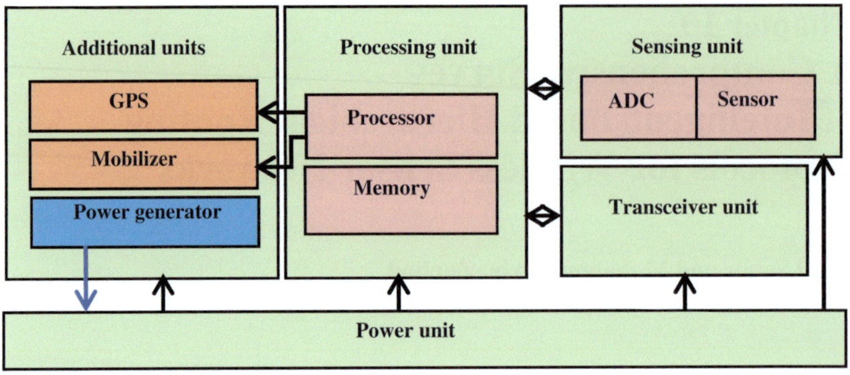

Fig. 1 Structure of a sensor node

saving energy and extending the lifetime of WSNs. The architecture of hierarchical-based routing consists of two layers, where the first layer is responsible for selecting the cluster head, and a selection of the second layer is responsible for constructing routes. In the hierarchical-based routing, the network is partitioned into multiple clusters. Each cluster has one head node called cluster head (CH) and many member nodes (MNs). The task of MN is to sensing data and forwarding it to its CH, while the CH is responsible for collecting and aggregating the data of its MNs and transferring the aggregated data to the sink. So, the hierarchical routing protocols can be called clustering protocols.

Hierarchical-based routing reduces the redundant data transmission and assigns a different task for each sensor node according to its capabilities which balance the network load. The hierarchical routing protocols can be non-intelligent or intelligent-based routing. In the former type, head nodes are selected randomly using the timer function, which causes an uneven traffic flow in different head nodes. Although they are suitable for applications of WSNs, they still have many challenges such as scalability, load balancing, connectivity, coverage, fault tolerance, and robustness. While in the intelligent-based routing, the head nodes are determined based on multi-criteria (i.e., fitness function) using different optimization algorithms to achieve the requirements of QoS. The fitness function is the most important issue in the evolutionary process of optimization algorithms because it plays a significant role in the performance of routing protocols [4–11].

So far, a large number of intelligent-based hierarchical routing protocols have been developed to improve the performance of static and mobile WSNs. In this chapter, we focus on reviewing the architecture and operation of the recently intelligent-based routing protocols that have designed based on utilizing Particle Swarm Optimization (PSO), Ant Colony Optimization (ACO), Fuzzy Logic (FL), Genetic Algorithm (GA), Neural Networks (NNs), and Artificial Immune Algorithm (AIA). Then, the surveyed protocols are classified based on different metrics as shown in Fig. 2. Moreover, the surveyed protocols are compared in terms of delay, network size, energy-efficient, scalability, advantages, and drawbacks.

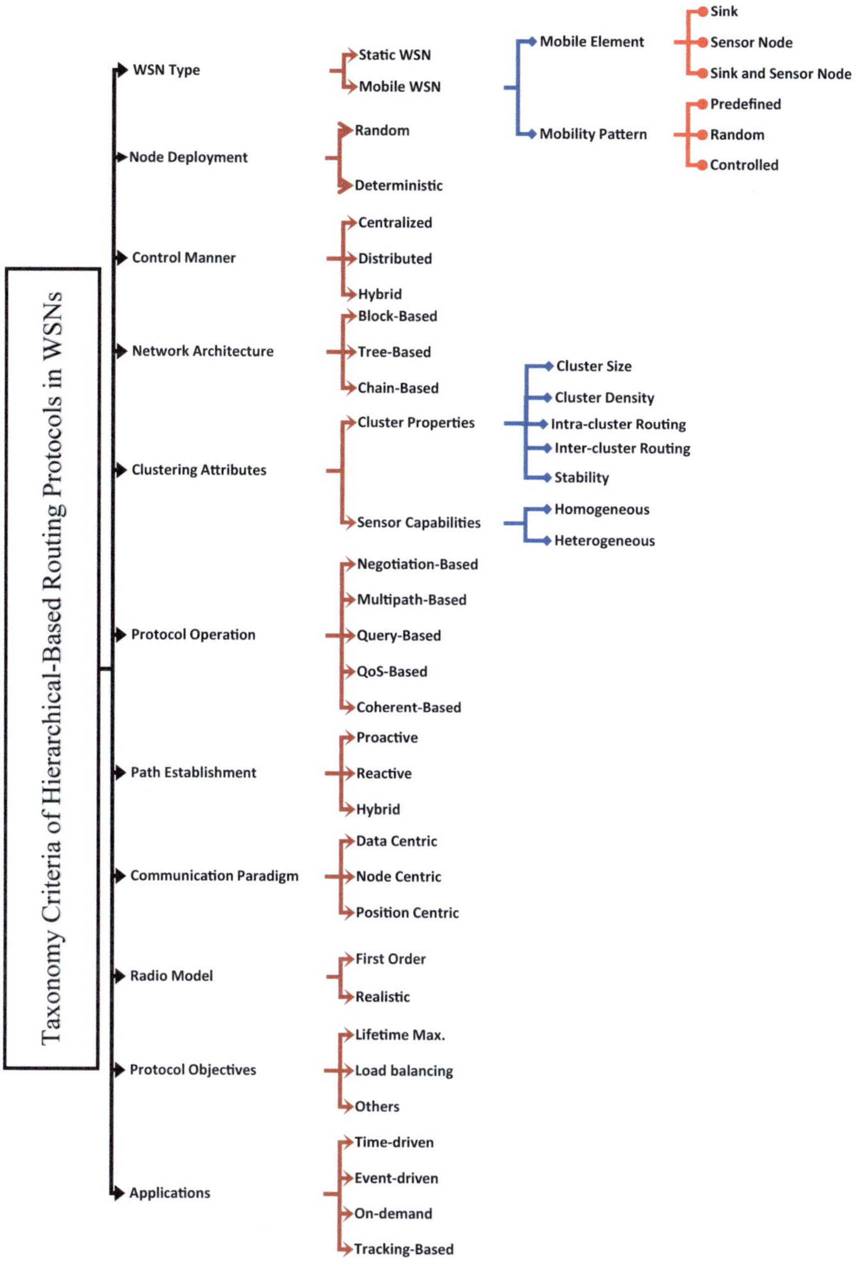

Fig. 2 Taxonomy of hierarchical-based routing protocols in WSNs

The chapter is organized as follows. Section 2 explains the different metrics that will use for classification of the intelligent hierarchical routing protocols. A review of the intelligent-based hierarchical routing protocols is presented in Sect. 3. In Sect. 4, a comparison between the reviewed protocols and discussion is given. Section 5 offers some conclusions and future work.

2 Taxonomy Metrics

In this section, we explain the different metrics that will use for classifying the intelligent hierarchical routing protocols.

2.1 WSN Types

WSN consists of a large number of the sensor nodes and one sink or multiple sinks depend on the applications. There are two types of WSNs, namely static WSNs (SWSNs) and mobile WSNs (MWSNs). In the first type, all elements of the network (i.e., sensor nodes and sink node(s)) are static. While in the second type, some or all sensor nodes can be mobile with static or mobile sink node(s). Therefore, the routing protocols of the MWSNs should be classified based on the mobile element and the mobility pattern.

Mobile Element: A WSN consists of a large number of the sensor nodes and one sink or multiple sinks. Depending on the applications, the sensor field can contain mobile sensor nodes or/and mobile sink node(s). Therefore, the routing protocols of MWSNs can be classified into three categories: protocol support sink node(s) mobility, protocol support mobility of the sensor nodes, and protocol support mobility of both the sensor nodes and the sink node(s). **Mobility pattern**: Determining the moving pattern of the mobile element (i.e., sensor nodes or sink node) is an important issue in the routing of the MWSNs. The predefined mobility pattern, the control mobility pattern, and the random mobility pattern are the famous mobility patterns in MWSNs. In **the predefined mobility pattern**, the path of the mobile element is priori-defined. So, the mobile element knows its path and sojourn positions along the path within the sensor field. This pattern is suitable for the mobile sink. While in **the controlled mobility pattern**, the mobility of the mobile element is controlled based on different factors such as improving lifetime, load balancing, avoiding hotspot problem [12], and improving connectivity. Finally, in **the random mobility pattern**, the mobile element moves with random velocity and random direction in the sensor field. This random mobility can be modeled and simulated using the Random Waypoint mobility model and the Reference Point Group Mobility Model [13–15].

i. **Random Waypoint Mobility Model**

Random Waypoint (RWP) mobility model has been widely used in the MWSNs due to its simplicity. In this model, the mobile element in the network selects a constant speed (V) uniformly and randomly from $\{[V_{\min}, V_{\max}], V_{\min} \geq 0$, where $V_{\min}$ and $V_{\max}$ are the minimum and the maximum allowable speed respectively$\}$ and random direction (θ) independently. Upon the selected V and θ, the mobile element moves from the initial location (x_0, y_0) toward the destination point (waypoint) (x_1, y_1). Upon reaching the destination, a mobile element pauses for a certain time called pause time (T_{pause}) and then selects another destination (x_2, y_2) based on randomly selected speed as shown in Fig. 3a. The mobile elements move in a linear direction, but reflect and change their directions sharply when reaching the boundary of the field.

ii. **Reference Point Group Mobility Model**

The Reference Point Group Mobility Model (RPGM) shown in Fig. 3b simulates the behavior of group mobility. Each node in the group follows the mobility direction

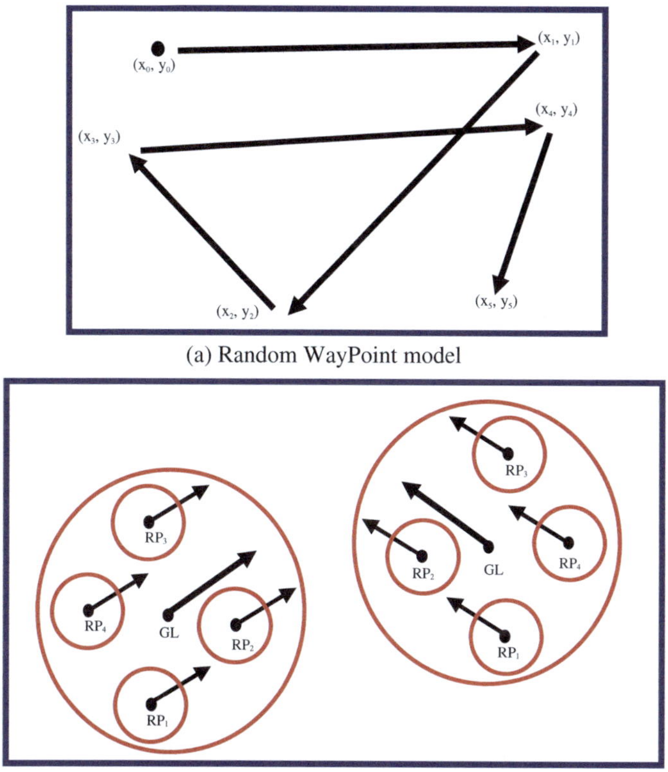

(a) Random WayPoint model

(b) Reference Point Group Mobility Model

Fig. 3 Mobility models

of the Group Leader (GL) that determines the group's motion behavior. The nodes in each group are distributed randomly around the Reference Point (RP) and added their mobility model to the RP, which drives them in the direction of the group. The Group Mobility Model is suitable for military battlefield communications, where soldiers may move together in a group.

2.2 Node Deployment

The deployment of sensor nodes affects the coverage and the connectivity of the WSNs. The sensor nodes can be deployed randomly or deterministically depending on the applications [16]. In the **random deployment**, the sensor nodes are scattered in the sensor field with uncertain locations. As a result of this, there is a tremendous change in node density because some nodes are placed closer to each other, and some are placed away from each other. Therefore, the random deployment does not ensure the connectivity among the nodes because it causes coverage holes in the field. Random deployment is preferred where the deployment area is inaccessible such as volcanoes and seismic zones. A random deployment can be cost-effective only if it provides the desired coverage. But generally, it doesn't work properly because the probability of throwing nodes on their exact locations is very less; therefore, an alternative approach is required.

While in the deterministic node deployment, positions of the nodes are predefined. These positions are calculated before deployment, and then, the sensor nodes are placed on their respective positions. Although the deterministic deployment satisfies the network connectivity, it is very complex in large networks and harsh environments. The deterministic deployment is used in those missions where the deployment area is physically reachable. As compared to the random deployment, deterministic deployment uses a fewer number of the sensor nodes to cover the given area. Therefore, it is more preferable over the random deployment.

2.3 Control Manner

Routing approaches in WSNs can be controlled in a centralized manner, distributed manner, or hybrid manner. In **centralized approaches**, global information about the network/group such as energy level and geographical position are used to control the clustering and routing processes. The centralized approach gives a better clustering architecture for the network, but it increases the overhead packets for establishing clusters. While in **distributed approaches**, the sensor nodes of the network collaborate with each other by exchanging information to control the routing process without the need for any global information about the network. Each sensor node decides locally to be the head node or not. The distributed approach does not guarantee the connectivity among the sensor nodes because it depends on the local information of the neighboring nodes, but it increases the network scalability. **Hybrid approaches** can be semi-distributed or semi-centralized approaches to optimize the features of centralized and distributed approaches.

2.4 Network Architecture

According to the network architecture, there are three categories, namely block-based routing, tree-based routing, and chain-based routing. In the **block-based hierarchical routing**, the network is partitioned into clusters, where each cluster has a head node called a cluster head (CH) and the other nodes in a cluster called member nodes (MNs). CH is responsible for collecting, aggregating, and forwarding data of its MNs to the sink. Construction of cluster and selecting CHs based on different QoS metrics is the aim of the block-based routing protocols. In the second type, **tree-based hierarchal routing**, a routing tree is constructed among the members of the network. Leaf nodes of the tree send the sensing data to their parent ones that aggregate and forward the received data to the upper-level parent nodes toward the sink. While in **chain-based hierarchical routing**, one or more chains are constructed to form a path for transmitting data. Each chain has a head node called Chain Leader (CL), which is responsible for collecting data of the chain members and forward it to the sink. This type of routing increases the packet delay because the data packet reaches to sink through multiple hops. Moreover, the chain can be broken and its data will be dropped due to the failure of one node in the chain, which increases the packet drop rate.

2.5 Clustering Attributes

The cluster properties and the sensor capabilities are the main two issues in the clustering attributes that affect the performance of the routing protocols [17, 18]. The cluster characteristics are used to compare between the routing protocols. These characteristics are:

(a) **Cluster Size**: The WSN can be partitioned into equal or unequal clusters. In the former type, all CHs have the same communication range. While in the latter type, heads of clusters have different communication ranges. In general, the unequal clustering is used to solve the energy hole problem [12] of the static sink.
(b) **Cluster Density**: Members of the cluster represent the cluster density or the cluster load. The load of the cluster can be fixed or variable according to the cluttering protocol.
(c) **Intra-cluster Routing**: Intra-cluster routing is the communication path between CH and its member nodes. This path can be a single-hop route, in which data

are directly sent from member nodes to CH, or a multi-hop route, in which relay nodes are used to forward data of member nodes to CH.

(d) **Inter-cluster Routing**: Inter-cluster routing is the communication outside the cluster, which may be from the sensor nodes to BS or CH to BS or CH to CH. Also, the inter-cluster communication can be a single-hop or a multi-hop routing.

(e) **Stability**: The cluster density controls the routing process stability. So, the stability can be fixed or variable based on the density of the clusters is fixed or vary throughout the routing process.

In term of **sensor capabilities**, WSNs can be classified into the homogeneous and heterogeneous networks according to the resources of sensor nodes. In the **homogeneous network**, the same resources such as battery level, computation, and communication are assigned to all nodes in the network. While in the **heterogeneous network**, different capabilities and resources are assigned to the sensor nodes. This is useful for clustering by assigning the role of CH to nodes that have more capabilities.

2.6 *Protocol Operation*

The routing protocols can be classified based on the protocol operation into five categories. The first category is **negotiation-based routing**, in which a high level of negotiation between the source and the destination is established before real data transmission to avoid the redundant data. **Query-based routing** is the second category, in which the source node does not transmit its data until receiving a query message from the destination node. The **multipath-based routing** category increases the robustness of routing and the packet delivery rate because multiple paths between the sensor nodes and the destination are constructed. The fourth category is called **coherent-based routing that** uses different processing mechanisms such as coherent and non-coherent methods. In the non-coherent technique, data are processed locally at the sensor node and then forwarded it to the aggregator. Then, the aggregator node reduces the redundancy of the received data from many sensor nodes and aggregates it into one packet before sending it to the sink. While in the coherent method, the processing and aggregation tasks are the responsibility of the aggregator. **QoS-based routing** is the last category, in which each routing protocol is designed to satisfy the QoS requirements such as reliability, delay or bandwidth.

2.7 *Path Establishment*

The path establishment mechanism has the ability to discover and establish routes from a source node to the intended receiver. This mechanism can be proactive, reactive, or hybrid. In the **proactive routing**, each node builds a routing table before real data transmission. This table contains a list of paths and their costs between a node and one or more next-hop neighbor. The proactive routing reduces the packet delay because the routing path is available and no need to wait for discovering it. This routing type can be used in periodic data monitoring applications like collecting data about temperature change over a particular area. While in the **reactive routing**, each node discovers its route path as a reaction to any sudden change in the event. The discovering process of a routing path adds some delay before transmitting the data. The reactive routing can be used in time-critical applications, e.g., explosion detection and intrusion detection. **Hybrid Routing** combines proactive and reactive routing types.

2.8 Communication Paradigm

There are three types of communication between the sink and the sensor nodes. The first type is the **node-centric**, in which destinations are identified using numerical addresses (ID). The sensor node uses the ID to forward its data to a specific destination. The second type is the **data-centric** technique, in which the sink sends a query message to a specific area within a sensor field and waits for data of the sensor nodes that are located within the range of sink. The sensor nodes that receive sink quires send their data to the sink via relay nodes. In the last type, **location-centric**, the location information of sensor nodes is required during the route construction process.

2.9 Radio Model

Since the energy dissipated in the communication process consumes the battery of sensor nodes, simulating and modeling the radio of nodes affect the performance of routing protocols. Thus, the routing protocols can be surveyed according to the used model of the sensor radio. According to literature, sensor radio was simulated as the first-order model [3, 19, 20] that shown in Fig. 4 or as the realistic radio model such as CC2420 [21–23]. In the **first-order radio model**, to transmit a k bit message over a distance (d), this radio model expends:

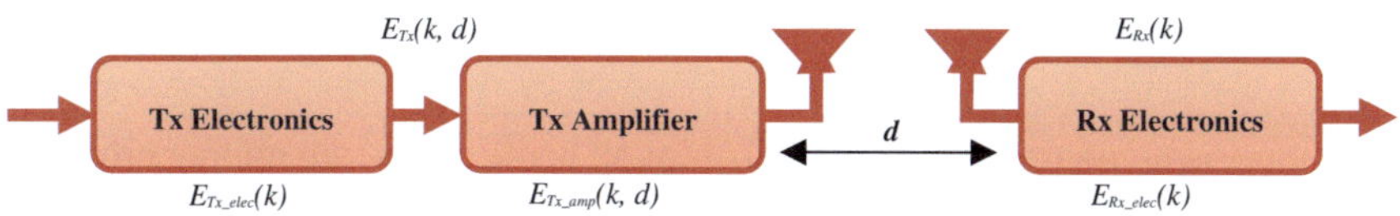

Fig. 4 First-order radio model

$$E_{Tx}(k,\ d) = E_{\text{Tx-elec}}(k) + E_{\text{Tx-amp}}(k, d) = k\, E_{\text{elec}} + E_{\text{Amp}}\, k\, d^p \quad (1)$$

and to receive this message, the radio expends:

$$E_{\text{Rx}}(k) = E_{\text{Rx-elec}}(k) = k\, E_{\text{elec}} \quad (2)$$

where E_{elec} and E_{Amp} are the dissipated energy in the electronic circuit and amplifier of the radio model, respectively, and p is the path loss. If distance d is less than the threshold distance $\left(d_o = \sqrt{E_{\text{fs}}/E_{\text{mp}}}\right)$, free space communication (fs) will be used by setting $p = 2$ and $E_{\text{Amp}} = E_{\text{fs}}$. Otherwise, the multipath (mp) fading communication should be adopted by setting $p = 4$ and $E_{\text{Amp}} = E_{\text{mp}}$.

The first-order radio model is simple, but it does not simulate the realistic model, while the **realistic radio model** enables sensor nodes to adjust their radio-transmitted power by selecting one of the four predefined power modes, i.e., 0, −5, −10, −15 dBm as in the CC2420. The data packets are sent by switching one of these modes depending on the distance between sender and receiver. However, the control packets are sent with maximum power. The total energy consumed (E_i) by a node i is calculated as:

$$E_i = \sum_{\text{state } j} P_{\text{state } j} \times t_{\text{state } j} + \sum E_{\text{transitions}} \quad (3)$$

where state j refers to the energy states: idle, transmit, receive, or sleep state. $P_{\text{state } j}$ is the consumption power in state j, and $t_{\text{state } j}$ is the time spent in the corresponding state. $E_{\text{transitions}}$ is the consumed energy in transitions between different states.

2.10 Protocol Objectives

The main objective of the hierarchical-based routing protocols is extending the lifetime of WSNs by reducing the dissipated energy of the communication process. However, the routing protocols can develop to achieve other objectives such as load balancing, improving stability period, enhancing network connectivity, fault tolerance, and QoS.

2.11 Applications

Since there is no routing protocol suitable for all applications, the routing protocols can be classified according to their applications. Applications of WSNs can broadly be divided into time-driven, event-driven, and on-demand applications. In the **event-driven applications**, the sensor nodes burst into activity only when an event is detected; otherwise, they stay in sleep mode. This type of applications can be found in forest fires' detection, volcanic eruptions' detection, grass fires' detection, etc. In the **time-driven applications**, sensor nodes periodically send their data to the sink. These applications are suitable for monitoring the environmental conditions like affecting crops, humidity, temperature, and lighting. While in **on-demand applications**, the sink is not interested in collecting data of sensor nodes periodically. In the on-demand applications, nodes send their data based on query messages from sink. This results in a more energy-efficient use of resources.

3 Intelligent-Based Hierarchical Routing Protocols

The objective of hierarchal-based routing protocols is prolonging the lifetime of WSN by finding the optimal CHs/Leaders, which reduce the dissipated energy during collecting data. The problem of finding the optimal CHs/Leaders is a Non-deterministic Polynomial (NP) optimization problem. Many robust hierarchical routing protocols are developed to solve the above-mentioned problem based on intelligent optimization algorithms such as Particle Swarm Optimization (PSO) [24], Genetic Algorithm (GA) [7, 25], Fuzzy Logic (FL) [26], Ant Colony Optimization (ACO) [27], and Artificial Immune Algorithm (AIA) [28]. These optimization algorithms are used to optimize the trade-off between energy consumption and the QoS requirements depending on the designed fitness function. The QoS metrics can be a delay, load balancing, scalability, connectivity, or fault tolerance. The fitness function is the most important issue in the evolutionary process of the optimization algorithms because it plays a significant role in the performance of a routing protocol. In this section, we review the operation and architecture of the recently intelligent-based hierarchical routing protocols. Taxonomy classification of surveyed protocols based on the above metrics is listed in Table 1.

3.1 Particle Swarm Optimization-Based Hierarchical Routing Protocols

Particle Swarm Optimization (PSO) is a randomly population-based optimization algorithm that was suggested by Kennedy et al. [24]. The behavior of PSO is inspired by the behavior of fish schooling or bird flocking. PSO comprises a swarm of S particles (potential solutions), where each particle i is defined by the position vector x_i and the velocity vector v_i. These particles fly through a D-dimensional space for searching for the global optimum position that produces the best fitness value of the given objective function. In each iteration t of the PSO algorithm, each particle changes its position based on the new value of its velocity as given by Eqs. 4 and 5.

$$v_i(t) = \omega v_i(t-1) + c_1 r_1 (x_{\text{best}}(t) - x_i(t-1)) + c_2 r_2 (g_{\text{best}}(t) - x_i(t-1)) \quad (4)$$

Table 1 Classification of intelligent-based hierarchical routing protocols

Protocol	Control manner	Node distribution	Mobile element	Mobility pattern	Network architecture	Clustering attributes				
						Cluster properties				Sensor capability
						Cluster Size	Cluster Density	Intra/inter-cluster routing	Stability	
EBUC 2010 [29]	C	Rn	SWSNs		Tree-Based	Uneq	V	SH/MH	V	Homo
PSO-DH 2011 [30]	C	Rn	SWSNs		Block-Based	Eq	V	MH/SH	V	Homo
PSO-SD 2012 [31, 32]	C/CH	Rn	SWSNs		Block-Based	Eq	V	MH/SH	V	Homo
AECRP 2013 [33]	C	Rn	SWSNs		Tree-Based	Eq	V	SH/MH	V	Homo
RCC-PSO 2014 [34]	C	Rn	SWSNs		Block-Based	Eq	V	SH/SH	V	Homo
PSO-HC 2014 [35]	C	Rn	SWSNs		Block-Based	Eq	V	MH/SH	V	Homo
E-OEERP 2015 [36]	C	Rn	SWSNs		Tree-Based	Eq	V	SH/MH	V	Homo
TPSO-CR 2015 [37]	C	Rn	SWSNs		Tree-Based	Eq	V	SH/MH	V	Homo/ Hetero
PSO-MBS 2011 [38]	C	Rn	One Sink	Ctr	Block-Based	Eq	V	SH/SH	V	Homo
CAGM 2015 [6]	C	Rn	One Sink	Rn	Block-Based	Eq	V	SH/SH	V	Homo

(continued)

Table 1 (continued)

Protocol	Control manner	Node distribution	Mobile element	Mobility pattern	Network architecture	Clustering attributes				
						Cluster properties				Sensor capability
						Cluster Size	Cluster Density	Intra/inter-cluster routing	Stability	
OQoS-CMRP 2017 [39]	C	Rn	SWSNs		Block-Based	Eq	V	MH/SH	V	Hetero
PSOBS 2019 [40]	C	Rn	One Sink	Ctr	Tree-Based	Eq	V	SH/SH	V	Homo
GABEEC 2012 [41]	C	Rn	SWSNs		Block-Based	Eq	F	SH/SH	F	Homo
Kumar et al. 2013 [8]	C	Rn	SWSNs		Block-Based	Eq	F	SH/SH	F	Homo
Kuila et al. 2013 [42]	C	Rn	SWSNs		Block-Based	Eq	V	SH/SH	V	Homo
GAHN 2014 [43]	C	Rn	SWSNs		Block-Based	Eq	V	SH/SH	V	Hetero
GAEEP 2014 [19]	H	Rn	SWSNs		Block-Based	Eq	V	SH/SH	V	Homo/ Hetero
GACR 2015 [44]	C	Rn	SWSNs		Tree-Based	Eq	V	SH/MH	V	Homo

(continued)

Table 1 (continued)

Protocol	Control manner	Node distribution	Mobile element	Mobility pattern	Network architecture	Clustering attributes				
						Cluster properties				Sensor capability
						Cluster Size	Cluster Density	Intra/inter-cluster routing	Stability	
GAECH 2015 [45]	C	Rn	SWSNs		Block-Based	Eq	V	SH/SH	V	Homo
GAROUTE 2011 [46]	C	Rn	Sensor nodes	Rn)RWP(	Block-Based	Eq	V	SH/MH	V	Homo
NSGAII-RP 2015 [47]	C	Rn	Sensor nodes	Ctr	Block-Based	Eq	V	SH/SH	V	Homo
GADA-LEACH 2016 [48]	C	Rn	SWSNs		Block-Based	Eq	V	SH/MH	V	Homo
MGAHP 2018 [49]	C	Rn	SWSNs		Block-Based	Eq	V	SH/SH	V	Homo
EAUCF 2013 [50]	D	Rn	SWSNs		Tree-Based	Uneq	V	SH/SH SH/MH	V	Homo
Mirsadeghi et al. 2014 [51]	D	Rn	SWSNs		Tree-Based	Eq	V	SH/MH	V	Homo
FBUC 2014 [52]	D	Rn	One Sink	Ctr	Tree-Based	Uneq	V	SH/SH SH/MH	V	Homo

(continued)

Table 1 (continued)

Protocol	Control manner	Node distribution	Mobile element	Mobility pattern	Network architecture	Clustering attributes				
						Cluster properties				Sensor capability
						Cluster Size	Cluster Density	Intra/inter-cluster routing	Stability	
OZEEP 2015 [53]	H	Rn	Sensor nodes	Rn (RWP)	Block-Based	Eq/ Uneq	V	SH/MH	V	Homo
SEPFL 2016 [54]	D	Rn	SWSNs		Block-Based	Eq	V	SH/SH	V	Homo
LEFUCMA 2018 [55]	D	Rn	SWSNs		Block-Based	Uneq	V	SH/MH	V	Homo
FMCR-CT 2019 [56]	D	Rn	SWSNs		Block-Based	Uneq	V	SH/MH	V	Homo
FCM-ACO 2016 [57]	C	Rn	SWSNs		Block-Based	Eq	V	SH/MH	V	Homo
ACOHC 2016 [58]	C	Rn	SWSNs		Block-Based	Eq	V	MH/MH	V	Homo
MH-GEER 2018 [59]	H	Rn	SWSNs		Block-Based	Eq	V	SH/MH	V	Homo
Jingyi et. al. 2016 [60]	C	Rn	SWSNs		Block-Based	Eq	V	SH/SH	V	Homo

(continued)

Table 1 (continued)

Protocol	Control manner	Node distribution	Mobile element	Mobility pattern	Network architecture	Clustering attributes				
						Cluster properties				Sensor capability
						Cluster Size	Cluster Density	Intra/inter-cluster routing	Stability	
Zahhad et al. 2015 [61]	C	Rn	Sensor nodes	Rn (RWP)	Block-Based	Eq	V	SH/SH	V	Homo
UMBIC 2016 [62]	H	Rn	Sensor nodes	Ctr	Block-Based	Uneq	V	SH/MH	V	Homo/ Hetero
ARBIC 2018 [9]	H	Rn	SWSNs		Block-Based	Uneq	V	SH/MH	V	Homo/ Hetero

Protocol	Protocol operation	Path establishment	Communication paradigm	Energy model	Protocol objectives	Applications
EBUC 2010 [29]	Coherent based	Proactive	Node/location-centric	First order	Lifetime Max., load balancing	Time-driven applications
PSO-DH 2011 [30]	Coherent based	Proactive	Node/location-centric	First order	Lifetime Max., load balancing	Time-driven applications
PSO-SD 2012 [31, 32]	Query based	Proactive	Data/location-centric	First order	Lifetime Max.	On-demand applications
AECRP 2013 [33]	Coherent based	Proactive	Node-centric	First order	Lifetime Max., load balancing, reliable data delivery	Large-scale monitoring applications

(continued)

Table 1 (continued)

Protocol	Protocol operation	Path establishment	Communication paradigm	Energy model	Protocol objectives	Applications
RCC-PSO 2014 [34]	Coherent based	Proactive	Node-centric	Realistic	Improving the lifetime, link quality and coverage	Habitats surveillance and area monitoring
PSO-HC 2014 [35]	Coherent based	Proactive	Node-centric	Realistic	Improving lifetime, link quality, scalability and coverage	Time-driven applications
E-OEERP 2015 [36]	Coherent based	Hybrid	Node/location-centric	NA	Improving lifetime, packet delivery ratio and throughput	Time-driven applications
TPSO-CR 2015 [37]	Coherent based	Proactive	Node-centric	Realistic	Improving lifetime, packet delivery ratio and coverage	Time-driven applications
PSO-MBS 2011 [38]	Coherent based	Proactive	Node/location-centric	First order	Maximizing lifetime; improving data delivery	Time-driven applications

(continued)

Table 1 (continued)

Protocol	Protocol operation	Path establishment	Communication paradigm	Energy model	Protocol objectives	Applications
CAGM 2015 [6]	Coherent based	Proactive	Node/location-centric	First order	Maximizing lifetime; energy consumption balancing	Time-driven applications
OQoS-CMRP 2017 [39]	Reactive	Node-centric	First order	Improving lifetime and solve the energy hole problem	Event-driven applications	
PSOBS 2019 [40]	Coherent based	Proactive	Node-centric	First order	Energy saving and improve packet delivery ratio	Time-driven applications
GABEEC 2012 [41]	Coherent based	Proactive	Node/location-centric	First order	Lifetime Max.	Time-driven applications
Kumar et al. 2013 [8]	Coherent based	Proactive	Node/location-centric	First order	Lifetime Max., energy saving	Time-driven applications
Kuila et al. 2013 [42]	Coherent based	Proactive	Node/location-centric	First order	Load balancing	Time-driven applications
GAHN 2014 [43]	Coherent based	Proactive	Node/location-centric	First order	Load balancing, lifetime Max.	Time-driven applications

(continued)

Table 1 (continued)

Protocol	Protocol operation	Path establishment	Communication paradigm	Energy model	Protocol objectives	Applications
GAEEP 2014 [19]	Coherent based	Proactive	Node/location-centric	First order	Lifetime Max., enhancing stability period, load balancing	Time-driven applications
GACR 2015 [44]	Coherent based	Proactive	Node/location-centric	First order	Load balancing, lifetime Max.	Time-driven applications
GAECH 2015 [45]	Coherent based	Proactive	Node/location-centric	First order	Improving lifetime and stability period	Time-driven applications
GAROUTE 2011 [46]	Coherent based	Proactive	Node-centric	First order	Improving lifetime, and Stability	Time-driven applications
NSGAII-RP 2015 [47]	Coherent based	Proactive	Node/location-centric	First order	Maximizing lifetime and coverage	Monitoring applications
GADA-LEACH 2016 [48]	Coherent based	Proactive	Node/location-centric	First order	Improving lifetime and communication stability	Time-driven applications
MGAHP 2018 [49]	Coherent based	Proactive	Node/location-centric	First order	Improving lifetime	On-demand applications

(continued)

Table 1 (continued)

Protocol	Protocol operation	Path establishment	Communication paradigm	Energy model	Protocol objectives	Applications
EAUCF 2013 [50]	Coherent based	Proactive	Node-centric	First order	Energy hole avoiding, Improving stability period	Time-driven applications
Mirsadeghi et al. 2014 [51]	Coherent based	Proactive	Node-centric	First order	Improving lifetime, energy consumption and coverage	Time-driven applications
FBUC 2014 [52]	Coherent based	Proactive	Node-centric	First order	Energy hole avoiding, improving lifetime	Time-driven applications
OZEEP 2015 [53]	Coherent based	Proactive	Node/location-centric	First order	Improving lifetime; load balancing, scalability	Time-driven applications
SEPFL 2016 [54]	Coherent based	Proactive	Node-centric	First order	First order	Time-driven applications
LEFUCMA 2018 [55]	Coherent based	Proactive	Node-centric	First order	First order	Time-driven applications
FMCR-CT 2019 [56]	Coherent based	Proactive	Node-centric	First order	First order	Time-driven applications

(continued)

Table 1 (continued)

Protocol	Protocol operation	Path establishment	Communication paradigm	Energy model	Protocol objectives	Applications
FCM-ACO 2016 [57]	Coherent based	Proactive	Node-centric	First order	First order	Time-driven applications
ACOHC 2016 [58]	Coherent based	Proactive	Node-centric	First order	First order	Time-driven applications
MH-GEER 2018 [59]	Coherent based	Proactive	Node-centric	First order	First order	Time-driven applications
Jingyi et. al. 2016 [60]	Coherent based	Proactive	Node-centric	First order	First order	Time-driven applications
Zahhad et al. 2015 [61]	Coherent based	Proactive	Node/location-centric	First order	Improving lifetime and stability period	Time-driven applications
UMBIC 2016 [62]	Coherent based	Proactive	Node/location-centric	First order	Energy hole avoiding; load balancing; stability period	Periodic and demand monitoring applications
ARBIC 2018 [9]	Coherent based	Proactive	Node/location-centric	Realistic	Improving lifetime and stability period, balancing the load	On-demand monitoring applications

C centralized, *D* distributed, *H* hybrid, *Eq* equal, *Uneq* unequal, *F* fixed, *V* variable, *SH* single-hop, *MH* multi-hop, *Homo* homogeneous, *Hetero* heterogeneous, *Rn* random, *Ctr* controlled

$$x_i(t) = x_i(t-1) + v_i(t-1) \quad (5)$$

where x_{best} and g_{Best} are the best particle position and best group position, respectively; ω is the weight factor, c_1 and c_2 are two positive constants within [0 1], and r_1 and r_2 two random parameters within [0, 1]. Recently, many hierarchical routing protocols are developed based on the PSO algorithm. Here, we will review some of them.

(1) **Energy-Balanced Unequal Clustering Protocol**

An Energy-Balanced Unequal Clustering (EBUC) protocol was suggested in [29] based on the PSO algorithm to prolong the lifespan of each sensor node. EBUC partitions the network into unequal clusters. The size of each cluster depends on its distance with reference to BS; the closest cluster to BS has a smaller size and vice versa as shown in Fig. 5. This solves the energy hole problem because CHs of the closer clusters can save some of the dissipated energy in intra-cluster traffic for the inter-cluster relay load. Moreover, the EBUC protocol reduces the dissipated energy of CHs by constructing an energy-aware routing tree among them. The operation of the EBUC is run on a centralized way at the BS. The EBUC protocol works in round manner, where each round starts with a setup phase, when the BS organizes the optimal clusters using the PSO algorithm and constructs a routing tree among them for the inter-cluster communication, followed by a steady-state phase when the sensory data are transferred periodically to BS. In the EBUC, BS uses the PSO algorithm for determining the best CHs based on minimizing the following cost function:

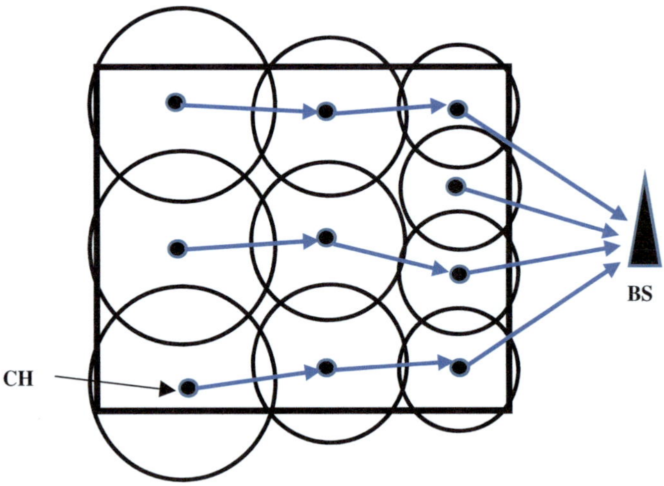

Fig. 5 An idea of EBUC protocol

$$F(p_i) = \alpha_1 f_1(p_i) + \alpha_2 f_2(p_i) + \alpha_3 f_3(p_i) \tag{6}$$

$$f_1(p_i) = \max_{k=1,2,\ldots,K} \sum_{\forall n_j \in C_{p_i,k}} \frac{d(n_j, \mathrm{CH}_{p_i,k})}{|C_{p_i,k}|} \tag{7}$$

$$f_2(p_i) = \frac{\sum_{j=1}^{N} E(n_j)}{\sum_{k=1}^{K} E(\mathrm{CH}_{p_i,k})} \tag{8}$$

$$f_3(p_i) = \frac{\sum_{i=1}^{K} d(\mathrm{BS}, \mathrm{CH}_{p_i,k})}{K \times d(\mathrm{BS}, \mathrm{NC})} \tag{9}$$

where f_1 is the maximum average Euclidean distance of nodes to their associated cluster heads related to the number of nodes that belong to cluster C_k of particle p_i $(C_{p_i,k})$, while the second objective f_2 is the ratio of the total initial energy of all nodes in the network to the total energy of the candidate CHs in the current round. Finally, f_3 is the ratio of the average Euclidean distance between CHs and BS to the Euclidean distance between the network center (NC) and BS. $\alpha_1, \alpha_2, \alpha_3$ are the weight factors within [0 1] where $\alpha_1 + \alpha_2 + \alpha_3 = 1$.

(2) **PSO-based Double Cluster -Heads Clustering Protocol**

A PSO-based Double cluster-Heads clustering protocol (PSO-DH) [30] was developed to balance the consumption energy and extend the lifetime of the WSN. In the PSO-DH, two nodes are selected in each cluster based on the location information and the energy levels of all nodes in the network to be CHs according to the fitness function given by Eq. 10.

$$F(i) = \omega f_1(i) + (1 - \omega) f_2(i) \tag{10}$$

$$f_1(p_i) = \frac{E(n_j)}{\sum_{q=1, q \neq j}^{|C_{p_i,k}|} E(n_q)} \tag{11}$$

$$f_2(p_i) = \frac{|C_{p_i,k}|}{\sum_{q=1, q \neq i}^{|C_{p_i,k}|} d(i, k)} \tag{12}$$

where $f_1(p_i)$ is the ratio of the energy of node j to the total energy of all members of a cluster k in particle p_i, $f_2(p_i)$ is the ratio of number of cluster members $|C_{p_i,k}|$ to the total Euclidean distance between node j and all other member nodes in the cluster, and ω is the weight factor within [0 1]. The first determined CH is called the Master CH (MCH) and is responsible for collecting data from members of the cluster and aggregating it. The aggregation data are transmitted to the second CH, which is called Vice CH (VCH). Then, the VCH forwards the aggregation data of the cluster directly to the sink, which saves the energy of MCH. Dividing tasks between MCH and VCH balances the load of the network and increases the lifetime of WSN.

(3) **PSO-based Semi-distributed Clustering Protocol**

In [31], authors developed a Semi-Distributed clustering protocol based on the PSO algorithm (PSO-SD), in which the PSO algorithm is performed within the cluster rather than BS. PSO-SD protocol is used for grouping the network into energy-aware clusters and selecting optimal CHs for these clusters. The PSO-SD protocol operates in round manner, where each round contains the CH selection phase and the data collection phase. At the beginning of CH selection, a sensor node in each cluster is chosen randomly to be a cluster assistant. Then, all sensor nodes in the cluster send their information, such as position and residual energy, to the cluster assistant node. The cluster assistant utilizes the PSO algorithm to find the optimal position of CH of its cluster based on the remaining energy $(E(p_i))$, the minimum average distance from the member nodes $\left(d\left(n_j, \mathrm{CH}_{p_i,k}\right)\right)$, and the head count of the probable CHs $(H(p_i))$ by optimizing the objective function of each particle p_i as follows:

$$F(p_i) = \alpha_1 f_1(p_i) + \alpha_2 f_2(p_i) + \alpha_3 f_3(p_i) \tag{13}$$

$$f_1(p_i) = \sum_{\forall n_j \in C_{p_i,k}} \frac{d\left(n_j, \mathrm{CH}_{p_i,k}\right)}{\left|C_{p_i,k}\right|} \tag{14}$$

$$f_2(p_i) = \frac{\sum_{i=1}^{N} E(p_i)}{\sum_{j=1}^{\left|C_{p_i,k}\right|} E\left(n_j\right)} \tag{15}$$

$$f_3(p_i) = \frac{1}{H(p_i)} \tag{16}$$

where α_1, α_2, α_3 are the weight factors within [0 1], where $\alpha_1 + \alpha_2 + \alpha_3 = 1$. In the data collection phase of the PSO-SD protocol, BS periodically collects the data of sensor nodes via CHs based on its request. The PSO-SD protocol improves the lifetime and the consumption energy of WSN, but it neglects the intra-cluster distance during the CH selection process. Authors in [32] solved this problem by modifying the objective function of the CH selection to employ the residual energy, node degree, intra-cluster distance, and number of the probable CHs.

(4) **Adaptive Energy-Efficient Clustering Routing Protocol**

An Adaptive Energy-efficient Clustering Routing Protocol (AECRP) [33] was designed to improve communication and data delivery of the large-scale WSNs. The operation of the AECRP protocol depends on combining the inter-cluster routing algorithm with the improved version of particle swarm-clustering algorithm. In the improved version of the particle swarm-clustering algorithm, PSO utilized to select the best CHs is based on the nodes' energy, the distribution among clusters, and the distribution within a cluster. While the inter-cluster routing algorithm combines the single-hop with the multi-hop in communication between the CH and the sink node to avoid long distances. Also, the inter-cluster routing algorithm adopts the threshold detection mechanism to decrease the load of the close CHs to the sink. Simulation results showed that the AECRP protocol balances the consumption energy of the network, prolongs the lifetime, and provides more reliable data delivery.

(5) **Realistic Centralized Clustering Protocol based on PSO**

A Realistic Centralized Clustering protocol for WSNs based on PSO (RCC-PSO) [34] was developed to enhance the coverage and the dissipated energy of a realistic network. PSO algorithm is utilized in the RCC-PSO to select the optimal CHs that improve the network coverage, the energy efficiency, and the link quality. The RCC-PSO protocol operates in round manner, where each round contains the setup period and the steady-state period. In the setup period, the network is configured and the optimal CHs are determined by BS. This phase starts with the neighbor discovery where each sensor node forms its neighbors' table that contains IDs of its neighbors. Then, each node uses the flooding method to transfer its ID, residual energy, and neighbors table to BS. Based on the received information, BS runs the PSO algorithm to find the best CHs from the candidate CHs in each particle p_i such that they optimize the energy-efficiency EE(p_i), cluster quality CQ(p_i), and network coverage NC(p_i) as follows:

$$F(p_i) = \alpha_1 \mathrm{EE}(p_i) + \alpha_2 \mathrm{CQ}(p_i) + \alpha_3 \mathrm{NC}(p_i) \tag{17}$$

$$\mathrm{EE}(p_i) = \sum_{k=1}^{K} \frac{E_{\mathrm{int}}\left(\mathrm{CH}_{p_i,k}\right)}{E\left(\mathrm{CH}_{p_i,k}\right)} \tag{18}$$

$$\mathrm{CQ}(p_i) = \max \frac{\sum_{\forall n_j \in C_{p_i,k}} \mathrm{LQ}(p_i)}{\left|C_{p_i,k}\right|}, \quad \text{where } \mathrm{LQ}(p_i) = \frac{\mathrm{RSSI}\left(n_j, \mathrm{CH}_{p_i,k}\right)}{\mathrm{minRSSI}} \tag{19}$$

$$\mathrm{NC}(p_i) = \frac{N - \sum_{k=1}^{K} \left|C_{p_i,k}\right|}{\sum_{k=1}^{K} \left|C_{p_i,k}\right|} \tag{20}$$

where $E_{\mathrm{int}}\left(\mathrm{CH}_{p_i,k}\right)$ and $E\left(\mathrm{CH}_{p_i,k}\right)$ are the initial and remaining energy of CH number k in particle p_i, LQ(p_i) is the link quality between node n_j and kth CH, $\mathrm{RSSI}\left(n_j, \mathrm{CH}_{p_i,k}\right)$ is the Received Signal Strength Indicator (RSSI) of the link between n_j and kth CH in particle p_i, and $\left|C_{p_i,k}\right|$ is the number of members in cluster k of particle p_i. While in the steady-state phase, each member node uses its TDMA schedule to transmit its data to its respective CH and then goes to the sleep state.

(6) **Realistic Centralized PSO Protocol for Hierarchical Clustering**

A realistic centralized PSO protocol for Hierarchical Clustering (PSO-HC) [35] is a modified version of the RCC-PSO protocol [34]. PSO-HC enhances the lifetime, coverage, and scalability of WSN by activating the best CHs and building two-hop

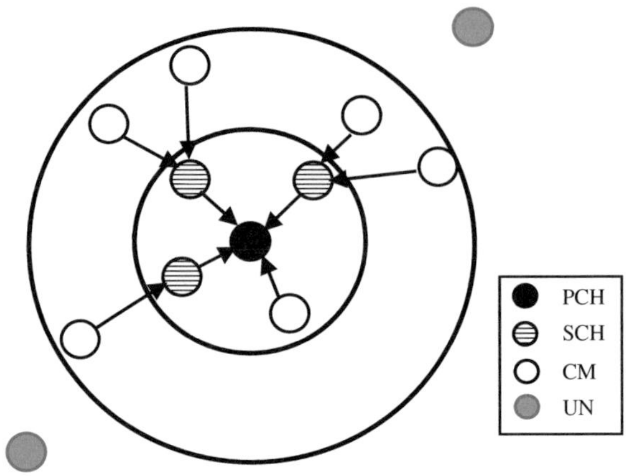

Fig. 6 PSO-HC protocol

routing between the nodes and their CHs. The optimal CHs are determined using the PSO as in RCC-PSO. Then, BS builds the first-tier clusters by allocating each un-clustered node (UN) to the nearest CH to it. The CH of the first tier called Primary CH (PCH) stays in active mode during the round time. After that, BS constructs the second tier by clustering all UNs (i.e., nodes that remained un-clustered from the first tier) to nodes in the first tier as shown in Fig. 6. A node that exists in the first tier and assigns cluster members (CM) from the second tier is named Secondary CH (SCH). SCH goes to sleep state after transmitting both its data packets and its members' data packets. Simulation results cleared that the PSO-HC outperforms the other cluster-based protocols in terms of dissipated energy and throughput.

(7) **Enhanced-Optimized Energy-Efficient Routing Protocol**

In [36], an Enhanced-Optimized Energy-Efficient Routing Protocol (E-OEERP) was developed to reduce the un-clustered or individual nodes to improve the lifetime of SWSNs. The individual nodes are nodes that not belong to any cluster in the network. The operation of E-OEERP protocol depends on combining the features of PSO algorithm and Gravitational Search Algorithm (GSA) for forming clusters and constructing a routing tree among these clusters, respectively. E-OEERP protocol forms the clusters by running the PSO algorithm for a certain number of iterations until all nodes become a member of any cluster. The fitness function that is used in E-OEERP protocol is based on the distance and energy of nodes as follows:

$$F(p_i) = \alpha_1 f_1(p_i) + \alpha_2 f_2(p_i) + \alpha_3 f_3(p_i) \tag{21}$$

$$f_1(p_i) = \frac{\sum_{j=1}^{m} d\left(n_j, p_i\right)}{\left|C_{p_i,k}\right|} \tag{22}$$

$$f_2(p_i) = \frac{E_{\text{Avg}}\left(C_{p_i,k}\right)}{E(p_i)} \tag{23}$$

$$f_3(p_i) = \frac{1}{|C_{p_i,k}|} \tag{24}$$

where $E_{\mathrm{Avg}}(C_{p_i,k})$ is the average energy of all members node of cluster k in particle p_i. The ability of E-OEERP protocol for assigning member nodes of each cluster improves the lifetime of WSN by reducing the individual node. Moreover, E-OEERP reduces the overhead of CH by electing a cluster assistant (CA) node that has maximum fitness value next to CH. Once clusters are constructed, the members of each cluster start to sense the data and send it to its respective CH that aggregates the collected data. Each CH finds the shortest available path with high reliability to transmit the aggregated data by finding the best next hop based on the distance and the force between the sensor nodes using GSA algorithm. Simulation results cleared the E-OEERP protocol which has the ability to reduce the individual nodes and improve the network lifetime.

(8) **Two-Tier PSO Centralized Routing Protocol**

A Two-tier PSO Centralized Routing protocol (TPSO-CR) [37] was suggested to solve the problem of clustering and routing in the homogeneous and the heterogeneous SWSNs. The operation of the TPSO-CR is broken into rounds, where each round consists of the setup stage and the steady-state stage. In the setup stage, BS finds the best CHs and relay nodes by running the PSO algorithm in two tiers. The first tier of PSO groups the network into optimal clusters and assigns the best CHs of these clusters based on the energy efficiency, cluster quality, and network coverage by optimizing Eq. 17 as the RCC-PSO protocol [34]. While the second tier constructs an efficient routing tree among the building clusters by finding the optimal relay node for each CH that saves energy by minimizing the number of relay nodes (R), balances energy consumption by selecting a relay node with a higher level of energy, and maximizes the link quality between relay nodes as follows:

$$F(p_i) = \alpha_1 \mathrm{EE}_a(p_i) + \alpha_2 \mathrm{EE}_b(p_i) + \alpha_3 \mathrm{LQ}(p_i) \tag{25}$$

$$\mathrm{EE}_a(p_i) = \frac{R}{C} \tag{26}$$

$$\mathrm{EE}_b(p_i) = \frac{\sum_{j=1}^{N} E(n_j)}{\sum_{r=1}^{R} E(\mathrm{RN}_{p_i,r})} \tag{27}$$

$$\mathrm{LQ}(p_i) = \max_{b=1,2,\ldots,R} \sum_{\forall rn_j \in r} \frac{\mathrm{RSSI}(rn_j, rn_{j+1})}{\mathrm{minRSSI}} \tag{28}$$

where R is the number of relay nodes and C is the total number of CHs. The second stage of TPSO-CR protocol is the steady-state stage, when data of the sensor nodes are transferred to CHs using the TDMA schedule that forwards the collected data to the BS via the relay nodes. The TPSO-CR protocol was developed and tested under a realistic network and energy consumption model CC2420. Simulation results proved that the TPSO-CR protocol significantly improves the packet delivery rate at both the CHs and the BS. Furthermore, it did not assume any unrealistic assumptions, for example, using GPS for location discovery.

(9) **PSO-based Routing Protocol with Mobile Base Station**

Authors in [38] presented a routing protocol based on PSO with Mobile Base Station (PSO-MBS) to enhance the network lifetime and the data delivery rate of WSNs. The task of PSO-MBS is dividing the network into clusters and determined pause sites of the mobile BS for gathering data of the sensor nodes as shown in Fig. 7. This task can be done in two phases, the setup phase and the steady-state phase. In the first phase, BS utilizes the PSO algorithm to find its optimal sojourn sites based on the location information of the sensors and the number of clusters in the network by minimizing the following fitness function.

$$F(p_i) = \sum_{k=1}^{K} \sum_{j=1}^{m} d(k, j) \tag{29}$$

where $d(k, j)$ is the distance between a node j and a BS location k, m is the number of member nodes of cluster k. K is the number of feasible sites of BS. While in the steady-state phase, each sensor node sends its data to its CH using the TDMA

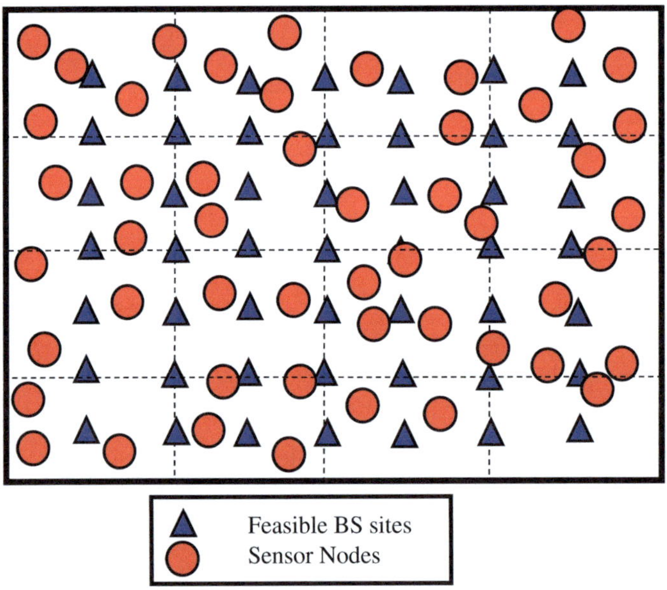

Fig. 7 Idea of PSO-MBS protocol

schedule. Then, CH aggregates the received data and waits for BS to visit the cluster to forward its data.

(10) Clustering Algorithm based on Glowworm Swarm Optimization

In [6], a Clustering Algorithm based on Glowworm swarm optimization (CAGM) was suggested to enhance the network lifetime and balance the network load by utilizing a mobile sink. The glowworm swarm optimization algorithm is used to partition the network into clusters and determine the best CHs for these clusters based on minimizing the consumption energy, while the mobile sink moves around the network to collect data from CHs which balances the load of CHs.

(11) An Optimized QoS-based Clustering with Multipath Routing Protocol

In [39], an Optimized QoS-based Clustering with Multipath Routing Protocol (OQoS-CMRP) is developed to reduce energy consumption and to solve the energy hole problem. The operation of OQoS-CMRP consists of four phases as shown in Fig. 8. In the first phase, OQoS-CMRP uses the PSO to form clusters and select head of each one based on optimizing the distance (d), residual energy (E), and the number of member nodes $\left(\left|C_{p_i,k}\right|\right)$ in each cluster k as follows:

$$F(p_i) = \alpha_1 f_1(p_i) + \alpha_2 f_2(p_i) + \alpha_3 f_3(p_i) \tag{30}$$

$$f_1(p_i) = \sum_{j=1}^{n} \frac{d\left(n_j, p_i\right)}{\left|C_{p_i,k}\right|} \tag{31}$$

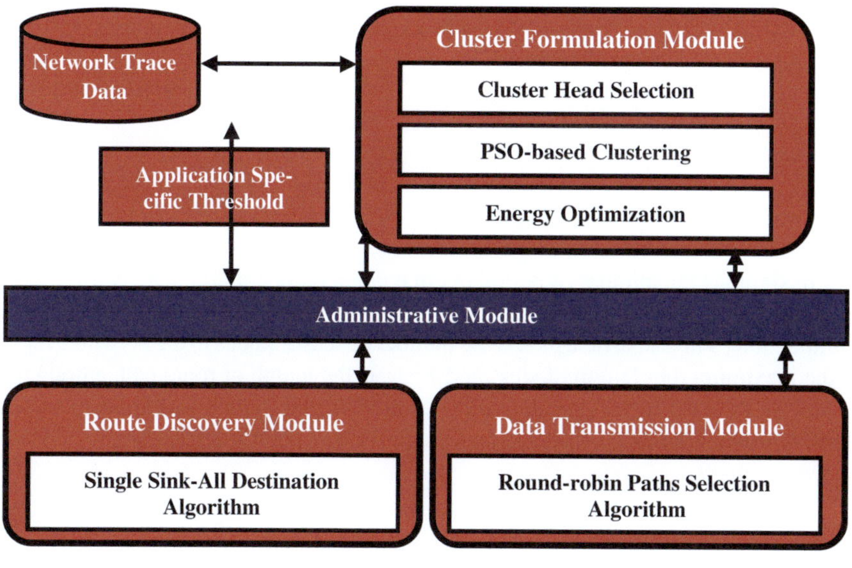

Fig. 8 System model of OQoS-CMRP protocol

$$f_2(p_i) = \frac{\sum_{j=1}^{n} E(n_j)}{E(p_i)} \tag{32}$$

$$f_3(p_i) = \frac{1}{\left|C_{p_i,k}\right|} \tag{33}$$

The second phase of OQoS-CMRP protocol is finding the shortest optimal multi-hop communication path from the sink to CHs using Single Sink-All Destination algorithm and selecting the next-hop neighbor nodes. While the third phase uses the round-robin path selection algorithm for transferring data of the sensor nodes toward the sink via selecting the optimal path with minimum cost, hop count, and maximum residual energy metrics. Finally, in the last phase of the OQoS-CMRP protocol, the sink initiates the process of re-clustering after a regular interval of time. Simulation results illustrated that the developed protocol enhances the network lifetime with better communication reliability and minimum delay.

(12) **Energy-Efficient Routing Mechanism for Mobile Sink**

An energy-efficient routing protocol based on the mobile sink is developed in [40] to reduce energy consumption and improve the packet delivery rate of WSNs. The authors utilized the PSO algorithm to develop a new clustering method called PSOBS. Based on the location information of all sensor nodes, sink uses the PSOBS method to select its rendezvous points (RP) depending on covering the entire network, minimizing the end-to-end delay, and saving the cost of the sink path as follows:

$$F(p_i) = \alpha_1 f_1(p_i) + \alpha_2 f_2(p_i) + \alpha_3 f_3(p_i) \tag{34}$$

$$f_1(p_i) = \frac{\sum_{j=1}^{\mathrm{NRP}_i} \mathrm{e}^{-(|\mathrm{NFD}_j - \mathrm{Sigma}|)}}{\mathrm{NRP}_i} \tag{35}$$

$$f_2(p_i) = \mathrm{e}^{-\left(\left|\frac{n}{\mathrm{Sigma}} - \mathrm{NRP}_i\right|\right)} \tag{36}$$

$$f_3(p_i) = \mathrm{e}^{-(|L_{\mathrm{max}i} - P_{\cos t_i}|)} \tag{37}$$

where sigma is the number of neighbors that can ideally communicate with node j, NFD_j is the number of packets that node j receives from its neighbors, NRP_i is the number of RPs in the particle p_i, n is the number of sensor nodes in the network, $L_{\mathrm{max}i}$ is the maximum tour length of sink, and $P_{\cos t_i}$ is the length of route path among the randomly selected RPs. After determining the RPs, the mobile sink moves to each point to collect the data of sensor nodes near to this point until visiting all RPs as shown in Fig. 9.

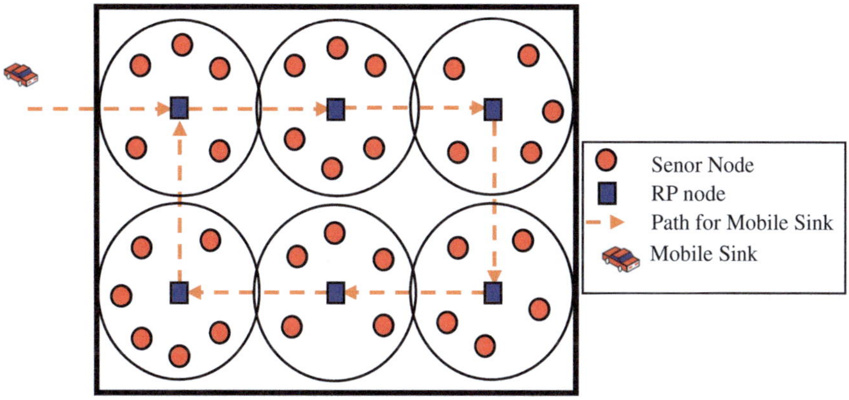

Fig. 9 Operation of PSOBS-based protocol

3.2 Genetic Algorithm-Based Hierarchical Routing Protocols

Genetic Algorithm (GA) [7, 25] is a randomized search and optimization technique which widely is used for solving the optimization problems that have a large number of possible solutions. The concept of GA depends on the survival of the fittest theory. GA starts with an initial population that contains a set of randomly generated solutions. Each individual solution in the initial population is named chromosome. A fitness function is used to evaluate each chromosome in the population. The chromosome with high fitness value is closer to the optimal solution. The main steps of GA are selection (reproduction), crossover, and mutation as shown in Fig. 10. The selec-

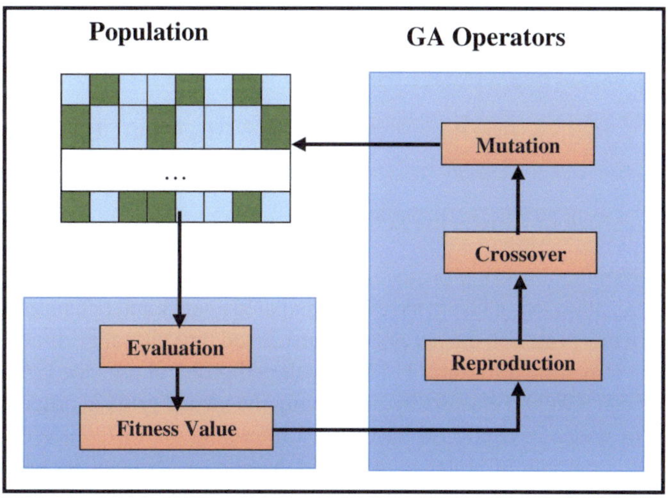

Fig. 10 Architecture of the Genetic Algorithm

tion step is used to select the best chromosomes, which will undergo the next steps. Crossover and mutation operations are used to generate offspring population from the parent population. In the crossover, a new offspring is generated by combining genes of two parents. While the mutation process makes a small change randomly to the current chromosome by mutating one or more genes based on the mutation rate. The entire process of GA is repeated in iteration manner until the stopping criteria are met. Many GA-based hierarchical routing protocols were designed to enhance the performance of WSNs. In this subsection, we survey some of these GA-based routing protocols.

(1) **Genetic Algorithm-Based Energy-Efficient Clustering Protocol**

A Genetic Algorithm-Based Energy-Efficient Clustering protocol (GABEEC) [41] was presented to save the dissipated energy and prolong the stability period of SWSNs. The GABEEC protocol operates in round manner, where each round contains the setup period and the steady-state period. In the setup period, the clusters are constructed and are not changed during the working of the network. In each round, the clusters are static, and CH of each cluster is dynamically changed based on the remaining energy of the current CH and its MNs. The clusters are constructed based on maximizing the network lifetime by minimizing the fitness function in Eq. 38, where R_{FND} and R_{LND} are the numbers of rounds which the first and the last nodes die, respectively, $d(j, \mathrm{CH}_h)$ and $d(\mathrm{CH}_h, \mathrm{BS})$ are the distances from node j to CH number h and from CH to BS respectively, and N is the number of nodes in the network. While in the steady-state period, MNs of each cluster sense the field and send data to their CHs using the TDMA schedule. Then, CH aggregates the received packets from its MNs into a fixed-length packet and forwards it to BS. Based on the received data, BS checks the remaining energies of CHs and the MNs at the end of each round. If the energy of a CH is less than the average energy of its cluster members, the highest energy node from its members is chosen to be the new CH, and the old CH becomes a cluster member.

$$F = \alpha_1 R_{\mathrm{FND}} + \alpha_2 R_{\mathrm{LND}} + \alpha_3 \sum_{j=1}^{N} -(d(j, \mathrm{CH}_h) + d(\mathrm{CH}_h, \mathrm{BS})) \tag{38}$$

(2) **Energy-Efficient Clustering Scheme based on Grid Optimization using GA**

An energy-efficient clustering protocol was developed by Kumar et al. [8] based on the grid optimization using GA to save the dissipated energy and enhance the lifetime of SWSNs. In this protocol, the network is partitioned into virtual grids and each grid acts as a cluster as shown in Fig. 11. The developed protocol uses the GA to balance the traffic load among the clusters by optimizing the virtual grids in order to equalize the number of nodes in each grid based on minimizing the difference between the ideal and actual number of nodes in each grid. Moreover, it considers the remaining energy of the nodes during the selection process of CHs in order to balance the

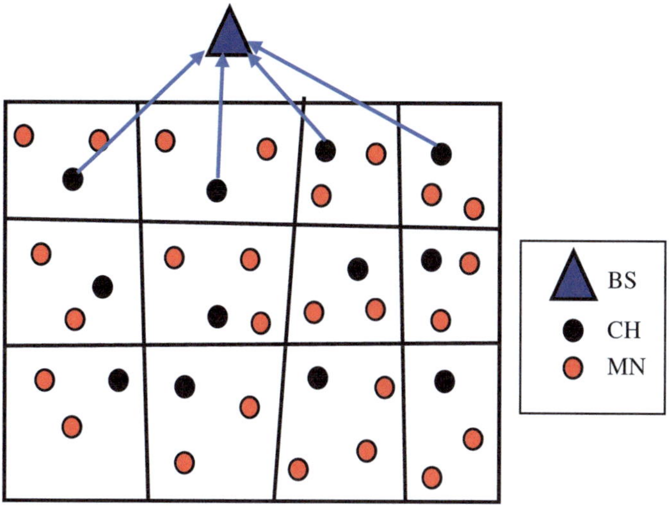

Fig. 11 Network model of optimized grid-based protocol

energy of the network. This protocol operates into round manner, where each round consists of five phases, namely grid partition phase, grids optimization phase using GA, CHs selection phase, TDMA announcement phase, and data transmission phase. Simulation results cleared that this protocol enhances the network lifetime.

(3) **Genetic Algorithm-based Load-Balanced Clustering Protocol**

A GA-based load-balanced clustering protocol was developed by Kuila et al. [42] to balance the load among CHs. This protocol groups the network into clusters based on minimizing the load of each CH. TCHs are determined priori, and the aim of the developed protocol is finding the optimal number of member nodes for each CH to form balanced clusters. Kuila's protocol uses the GA algorithm to balance the load among CHs based on optimizing the standard deviation of the CH load as illustrated in Eq. 39, where K is the number of CHs, n is the number of sensor nodes in cluster k, Ld_j is the load of node j, and W_k is the overall load of cluster k. Here, the authors modified the initial population generation step and the mutation step to fast the GA algorithm. They considered the connectivity between CHs and the sensor nodes during the generation of the initial population. While in the mutation step, a gene is selected for the mutation process based on ensuring better load balancing.

$$F = \frac{1}{\delta} = \frac{1}{\sqrt{\frac{1}{k}\sum_{k=1}^{K}\left(\left(\sum_{j=1}^{n} Ld_j/n\right) - W_k\right)^2}} \tag{39}$$

(4) Genetic Algorithm-based Clustering Protocol for Heterogeneous Networks

A GA-based clustering protocol for Heterogeneous Networks (GAHNs) [43] was designed to balance the energy consumption and improve the lifetime of the heterogeneous WSNs. GAHN is a centralized protocol, in which BS runs the GA algorithm after each round to find the optimal dynamic network clustering structure for the heterogeneous WSN. The fitness function of GA employs the remaining energy of CHs at round $t(E(t))$ related to the initial energy $(E(0))$, the consumed energy of direct transmission from the sensor node to BS $(\widetilde{E})$ related to the expected consumed energy according to the clustering structure $(\widehat{E})$, the local density of each non-CH node j within the δ-vicinity $\left(G_j(\delta)\right)$, and the distance from CHs to the BS $(d(\mathrm{CH}_k, \mathrm{BS}))$ as given by Eq. 40, where K is the number of CHs and N is the total number of sensor nodes in the network. These all factors of the fitness function ensure the load balancing among all clusters in the network, as authors explained in the simulation results.

$$F = \sum_{k=1}^{K} \frac{E_j(t)}{E_j(0)} + \frac{\widetilde{E}}{\widehat{E}} + \frac{1}{\sum_{k=1}^{K} d(\mathrm{CH}_k, \mathrm{BS})} + \frac{1}{N-K} \sum_{j=1}^{N-K} G_j(\delta) \tag{40}$$

(5) Genetic Algorithm-based Energy-Efficient Adaptive Clustering Protocol

A Genetic Algorithm-based Energy-Efficient adaptive clustering hierarchy Protocol (GAEEP) [19] was suggested to efficiently prolong the stability period and the lifetime of WSNs. The idea of GAEEP is determining the locations of the optimum CHs based on minimizing the communication cost of the sensor nodes using the GA algorithm. The framework of GAEEP protocol consists of the setup phase and the steady-state phase in each round. In the setup phase, BS constructs the optimum clusters by using GA based on minimizing the consumption energy $(E_d(t))$ in overhead control packets and data collection according to the clustering structure at round t and controlling number of $\mathrm{CH}(L)$ as follows:

$$F = \alpha\left(\frac{E_d(t)}{E(0)}\right) + (1-\alpha)\left(\frac{L}{N}\right) \tag{41}$$

where N is the number of sensor nodes in the network and $E(0)$ is the initial energy of the network at round 0. While in the steady-state phase, CHs gather data of their MNs and transfer it to BS in a frame format. In the last frame, sensor nodes send their energy information besides their data packets to BS for the next round. Simulation results showed that the GAEEP protocol outperforms the other protocols in terms of stability period and lifetime for both the heterogeneous and the homogeneous networks. Moreover, the GAEEP protocol enhances the clustering process reliability and balances the load among the nodes.

(6) Genetic Algorithm-based Clustering and Routing Algorithms

Two algorithms based on the GA algorithm for Clustering and Routing (GACR) were developed in [44] to conserve the overall energy of WSNs. GACR algorithms prolong the lifespan of CHs, thereby prolonging the lifetime of the network. The clustering algorithm determines CHs based on the residual energy of CHs (E_r(CH)) and distances from the sensor nodes to their corresponding CHs ($d(n_j$, CH)) as illustrated by Eq. 42, where m_k is the members of cluster k and, while the routing algorithm finds out the routes from all CHs to BS that have short distances and minimum number of hops with maximum energy relay nodes. So, the chromosomes that represent paths are evaluated by the fitness function in Eq. 43, where MaxDist and MaxHop are the maximum total covered distance and the total number of hop of a chromosome, respectively.

$$F_{\text{clustering}} = \frac{1}{\sqrt{\frac{1}{K}\sum_{k=1}^{K}\left(\left(\sum_{k=1}^{K} E_{\text{r}}(\text{CH}_k)/m_k\right) - \left(E_{\text{r}}(\text{CH}_k)/m_k\right)\right)^2}} \times \frac{1}{\frac{1}{N}\sum_{j=1}^{N} d\left(n_j, \text{CH}\right)} \tag{42}$$

$$F_{\text{routing}} = \alpha \times \text{MaxDist} + (1-\alpha) \times \text{MaxHop} \tag{43}$$

(7) Genetic Algorithm-Based Energy-Efficient Clustering Hierarchy Protocol

A Genetic Algorithm-based Energy-efficient Clustering Hierarchy (GAECH) protocol [45] was designed to extend the stability period and the lifetime of WSNs with a novel fitness function. GAECH uses GA algorithm to divide the network into clusters and select the head of each cluster based on balancing the load among the sensor node to extend the stability period. So, the GAECH considered the dissipated energy in data collection of the round $t(E_d(t))$, standard deviation of dissipated energy between clusters (SD), CH dispersion (CH_{disp}), and energy consumption of CHs ($E_{d\text{CH}}(t)$) in the fitness function as given by Eq. 44, where α_1, α_2, α_3 and α_4 are the weight coefficients within [0 1]. The values of weight coefficients in the fitness function of GAECH will be varied based on the required application in order to get better results. The results cleared that the GAECH outperforms the other algorithms in all the necessary aspects.

$$F = \alpha_1 E_d(t) + \alpha_2 \text{SD} + \alpha_3 \frac{1}{\text{CH}_{\text{disp}}} + \alpha_4 E_{d\text{CH}}(t) \tag{44}$$

(8) Genetic Algorithm-based Routing Protocol with Mobile Sensor Nodes

A routing protocol based on GA (GAROUTE) [46] was designed for MWSNs to build a stable clustering network. GAROUTE works in a centralized manner, in which BS broadcasts a request message to ask all mobile nodes in the network to send a list of its neighboring nodes within its transmission range. Then, each node broadcasts a message to find the list of its neighbors. After that, a node sends the neighbors' list along with its own energy information and speed to BS. Based on the obtained information, GAROUTE uses GA algorithm to construct stable clusters with best CHs based on minimizing the following function:

$$F = \alpha_1 E_{\text{avg}}(t) + \alpha_2 K + \alpha_3 \text{CH}_{\text{speed}} \tag{45}$$

where $E_{\text{avg}}(t)$ is the average dissipated energy in around t, K is the number of CHs, and CH_{speed} is the sum of the speed of all CHS.

(9) **Multi-Objective Evolutionary Routing Protocol**

A Routing Protocol based on Non-dominated Sorting Genetic Algorithm-II (NSGAII-RP) was developed in [47] to increase the lifetime of the MWSNs by building efficient routes. NSGAII-RP protocol considers the coverage and the routing problem. NSGAII-RP uses the NSGAII algorithm to find the best CHs and optimize the coordinates of the mobile nodes based on maximizing the network coverage and minimizing the dissipated energy. Simulation results cleared that NSGAII-RP has the ability to build energy-efficient routes with guarantee the network coverage, as a result the network lifetime is improved.

(10) **Genetic Algorithm-based Distance-aware Routing Protocol**

In [48], a Genetic Algorithm-based Distance-Aware routing protocol called GADA-LEACH was developed to improve the performance of LEACH protocol. The idea of GADA-LEACH protocol depends on using the GA algorithm and the distance-aware routing concept. GADA-LEACH considers three objectives in the fitness function of GA as given by Eq. 46. These objectives are the ratio of consumption energy of all nodes to consumption energy of CHs $\left(\frac{E_{\text{nodes}}}{E_{\text{CHs}}}\right)$, the ratio of distances between CH and its members to the cluster size $\left(\frac{d(\text{SN,CH})}{m}\right)$, and the ratio of the distance between CHs and BS to the number of CHs $\left(\frac{d(\text{BS,CH})}{K}\right)$. Also, GADA-LEACH introduces relay nodes between CHs and sink to forward data of CHs to sink especially if the separation distance between CH and sink is large. This preserves the connectivity between CH and sink and also decreases the load of CHs.

$$F = 0.3\frac{E_{\text{nodes}}}{E_{\text{CHs}}} + 0.35\frac{d(\text{SN, CH})}{m} + 0.35\frac{d(\text{BS, CH})}{K} \tag{46}$$

(11) **Mobility-Based Genetic Algorithm Hierarchical Routing Protocol**

In [49], the authors developed a new routing protocol for MWSNs called Mobility-based Genetic Algorithm Hierarchical Routing Protocol (MGAHP). MGAHP uses GA algorithm for determining the optimum CHs that have high energy based on minimizing the consumption energy in overhead control packets and data collection. The framework of MGAHP protocol consists of the setup phase and the steady-state phase in each round. But, the information collection phase runs only at the beginning of the first round to collect information about the network such as location and initial energy of sensor nodes. The MGAHP protocol considers the dissipated energy in both data and overhead packets ($E_d(t)$) and the number of formed CHs (L) in the fitness function during CH selection process as given by (47).

$$F = \alpha E_d(t) + (1 - \alpha)\left(\frac{L}{N}\right) \tag{47}$$

where N is the number of sensor nodes in the network. While in the steady-state phase, the sensor nodes forward their data to CHs based on request data message from them. If a sensor node moves outside its cluster, its packet will drop because it does not receive the request message.

3.3 *Fuzzy Logic-Based Hierarchical Routing Protocols*

Fuzzy Logic (FL) [26] is a logical method that is similar to human reasoning. The behavior of FL mimics the way of decision making in humans. In FL, the relationship of input–output is defined by a set of relational expressions. FL contains four important parts, including fuzzification, defuzzification, inference engine, and a rule base as shown in Fig. 12. Rule base contains the set of rules and conditions that given the decision-making system. The fuzzification step is used to convert the inputs into fuzzy sets. Then, the inference engine determines the matching degree of the current fuzzy input with respect to each rule and decides which rules are to be fired according to the input field. Finally, the defuzzification step converts the fuzzy sets obtained by the inference engine into a crisp value. The less computational complexity of FL makes it is most suitable for WSN, and various areas of WSNs have been effectively covered by the rules of FL. Recently, FL-based hierarchical routing protocols will be surveyed below.

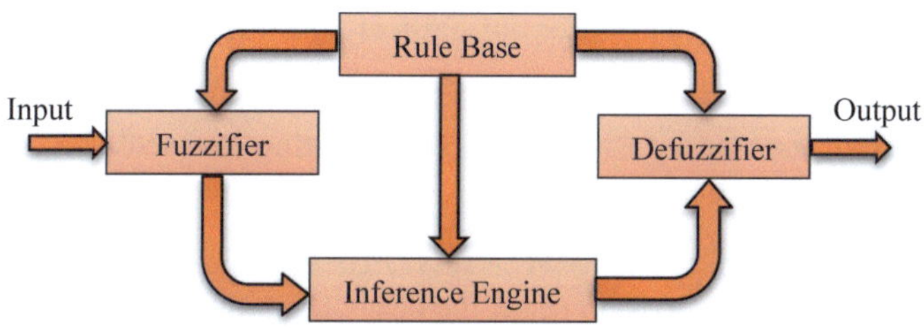

Fig. 12 Fuzzy Logic architecture

(1) Fuzzy Energy-Aware Unequal Clustering Algorithm

A Fuzzy Energy-Aware Unequal Clustering algorithm (EAUCF) was presented in [50] to solve the energy hole problem by using an unequal clustering mechanism. EAUCF is a distributed competitive unequal clustering algorithm, which aims to reduce the intra-cluster work of CHs that are either close to BS or have low remaining battery power. In EAUCF, the CHs are determined randomly based on a predefined threshold. Then, EAUCF uses the FL approach to calculate the competition radius of each CH bases on two fussy input variables, namely the distance to BS and residual energy of the determined CHs. Varying the competition radius of CHs distributes the workload among all nodes. Simulation results showed that the EAUCF protocol is energy-efficient clustering algorithm and stable as compared to the other protocols.

(2) A Novel Distributed Clustering Protocol Using Fuzzy Logic

Mirsadeghi et al. [51] developed a novel distributed clustering protocol using FL to improve the energy-efficient and coverage of WSNs. The developed protocol uses FL to calculate the node CH's chance based on three fussy variables: node residual energy, local node density, and node centrality. Nodes with high change have high probabilities to be CHs. Each node computes its delay-time based on its chance. Then, the node declares itself a CH if it does not receive any message from other CHs during its delay-time. After that CH node broadcast a message within its communication range to announce its state. Simulation results showed that the developed protocol has a high coverage ratio, a low number of orphan nodes, and low energy consumption that prolong the network lifetime.

(3) Fuzzy Logic-Based Unequal Clustering Algorithm

A Fuzzy Logic-Based Unequal Clustering algorithm (FBUC) was suggested in [52] to enhance the performance of WSNs by solving the energy hole problem. FBUC is developed to overcome the drawbacks of Fuzzy Energy-Aware Unequal Clustering algorithm (EAUCF) that is considered in [50]. Besides the distance to BS and residual energy of a node that used in EAUCF as fuzzy input variables for calculating the CH competition, FBUC adds a node degree as a third-input fuzzy variable. After computing the competition radius, nodes with maximum competition radius and high residual energy are elected as CHs. Also, FBUC protocol considers the distance to CH and CH degree in the process of clusters construction to balance the load among the CHs.

(4) Optimized Zone-based Energy-Efficient Routing Protocol

An Optimized Zone-based Energy-Efficient routing Protocol (OZEEP) [53] was suggested to enhance the lifetime of MWSNs. OZEEP uses the Genetic Fuzzy System (GFS) in the clustering process to select the best CHs in two steps. The first step is called the screening process, in which the fuzzy system is used to elect the candidate nodes for the role of CH based on residual energy, neighbors, distance from BS, and mobility. The candidate CHs declare themselves the BS. While in the second step,

BS runs the GA algorithm to take the final decision and select CH from the candidate CHs based on mobility factor, consumption energy, and number of CHs. Combining the fuzzy system with the GA algorithm helps for selecting the best CH and eases the task of the genetic system in the second step. Simulation results concluded that the OZEEP protocol has the ability to enhance the packet delivery ratio and the network lifetime.

(5) **Stable Election Protocol based on Fuzzy Logic**

A Stable Election Protocol based on Fuzzy Logic (SEPFL) was designed in [54] to prolong the lifespan and the throughput of heterogeneous WSNs. SEPFL utilizes the FL control to enhance the CH selection and prolong the lifetime of the SEP protocol [63]. In SEPFL protocol, two probabilities are used for selecting CHs. The first probability is calculated using FL algorithm based on the distance of the nodes from BS, the nodes' density, and the battery level of the nodes, while the second probability is calculated using threshold equations. Based on the mean value of the obtained probabilities, the weighted probability for each node is computed. Nodes with high probability and requiring less energy for communicating with other nodes as well as BS are elected as CHs.

(6) **Low-Energy Fuzzy-based Unequal Clustering Multihop Architecture**

In [55], a Low-Energy Fuzzy-based Unequal Clustering Multihop Architecture (LEFUCMA) protocol was developed to enhance the lifetime and balance the load of WSNs. The operation of LEFUCMA consists of neighbor-finding phase and steady-state phase. The neighbor-finding phase is used to find the neighbors of each node and repeats after completion of a certain number of rounds of the steady-state phase depending on the node death rate. While the steady-state of LEFUCMA is divided into rounds: CH selection algorithm, CHs distribution algorithm, unequal clustering mechanism, and multihop routing algorithm. In CH selection algorithm, FL is used to select CHs based on residual energy, number of neighboring nodes, packet reception rate, and distance between node and BS. To avoid the hotspot problem, the number of CHs in a sector is decided based on the node density and distance of sector from BS. After selecting CHs, the communication range of each CH is computed based on the residual energy and the distance to BS. Finally, an efficient multihop routing tree is constructed by selecting the next hop depending on distance form BS and current CH, residual energy, cluster density, and inter-cluster traffic.

(7) **Fuzzy Multi Cluster-Based Routing with a Constant Threshold**

Since clustering the network in each round increases the overhead control messages and the possibility of collision, so a Fuzzy Multi Cluster-Based Routing with a Constant Threshold (FMCR-CT) is designed in [56] to avoid the clustering process in each round. FMCR-CT utilizes FL to select the best CHs based on two inputs fuzzy variables: the residual energy and the number of neighbors of each node. The selected CHs will be used until the residual energy of any CH becomes less than a threshold value. After that a multihop routing is constructed based on the distance to

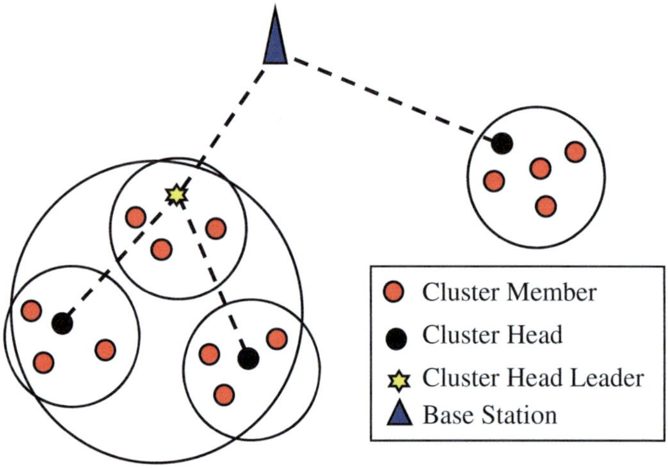

Fig. 13 Routing of FMCR-CT protocol

BS and the residual energy for selecting the CH leader. Then, the competition radius of the selected CH leader is calculated based on the distance of CH to BS. If there are CHs within the competition radius of the CH leader, these CHs use a multihop routing to forward their data to BS via the CH leader as shown in Fig. 13. Otherwise, a single routing is used by forwarding data to BS directly.

3.4 *Ant Colony Optimization-Based Hierarchical Routing Protocols*

Ant Colony Optimization (ACO) algorithms were introduced in the early 1990s [27]. The behavior of ACO algorithms was inspired from ants, in how ants can find the shortest path between the food source and their nest. Initially, they explore the area surrounding their nest in a random manner and leave a chemical pheromone trail on the ground. When an ant finds a food source, it evaluates the quantity and the quality of the food and takes some of it back to its nest. During the return trip, the ant leaves pheromone trails on the ground depending on the quantity and quality of the food to guide the other ants. After a number of iterations, all ants follow the short path that has strong pheromone concentrations because the pheromone of the long paths will evaporate as shown in Fig. 14. So, the ACO algorithm has the ability to construct routes among the sensor nodes by selecting short paths. Different routing protocols for WSNs were developed based on the principle of the ACO algorithm as follows.

(1) **Fuzzy C Means-based Routing Protocol with Ant Colony Optimization**

A hierarchical routing protocol based on both the fuzzy C means and ACO algorithm (FCM-ACO) [57] was designed to gather the data of the sensor nodes in an efficient

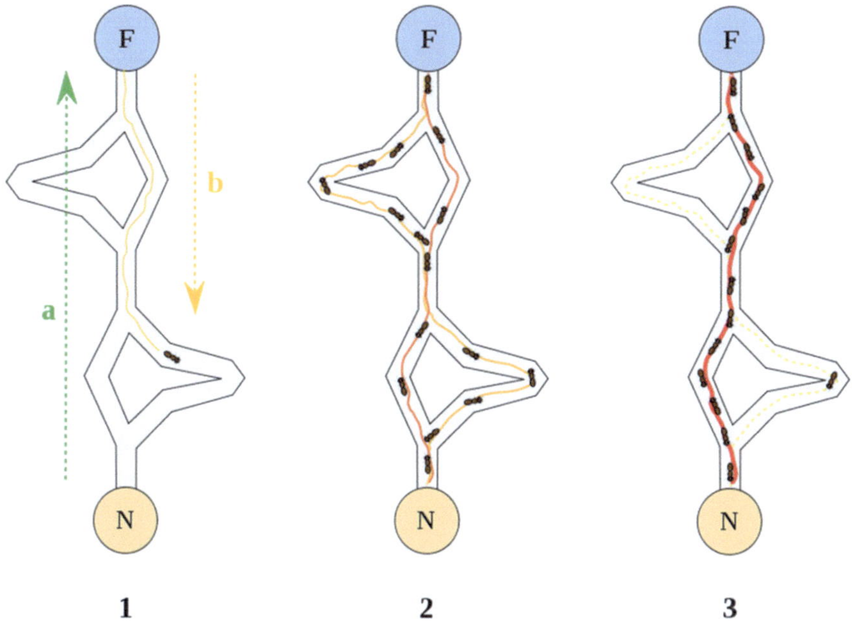

Fig. 14 Behavior of the ACO algorithm [23]

way. BS uses the FCM algorithm to divide the network into an optimal number of clusters. These clusters will be fixed during the lifetime of the network. The value of the optimal cluster is the value that gives minimum consumed energy per round. The head of each cluster is determined locally based on the proximity to BS and remaining energy. Then, BS uses the ACO algorithm to construct a chain among the CHs for collecting data of sensor nodes, where the state transition probability of kth ant at CH_i to move to CH_j is given by:

$$p_{ij}^{k} = \frac{(\tau_{ij})^{\alpha}(\eta_{ij})^{\beta}}{\sum_{l \in N_i^k} (\tau_{il})^{\alpha}(\eta_{il})^{\beta}}; \quad \text{if } j \in N_i^k \tag{48}$$

where τ_{ij} is the pheromone level of the direct edge between CH_i and CH_j, η_{ij} is the inverse of the Euclidean distance between CH_i and CH_j, N_i^k is the list of nodes is not visited yet by kth ant, α and β are relative importance parameters (>0).

(2) Hybrid Routing Protocol Based on the Artificial Fish Swarm and Ant Colony

In [58], an ant colony-based hierarchical clustering protocol called ACOHC was developed to gather data of sensor nodes in an efficient way. The operation of ACOHC protocol consists of two steps. In the first step, BS uses the K-means algorithm to partition the network into an optimal number of clusters. These clusters still fixed during the lifetime of the network. While in the second step, BS utilizes the ACO algorithm to form a cluster chain among the member nodes of each cluster; the chain leader is selected based on residual energy and distance to BS. Then, an upper chain is constructed among the chain leaders of clusters using the ACO algorithm, and the super leader is selected based on residual energy and distance to BS. The ACO calculates the state transition probability of kth ant using Eq. 48 to select the next hop in its path.

(3) **Multi-Hop Graph-Based Energy-Efficient Routing Protocol**

A multi-hop graph-based energy-efficient routing protocol, known as MH-GEER, was designed in [59] to improve the network lifetime and balance the network load. The operation of MH-GEER consists of three phases: clustering; data collection and aggregation; and constructed routing between CHs. The data collection and aggregation phase within each cluster are similar to LEACH protocol [64] in its steady-state phase. In the clustering phase, BS uses the K-means algorithm to partition the network into fixed clusters, and the role of CH is rotated among all sensor nodes within each cluster based on the residual energy. Then, BS uses the ACO algorithm to construct inter-cluster routing among CHs. In ACO, kth ant at CH_i calculates its state transition probability to move to CH_j as follows:

$$p_{ij}^{k} = \begin{cases} \dfrac{\left(\frac{1}{\tau_{ij}}\right)^{\alpha}(\eta_{ij})^{\beta}}{\sum_{\substack{l \in \{1,2,3,\ldots,N\} \\ l \notin \text{pth}\, k}} \left(\frac{1}{\tau_{ij}}\right)^{\alpha}(\eta_{ij})^{\beta}}; & \text{if } j \notin \text{pth}\, k \\ 0; & \text{otherwise} \end{cases} \tag{49}$$

where τ_{ij} is the pheromone level of the direct edge between CH_i and CH_j, η_{ij} represents the visibility of the edge between CH_i and CH_j, $\text{pth}\, k$ is the constructed path, α and β are relative importance parameters (>0).

3.5 *Artificial Immune Algorithm-Based Hierarchical Routing Protocols*

Artificial Immune algorithm (AIA) [28] is a recently evolutionary algorithm that mimics the antigen-antibody reaction in the mammal's immune system. The antibody and the antigen in the AIA represent the feasible solution and the objective function for a traditional optimization method. AIA has much less computational cost and produces the solution sets that are highly competitive in terms of convergence, diversity, and distribution. The framework of AIA consists of selection, replication, clonal with hypermutation, and mutation step as shown in Fig. 15. Initially, antibodies population is generated randomly, where the length of each antibody depends on the decision variables vector. Then, the initial population goes through the evolutionary

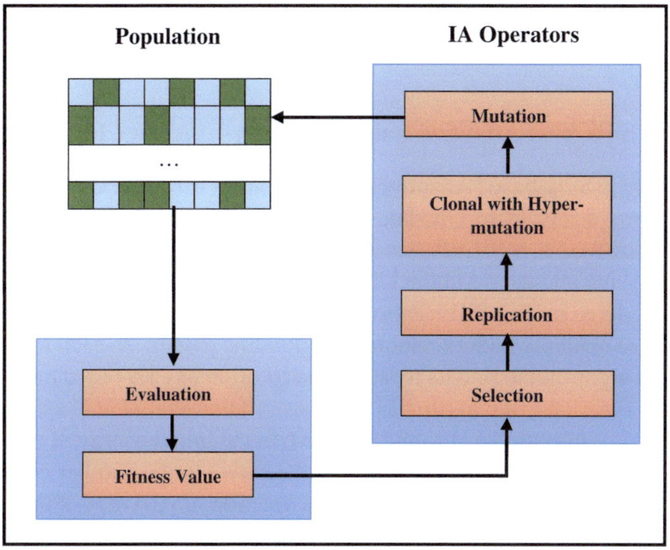

Fig. 15 Architecture of the Artificial Immune Algorithm

steps of the AIA algorithm. The first step is the selection step which selects the best antibodies based on the antibodies' probabilities. Some of the selected antibodies are chosen using the replication step to join the clonal with hypermutation step. The hypermutation step is used to increase the antibodies' exploitation by proliferating the parent antibodies to produce new offsprings. Finally, the mutation step is used to provide exploration and prevent the algorithm from dropping at a local minimum by altering one or more entries depending on the mutation rate. The entire process of AIA has repeated in generation manner until the stopping criteria are met. Due to features of AIA compared to the other evolutionary algorithms, it used in many hierarchical clustering protocols to improve the performance of WSNs.

(1) **Clustering Protocol based on Immune Optimization Algorithms**

A clustering protocol based on AIA algorithm was suggested in [60] to prolong the network lifetime and enhance the performance of WSNs. The suggest protocol considers the energy (E), distance (D), cluster size (Cz), and information volume (I) in the objective function of AIA for selecting CHs as given by Eq. 50. In this protocol, authors use arbitrary weight (w_i) for each objective in the function. The value of each weight (w_i) is updated based on the value of its objective at the previous and current generations and its value at the previous generation as explained in Eq. 51.

$$F = \sum_{i=1}^{4} w_i^g \times f_i^g, \ \forall f_i^g \in \{E, D, Cz, I\} \tag{50}$$

$$w_i^g = w_i^{g-1} + \frac{1}{1 + e^{-f_i^{g-1}}} \times \left(f_i^g - f_i^{g-1}\right) \tag{51}$$

(2) **Mobile Sink-base Adaptive Immune Energy-Efficient Clustering Protocol**

In [61], a Mobile Sink-based Adaptive Immune Energy-Efficient Clustering Protocol (MSIEEP) was designed to alleviate the energy hole problem and improves the performance of WSNs. MSIEEP uses the AIA algorithm to find the sojourn locations of the mobile sink and the optimum CH. Three mobility path patterns for the mobile sink are considered in the MSIEEP protocol, namely four-region rectangular pattern, eight-region rectangular pattern, and four-region line pattern. In each region (r), BS runs the AIA algorithm to find its location and location of CHs based on minimizing the objective function given in Eq. 52. The consumption energy in data and overhead packets ($E_{d-r}(t)$) of region r at round t relative to the initial energy of all live nodes in the same region and number of CHs ($L_r(t)$) relative to the number of alive nodes in region r are considered as objectives in the function of AIA algorithm. Moreover, MSIEEP enhances the lifetime and balances the load among the nodes in WSN via selecting CHs from nodes that have high residual energy.

$$F_r = \omega\left(\frac{E_{d-r}(t)}{E_r(0)}\right) + (1-\omega)\left(\frac{L_r(t)}{N_r}\right) \tag{52}$$

(3) **Unequal Multi-hop Balanced Immune Clustering Protocol**

An Unequal Multi-hop Balanced Immune Clustering protocol (UMBIC) [62] was suggested to alleviate the energy hole problem and improve the lifetime of small and large-scale/homogeneous and heterogeneous WSNs with different nodes density. UMBIC protocol considers the Unequal Clustering Mechanism (UCM) and AIA to balance the load among nodes by saving the intra-cluster load of the close clusters for the inter-cluster relay traffic. The UCM is used to partition the network into clusters of unequal size based on distance with reference to BS and residual energy. While the AIA constructs optimum clusters based on covering the entire sensor field (R_{Cov}), reducing the communication cost of all nodes ($E_d(t)$) and controlling the number of CHs ($L_r(t)$) as illustrated in Eq. 53. The UMBIC protocol rotates the role of CHs among the nodes only if the residual energy of one of the current CHs less than a predefined energy threshold, as a result the computational time and overheads of selecting new CHs are saved. Also, the UMBIC protocol constructs a routing tree among the determined CHs to reduce the dissipated energy and avoid the energy hole problem. Each CH selects its next hop that has high residual energy, close to BS and in the communication range of CH.

$$F = \omega_1\left(\frac{E_{\mathrm{d}}(t)}{E_{\mathrm{r}}(0)}\right) + \omega_2(1 - R_{\mathrm{Cov}}) + (1 - \omega_1 - \omega_2)\left(\frac{L_{\mathrm{r}}(t)}{N_{\mathrm{r}}}\right) \tag{53}$$

(4) **Adjustable Range-Based Immune Hierarchy Clustering protocol for MWSNs**

In order to gather data of the mobile sensor nodes in an efficient way, an Adjustable Rang-Based Immune hierarchy Clustering protocol (ARBIC) was developed in [9]. The ARBIC protocol organizes the network into optimum clusters, and the size of each cluster is adjusted based on the speed of its mobile sensor nodes to preserve the cluster connectivity. In order to establish stable clusters, ARBIC protocol uses the AIA algorithm to select nodes that have high residual energy and less mobility to be CHs based on optimizing the trade-off among the energy consumption, coverage, and number of CHs as given in Eq. 53. Also, a close high energy node to CH is selected to be a Vice CH (VCH) in each cluster. The VCH takes the responsibility of a cluster when the main CH is failed. Then, the member nodes join to CH depend on the residual energy, distance, and link connection time. The clustering process is run if and only if the residual energy of any CHs is less than a predefined energy threshold to reduce the overhead packets and the computational time. Moreover, the stability of links between CHs and their member nodes is maintained by running a fault tolerance mechanism after sending each frame to reduce the packets' drop rate by maintaining.

4 Comparison and Discussion

A comparison among the above-surveyed protocols based on packet delay, network scale, energy-efficiency and scalability, advantages, and drawbacks is listed in Table 2. It is noticed that the existence routing protocols had the ability to improve the network lifetime and ensure network connectivity. Figure 16 shows the distribution of the surveyed protocols according to the taxonomy metrics. It is observed that 70% of the surveyed protocols are centralized; however, the centralized manner adds more overheads that increase the consumption energy and limit the network scalability.

However, the distributed routing protocols do not guarantee the network connectivity because they depend on the local information of the neighboring nodes. Therefore, researchers should optimize the features of centralized protocols and distributed protocols by designing semi-distributed or semi-centralized routing protocols. Utilizing the mobile sink in MWSN solves the hotspot problem of the static sink and balances the load among the sensor nodes, but it requires more time to collect data from all sensor nodes in the network which increases the packet delay. Thus, the mobility of the sink node should be controlled based on optimizing the consumption energy and the packet delay. Also, 90% of the surveyed protocols simulate the radio model as the first-order model, which is not a realistic assumption. Thus, the researchers should consider the realistic radio model for simulating the developed routing protocols.

Table 2 Comparison of the intelligent-based hierarchical routing protocols

Protocol	Delay	Network size	Energy efficiency	Scalability	Advantages	Disadvantages
EBUC 2010 [29]	M	Large scale	H	Ltd.	• Considers the cost of both the inter-cluster and intra-cluster communications	• Requires location information • Assumes that the CHs can communicate with each other regardless of their connectivity
PSO-DH 2011 [30]	M	Small scale	M	Ltd.	• Considers the distance between CH and its member, and residual energy the CH selection process	• Requires location information • Requires more computations in the selection of double CHs for each cluster
PSO-SD 2012 [31, 32]	M	Small/large scale	M	G	• Prolongs the lifetime of the network • Considers the residual energy, minimum average distance from the member nodes and count of the probable head nodes in CHs selection process	• The cluster assistant drains energy in computations of CHs selection • The direct communication between CHs and BS is an unrealistic assumption
AECRP 2013 [33]	M	Large scale	L	Ltd.	• Balances the energy consumption of the overall network • Delays the death time of the node • Provides more reliable data delivery	• The coverage of the monitoring area is not considered • The multi-hop routing increases the packet delay

(continued)

Table 2 (continued)

Protocol	Delay	Network size	Energy efficiency	Scalability	Advantages	Disadvantages
RCC-PSO 2014 [34]	L	Small scale	M	G	• It is developed for a realistic network and energy model • Considers link quality and network coverage in CHs selection	• Neighbor discovery adds extra overheads • The direct communication between CHs and BS is an unrealistic assumption
PSO-HC 2014 [35]	M	Small/large scale	M	VG	• It is developed for a realistic network and energy model • Constructs two-tier clusters to maximize scalability • Considers link quality and network coverage in CHs selection	• Neighbor discovery adds extra overheads • The direct communication between CHs and BS is an unrealistic assumption
E-OEERP 2015 [36]	H	Small scale	H	Ltd.	• Reduces the individual nodes • Eliminates the control overheads messages • Avoids long-distance transmission	• Finding the best next hop adds some delay before transmitting the packets to BS • Requires location information

(continued)

Table 2 (continued)

Protocol	Delay	Network size	Energy efficiency	Scalability	Advantages	Disadvantages
TPSO-CR 2015 [37]	H	Small/large scale	M	VG	• Improves the packet delivery rate at both CHs and BS • Improves the network coverage and energy consumption • It does not assume any unrealistic assumptions, for example, using GPS for location discovery	• The packet latency is high because the MAC layer of each relay node buffers each packet before transmitting it to the next relay node • Running PSO in two-tiers increases the complexity
PSO-MBS 2011 [38]	H	Small scale	M	Ltd.	• Improves the data delivery rate and the network lifetime • Considers the distances between sink and the sensor nodes in the fitness function of PSO	• Requires locations information of sensor nodes • The packets delay is increased due to waiting CH for BS visit
CAGM 2015 [6]	H	Small scale	M	Ltd.	• Balances the network load • Utilizes the swarm optimization algorithm to find the best CHs	• The packets delay is increased due to waiting CH for BS visit • It requires information of sensor nodes for optimization which increases the protocol overhead packets

(continued)

Table 2 (continued)

Protocol	Delay	Network size	Energy efficiency	Scalability	Advantages	Disadvantages
OQoS-CMRP 2017 [39]	L	Small scale	M	Ltd.	• Solves the energy hole problem • Achieves better communication reliability with minimum delay	• Uses the simple order radio model • Requires location information during the clustering process
PSOBS 2019 [40]	NA	Small scale	H	NA	• Reduces the packet loss rate • Solves the energy hole problem by using a mobile sink	• Uses the simple order radio model • Requires location information during the clustering process • Delay will be high especially for large-scale networks
GABEEC 2012 [41]	L	Small scale	L	Ltd.	• The clusters are not recreated for each round, and this reduces the computational complexity	• Needs load balance • Requires location information
Kumar et al. 2013 [8]	L	Small scale	M	Ltd.	• Considers the residual energy of nodes in CHs selection • Prolongs the network lifetime	• Requires location information • It cannot be used in large-scale network because it uses single-hop communication

(continued)

Table 2 (continued)

Protocol	Delay	Network size	Energy efficiency	Scalability	Advantages	Disadvantages
Kuila et al. 2013 [42]	L	Small scale	M	Ltd.	• Balances the load among CHs • Considers the standard deviation of the CH load in the objective function	• Requires location information • The CHs selection process is not stated • It cannot be used in large-scale network because it uses single-hop communication
GAHN 2014 [43]	L	Small scale	H	G	• Employs remaining energy, expected energy expenditure, network locality, and distance to BS in the clustering process • Balances the energy consumption in the heterogeneous network	• It cannot be used in a large-scale network • Gathering the information of nodes increases the overhead packets.
GAEEP 2014 [19]	L	Small scale	H	Ltd.	• Improves the network lifetime and stability period of homogeneous and heterogeneous networks • Increases the reliability of clustering process	• Requires location information • It cannot be used in large-scale network because it uses single-hop communication

(continued)

Table 2 (continued)

Protocol	Delay	Network size	Energy efficiency	Scalability	Advantages	Disadvantages
GACR 2015 [44]	M	Small/large scale	H	VG	• Deals with clustering as well routing • Performs energy balancing based on residual energy of the CHs and the distance between the member nodes and their CHs	• The CHs selection process is not stated • Requires location information • It did not consider the bandwidth and latency
GAECH 2015 [45]	L	Small scale	H	Ltd.	• Improves the network lifetime and stability period • Conserves energy by balancing the energy consumption among the clusters	• It did not consider the node degree and residual energy of a node in the fitness function • Requires location information
GAROUTE 2011 [46]	H	Small scale	L	Ltd.	• Consider energy consumption and speed for selecting CHs • It does not require the location information of nodes in the clustering process	• It does not ensure that all mobile nodes can participate in the clustering process • Requires the list of neighbors and the energy information of each node, which increases the overhead packets

(continued)

Table 2 (continued)

Protocol	Delay	Network size	Energy efficiency	Scalability	Advantages	Disadvantages
NSGAII-RP 2015 [47]	M	Small scale	M	Ltd.	• Considers both the routing and coverage problems in MWSNs • Controls the mobile nodes in the network to increase coverage and lifetime	• The dissipated energy in the movement of sensor nodes is ignored during the clustering process • Requires the location information of all sensor nodes in the networks
GADA-LEACH [48]	NA	Small scale	H	Ltd.	• Improves the network lifetime • Eases the communication between CHs and sink by introducing relay nodes	• It ignores the balancing problem • It did not explain how the relay nodes are selected
MGAHP [49]	NA	Small scale	H	Ltd.	• Improves the network lifetime	• It did consider the speed of sensor nodes during the selection of CHs • It did not study the delay and load-balancing problems
EAUCF 2013 [50]	M	Small/large scale	H	VG	• Solves the energy hole problem using unequal clusters • Improves the network lifetime	• It does not use for applications when BS is located far away from the sensing region • It is developed for homogeneous networks

(continued)

Table 2 (continued)

Protocol	Delay	Network size	Energy efficiency	Scalability	Advantages	Disadvantages
Mirsadeghi et al 2014 [51]	L	Small/large scale	H	VG	• Improves the coverage and lifetime of WSNs • Considers the residual energy, local density, and node centrality to calculate the node cluster head's chance	• It assumed that BS should know global network information • It is developed for homogeneous networks
FBUC 2014 [52]	L	Small/large scale	H	VG	• Solves the energy hole problem using unequal clusters • Considers the separation to BS, current energy level of the node and CH node degree in the election of CHs	• It does not use for applications when BS is located far away from the sensing region • It is developed for homogeneous networks
OZEEP 2015 [53]	NA	Small/large scale	H	G	• Selects the best CHs by combining fussy system and GA • Considers energy, mobility, density, and distance in the CH selection	• Authors ignore the packets delay • The disconnected nodes wait for the next round to join with a new CH during the re-clustering process

(continued)

Table 2 (continued)

Protocol	Delay	Network size	Energy efficiency	Scalability	Advantages	Disadvantages
SEPFL 2016 [54]	L	Small scale	M	Ltd.	• All sensors have almost an equal lifetime • It does not require the knowledge of the global network	• It cannot be used in large-scale network because it uses single-hop communication • Improving in the stability period is limited • Needs load balance
LEFUCMA [55]	NA	Small/large scale	H	VG	• Avoids the energy hole problem by using unequal clustering mechanism • Balances the network load by distributing CHs based on sector density and distance to BS	• It did not consider the delay problem
FMCR-CT [56]	NA	Small/large scale	H	G	• Avoids the problem of CH selection in each round • Decreases the overhead control packets	• It did not improve the stability period • It did not consider the delay problem
(FCM-ACO) [57]	NA	Small scale	H	Ltd.	• Performs the clustering process only one times • Uses FCM for clustering the network and uses ACO for constructing the routing chain	• It did not explain how the CH of each cluster is selected • The constructed chain increases the delay especially for large-scale network

(continued)

Table 2 (continued)

Protocol	Delay	Network size	Energy efficiency	Scalability	Advantages	Disadvantages
ACOHC [58]	M	Small scale	H	Ltd.	• Performs the clustering process only one times • Uses *K*-means algorithm for clustering the network and uses ACO for constructing the routing chain.	• Repetition of constructing cluster chain and upper chain each round consumes more computation time • The constructed chain increases the delay especially for large-scale network
MH-GEER [59]	NA	Small scale	H	G	• Performs the clustering process only one times • The CHs of clusters are selected locally based on residual energy • Uses *K*-means algorithm for clustering the network and uses ACO for constructing the routing chain	• It cannot be used for large-scale networks • Requires location information
Jingyi et. al. [60]	L	Small scale	M	Ltd.	• Improves the clustering topology by considering the node energy and distance	• It cannot be used when BS is located far because it used a single-hop routing • May lead to non-uniform CHs distribution

(continued)

Table 2 (continued)

Protocol	Delay	Network size	Energy efficiency	Scalability	Advantages	Disadvantages
MSIEEP (Fixed BS) 2015 [61]	L	Small scale	H	Ltd.	• Enhances the lifetime and the stability period of the network • Balances the load among the nodes • It is reliable and energy-efficient	• The computational time increases as the node degree increases • Requires location information • It cannot be used in a large-scale network
UMBIC 2016 [62]	M	Small/large scale	H	H	• Enhances the lifetime and the stability period of the network • Balances the load among the nodes and solves the energy hole problem • Saves the CPU time and overhead packets of selecting new CHs	• Requires location information • Requires sensor nodes with power control
ARBIC 2018 [9]	L	Small/large scale	H	H	• Adjusts communication range of CHs based on the speed of MSN • Runs the clustering process if the energy of any CH less than a threshold • Considers, the energy, distance, mobility factor, coverage and link connection time during the cluster construction process	• Requires location and speed information • Requires sensor with adjustable communication ranges

L low, *M* medium, *H* high, *VH* very high, *Ltd.* limited, *G* good, *VG* very good and *NA* not available

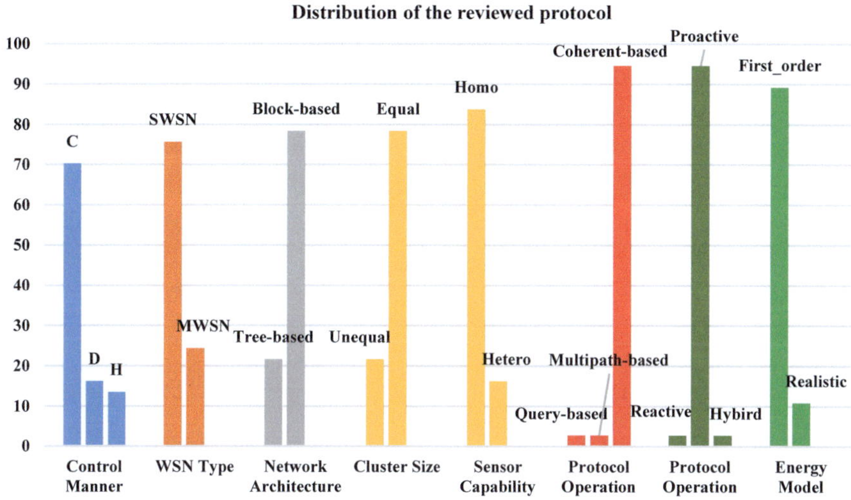

Fig. 16 Distribution of the surveyed protocols

Sometimes real-world applications such as outdoor medical applications and wildlife applications require mobile sensors environments. Data gathering in the mobile sensor environments is a difficult task as compared to static one due to the frequency topology change of the network. So, the work in this area should be continued to develop stable routing protocols for the mobile sensor environments because the work of this area is still limited. Moreover, the real-work implementation of the designed protocol is an important issue for facing the problems and challenges of the designed protocol that can hardly be discovered in simulation.

5 Conclusion and Future Directions

In the last decade, a large number of intelligent-based hierarchical routing protocols have been developed for WSNs based on a variety of different optimization algorithms. This chapter focuses on reviewing the recently intelligent routing protocols that have designed based on Particle Swarm Optimization, Ant Colony Optimization, Fuzzy Logic, Genetic Algorithm, and Artificial Immune Algorithm. Also, a detailed taxonomy is presented to classify the surveyed protocols according to different metrics. Moreover, the reviewed protocols are evaluated and compared on the basis of delay, network size, energy-efficient, features, and drawbacks. This chapter can be a guide for designers of WSNs to select the appropriate hierarchical routing protocol for a specific application. The much research works have been done to solve the drawback of the hierarchical routing protocols, but still there are some open issues should be considered. The first issue is the real-work implantations. Experiments with real test beds force the experimenter to discover a wide set of problems and challenges that can hardly be noticed in simulations. The second issue is the overheads and computational time of clustering the network. Most of the intelligent routing protocols operate in a centralized manner which increases the overheads and computational time, especially in a large-scale network. Designers should be developed semi-distributed protocols that run within the cluster head rather than sink.

References

1. Rahman MA, Anwar S, Pramanik MI, Rahman MF (2013) A survey on energy efficient routing techniques in wireless sensor network. In: 2013 15th international conference on advanced communication technology (ICACT), pp 200–205
2. Yu J, Qi Y, Wang G, Gu X (2012) A cluster-based routing protocol for wireless sensor networks with nonuniform node distribution. AEU—Int J Electron Commun 66:54–61
3. Sabet M, Naji HR (2015) A decentralized energy efficient hierarchical cluster-based routing algorithm for wireless sensor networks. AEU—Int J Electron Commun 69:790–799
4. Do sung K, Hyun soo C, Seungwha Y (2012) Improve far-zone LEACH protocol for energy conserving. In: 2012 8th international conference on wireless communications, networking and mobile computing (WiCOM), Shanghai, pp 1–4
5. Hui L, Xiaoguang Z, Lijun L (2013) A hybrid deployment algorithm based on clonal selection and artificial physics optimization for WSN. Inf Technol 12:917–925
6. Wang J, Cao Y-Q, Li B, Lee S-Y, Kim J-U (2015) A glowworm swarm optimization based clustering algorithm with mobile sink support for wireless sensor networks. J Internet Technol 16:825–832
7. Goldberg DE (1989) Genetic algorithms in search, optimization and machine learning. Addison-Wesley Longman Publishing Co., Inc.
8. Kumar G, Singh J (2013) Energy efficient clustering scheme based on grid optimization using genetic algorithm for wireless sensor networks. In: 2013 fourth international conference on computing, communications and networking technologies (ICCCNT), Tiruchengode, pp 1–5
9. Sabor N, Ahmed SM, Abo-Zahhad M, Sasaki S (2018) ARBIC: an adjustable range based immune hierarchy clustering protocol supporting mobility of wireless sensor networks. Pervasive Mob Comput 43:27–48
10. Das SK, Tripathi S (2019) Energy-efficient routing formation algorithm for hybrid ad-hoc network: a geometric programming approach. Peer-to-Peer Netw Appl 12:102–128
11. Das SK, Tripathi S (2017) Energy efficient routing formation technique for hybrid ad hoc network using fusion of artificial intelligence techniques. Int J Commun Syst 30:e3340
12. Xiaobing W, Guihai C, Das SK (2006) On the energy hole problem of nonuniform node distribution in wireless sensor networks. In: 2006 IEEE international conference on mobile adhoc and sensor systems (MASS), Vancouver, BC, pp 180–187
13. Camp T, Boleng J, Davies V (2002) A survey of mobility models for ad hoc network research. Wirel Commun Mob Comput 2:483–502
14. Pushpa RA, Vallimayil A, Dhulipala VRS (2011) Impact of mobility models on mobile sensor networks. In: 3rd international conference on electronics computer technology (ICECT), Kanyakumari, pp 102–106
15. Jayakumar G, Ganapathi G (2008) Reference point group mobility and random waypoint models in performance evaluation of MANET routing protocols. J Comput Syst Netw Commun 2008
16. Abdollahzadeh S, Navimipour NJ (2016) Deployment strategies in the wireless sensor network: a comprehensive review. Comput Commun 91–92:1–16

17. Liu X (2012) A survey on clustering routing protocols in wireless sensor networks. Sensors (Basel) 12:11113–11153
18. Gupta SK, Jain N, Sinha P (2013) Clustering protocols in wireless sensor networks: a survey. Int J Appl Inf Syst 5:41–50
19. Abo-Zahhad M, Ahmed SM, Sabor N, Sasaki S (2014) A new energy-efficient adaptive clustering protocol based on genetic algorithm for improving the lifetime and the stable period of wireless sensor networks. Int J Energy Inf Commun 5:47–72
20. Heinzelman WB, Chandrakasan AP, Balakrishnan H (2002) An application-specific protocol architecture for wireless microsensor networks. IEEE Trans Wirel Commun 1:660–670
21. Do-Seong K, Yeong-Jee C (2006) Self-organization routing protocol supporting mobile nodes for wireless sensor network. In: First international multi-symposiums on computer and computational sciences, 2006, IMSCCS'06, Hanzhou, Zhejiang, pp 622–626
22. Awwad SAB, Ng CK, Noordin NK, Rasid MFA (2011) Cluster-based routing protocol for mobile nodes in wireless sensor network. Wireless Pers Commun 61:251–281
23. Cakici S, Erturk I, Atmaca S, Karahan A (2014) A novel cross-layer routing protocol for increasing packet transfer reliability in mobile sensor networks. Wireless Pers Commun 77:2235–2254
24. Kennedy J, Eberhart R (1995) Particle swarm optimization. In: Proceedings of IEEE international conference on neural networks, vol 4, pp 1942–1948
25. Yarushkina NG (2002) Genetic algorithms for engineering optimization: theory and practice. In: 2002 IEEE international conference on artificial intelligence systems, 2002 (ICAIS 2002), pp 357–362
26. Mamdani EH, Assilian S (1975) An experiment in linguistic synthesis with a fuzzy logic controller. Int J Man Mach Stud 7:1–13
27. Blum C (2005) Ant colony optimization: introduction and recent trends. Phys Life Rev 2:353–373
28. Zhang Z (2007) Immune optimization algorithm for constrained nonlinear multiobjective optimization problems. Appl Soft Comput 7:840–857
29. Jiang C-J, Shi W-R, Xiang M, Tang X-L (2010) Energy-balanced unequal clustering protocol for wireless sensor networks. J China Univ Posts Telecommun 17:94–99
30. Zhang R, Jia Z, Li X, Han D (2011) Double cluster-heads clustering algorithm for wireless sensor networks using PSO. In: 2011 6th IEEE conference on industrial electronics and applications (ICIEA), Beijing, pp 763–766
31. Singh B, Lobiyal DK (2012) Energy-aware cluster head selection using particle swarm optimization and analysis of packet retransmissions in WSN. Procedia Technol 4:171–176
32. Singh B, Lobiyal D (2012) A novel energy-aware cluster head selection based on particle swarm optimization for wireless sensor networks. Hum-Centric Comput Inf Sci 2:1–18
33. Xia L, Wang G, Liu Z, Zhang Y (2013) An energy-efficient routing protocol based on particle swarm clustering algorithm and inter-cluster routing algorithm for WSN. In: 2013 25th Chinese control and decision conference (CCDC), Guiyang, pp 4029–4033
34. Elhabyan RS, Yagoub MCE (2014) Particle swarm optimization protocol for clustering in wireless sensor networks: a realistic approach. In: 2014 IEEE 15th international conference on information reuse and integration (IRI), Redwood City, CA, pp 345–350
35. Elhabyan RS, Yagoub MCE (2014) PSO-HC: particle swarm optimization protocol for hierarchical clustering in wireless sensor networks. In: 2014 international conference on collaborative computing: networking, applications and worksharing (CollaborateCom), Miami, FL, pp 417–424
36. RejinaParvin J, Vasanthanayaki C (2015) Particle swarm optimization-based clustering by preventing residual nodes in wireless sensor networks. IEEE Sens J 15:4264–4274
37. Elhabyan RSY, Yagoub MCE (2015) Two-tier particle swarm optimization protocol for clustering and routing in wireless sensor network. J Netw Comput Appl 52:116–128
38. Latiff NAA, Latiff NMA, Ahmad RB (2011) Prolonging lifetime of wireless sensor networks with mobile base station using particle swarm optimization. In: 2011 4th international conference on modeling, simulation and applied optimization (ICMSAO), Kuala Lumpur, pp 1–6

39. Deepa O, Suguna J (2017) An optimized QoS-based clustering with multipath routing protocol for wireless sensor networks. J King Saud Univ—Comput Inf Sci
40. Tabibi S, Ghaffari A (2019) Energy-efficient routing mechanism for mobile sink in wireless sensor networks using particle swarm optimization algorithm. Wirel Pers Commun 104:199–216
41. Bayraklı S, Erdogan SZ (2012) Genetic algorithm based energy efficient clusters (GABEEC) in wireless sensor networks. Procedia Comput Sci 10:247–254
42. Kuila P, Gupta SK, Jana PK (2013) A novel evolutionary approach for load balanced clustering problem for wireless sensor networks. Swarm Evol Comput 12:48–56
43. Elhoseny M, Yuan X, Yu Z, Mao C, El-Minir H, Riad A (2014) Balancing energy consumption in heterogeneous wireless sensor networks using genetic algorithm. IEEE Commun Lett 19:2194–2197
44. Gupta S, Jana P (2015) Energy efficient clustering and routing algorithms for wireless sensor networks: GA based approach. Wireless Pers Commun 83:2403–2423
45. Baranidharan B, Santhi B (2015) GAECH: genetic algorithm based energy efficient clustering hierarchy in wireless sensor networks. J Sens 2015:8
46. Sarangi S, Kar S (2011) Genetic algorithm based mobility aware clustering for energy efficient routing in wireless sensor networks. In: 2011 17th IEEE international conference on networks, Singapore, pp 1–6
47. Attea BA, Khalil EA, Cosar A (2015) Multi-objective evolutionary routing protocol for efficient coverage in mobile sensor networks. Soft Comput 19:2983–2995
48. Bhatia T, Kansal S, Goel S, Verma AK (2016) A genetic algorithm based distance-aware routing protocol for wireless sensor networks. Comput Electr Eng 56:441–455
49. Rady A, Sabor N, Shokair M, El-Rabaie EM (2018) Mobility based genetic algorithm hierarchical routing protocol in mobile wireless sensor networks. In: 2018 international Japan-Africa conference on electronics, communications and computations (JAC-ECC), pp 83–86
50. Bagci H, Yazici A (2013) An energy aware fuzzy approach to unequal clustering in wireless sensor networks. Appl Soft Comput 13:1741–1749
51. Mirsadeghi M, Mahani A, Shojaee M (2014) A novel distributed clustering protocol using fuzzy logic. Procedia Technol 17:742–748
52. Logambigai R, Kannan A (2015) Fuzzy logic based unequal clustering for wireless sensor networks. Wireless Netw 1–13
53. Srivastava JR, Sudarshan TSB (2015) A genetic fuzzy system based optimized zone based energy efficient routing protocol for mobile sensor networks (OZEEP). Appl Soft Comput 37:863–886
54. Tamandani YK, Bokhari MU (2016) SEPFL routing protocol based on fuzzy logic control to extend the lifetime and throughput of the wireless sensor network. Wireless Netw 22:647–653
55. Arjunan S, Sujatha P (2018) Lifetime maximization of wireless sensor network using fuzzy based unequal clustering and ACO based routing hybrid protocol. Appl Intell 48:2229–2246
56. Mazinani A, Mazinani SM, Mirzaie M (2019) FMCR-CT: an energy-efficient fuzzy multi cluster-based routing with a constant threshold in wireless sensor network. Alex Eng J 58:127–141
57. Ghosh S, Mondal S, Biswas U (2016) Fuzzy C means based hierarchical routing protocol in WSN with ant colony optimization. In: 2016 2nd international conference on applied and theoretical computing and communication technology (iCATccT), pp 348–354
58. Mondal S, Ghosh S, Biswas U (2016) ACOHC: ant colony optimization based hierarchical clustering in wireless sensor network. In: 2016 international conference on emerging technological trends (ICETT), pp 1–7
59. Rhim H, Tamine K, Abassi R, Sauveron D, Guemara S (2018) A multi-hop graph-based approach for an energy-efficient routing protocol in wireless sensor networks. Hum-Centric Comput Inf Sci 8:1–21
60. Jingyi W, Yuhao J, Xiaotong Z, Hongying B (2016) Clustering protocol based on immune optimization algorithms for wireless sensor networks. In: 2016 2nd IEEE international conference on computer and communications (ICCC), pp 2272–2276

61. Abo-Zahhad M, Ahmed SM, Sabor N, Sasaki S (2015) Mobile sink-based adaptive immune energy-efficient clustering protocol for improving the lifetime and stability period of wireless sensor networks. IEEE Sens J 15:4576–4586
62. Sabor N, Abo-Zahhad M, Sasaki S, Ahmed SM (2016) An unequal multi-hop balanced immune clustering protocol for wireless sensor networks. Appl Soft Comput (in Progress)
63. Smaragdakis G, Matta I, Bestavros A (2004) SEP: a stable election protocol for clustered heterogeneous wireless sensor networks. In: Second international workshop on sensor and actor network protocols and applications (SANPA 2004), pp 1–11
64. Mahapatra RP, Yadav RK (2015) Descendant of LEACH based routing protocols in wireless sensor networks. Procedia Comput Sci 57:1005–1014

Chapter 11
Qualitative Survey on Sensor Node Deployment, Load Balancing and Energy Utilization in Sensor Network

Ayan Kumar Panja and Arka Ghosh

1 Introduction

The Internet of Things (IoT) [35] we know of today is much more than connecting devices over the internet, it is greatly used in the fields of modern intelligent services around the world. Internet of things has already percolated in our everyday usage like smartphones, smartwatches, smart healthcare systems, etc. Sharing of information is very necessary to tackle various real-world problems related to business, science or taking preventative measures against a catastrophic event. Although in the lower level of the services provided by this technology are sensor nodes. The sensors are the prime components in any wireless sensor network application [1, 2]. The sensors are composed of battery-operated devices capable of gathering information such as temperature, proximity, motion of objects according to application-specific need. Sensors are also used for relaying information from one sensor node to another, for example, Zigbee module. Zigbee uses IEEE 802.15.4 protocol and is used for creating Personal Area Networks (PAN) which communicates over low power high-level radio transmission. In any wireless sensor network architecture be it a centralized or an ad hoc network [3–6] or opportunistic information relaying network there are three prime node components one is the main sink node called the base station, the coverage nodes for information gathering and the relay nodes for the transmission of the data towards the base station.

The overall work of a sensor node can be generalized into four categories sending, receiving, aggregating and processing. Although modern-day sensors are highly energy-efficient and communicate using very low power still we have to solely rely

A. K. Panja (✉)
Department of Computer Science and Engineering/BSH, Institute of Engineering and Management, Sector V Salt Lake, Kolkata 700091, India

A. Ghosh
Department of Nano Technology, University of Siegen Germany, Siegen, Germany

D. De et al. (eds.), *Nature Inspired Computing for Wireless Sensor Networks*, Springer Tracts in Nature-Inspired Computing,
https://doi.org/10.1007/978-981-15-2125-6_11

on the battery. Thus one of the prime important concerns is the lifetime of the sensor node hence one of the biggest challenges in energy efficiency for which researchers have worked and provided various models and algorithms for energy-efficient gathering and routing [37] of information and data. Node deployment [27] plays one of the crucial roles in energy management. Nodes can be stationary or mobile but the information gathered has to be relayed back to its respective base station either periodically or with respect to some occurrence of an event. As the sensors are battery-operated devices it becomes quite difficult to replenish the battery power when there are deployed in places where manual intervention is quite difficult to achieve. Furthermore, the sensors are usually low powered devices and there range of transmission is quite small hence the sensor nodes are usually group into clusters where a tree-like route is formed the coverage nodes relay the information hop by hop with the help of the relay nodes and channels the information towards a base station. Base stations are usually are powerful than the coverage sensors. If we consider an environment where numerous sensors are deployed along with more than one base station to which the sensors subscribes to, then identifying the base station to which a particular coverage sensor will relay its information is important. The distribution of coverage sensors is done based on various criteria such as the routing distance, number of hops, etc. The communication protocol in a wireless sensor network can be considered as a three-dimensional protocol stack which consists of standard layer of TCP/IP suite. There is numerous methodology of data transmission in a wireless sensor network such as Unicast, multicast, broadcast, but the objective of deploying sensor nodes is to collect data's from numerous target points and relay that very information towards a sink node. Collecting and relaying the sensed data towards a sink node is called converge casting which is actually the inverse of broadcasting. A process of converge cast in a small network with node deployed is shown in Fig. 1.

2 Overview of Sensor Node Deployment

In order to increase the lifetime metric of the wireless sensor network, it is very crucial to take into consideration various criteria into accounts such as the area of coverage [36], distance to its base station, and a number of sensors used, etc. Proper sensor node deployment is one of the most important criteria for increasing the lifetime of the sensor nodes. It can be defined as the optimal positioning of coverage nodes and the base station. Coverage nodes are the sensor nodes used for data gathering purposes, while base stations are the sink nodes where the deployed sensors send their gathered data. Usually the deployment is done in two ways. The first procedure involves determining the exact position of the target points that cover a given area and placing the sensor nodes in the required vicinity to carry out the sensing operation. Another procedure involves a non-deterministic approach where the actual position or area of coverage is not known or prior prediction of the location of sensing is not possible. Deployment plays a very important role in sustaining the wireless sensor network, effective deployment increases the network lifetime.

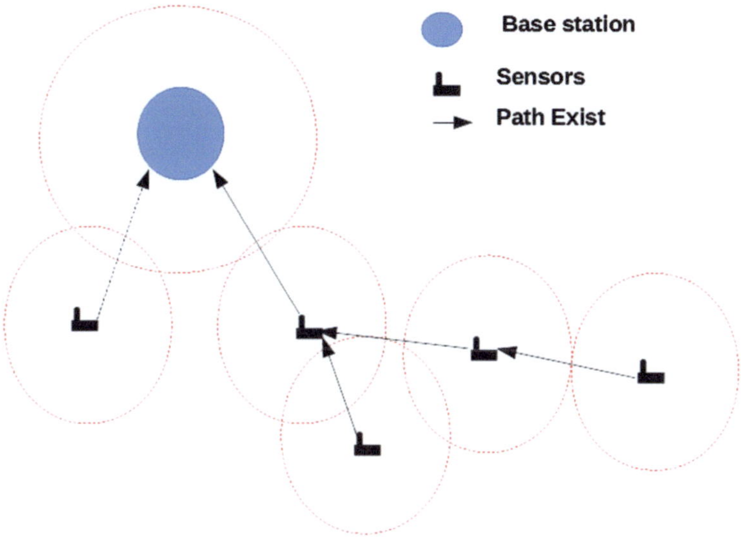

Fig. 1 Convergecasting

Node deployment is one of the fundamental issues in WSN that reduces the complexity of various problems related to energy management, routing [7, 8], etc. There are many challenges pertaining to node deployment. Sensor nodes are prone to failure, malfunctioning of sensor nodes can occur at any time this can cause significant re-routing of packets and reorganization of the network. Furthermore, there might be a case where same sensing area is covered by multiple sensors due to which redundant data's are relayed towards the base station which in turn reduces the efficiency of the network and its lifetime, Fig. 2 depicts the problems related to improper node deployment and redundant data collection.

2.1 IPP Based Approach for Ensuring Coverage

Integer programming is an NP-Complete problem practically used in every sphere of an optimization problem. The main motive lies in maximizing or minimizing an objective function based on certain constraints that are linear in nature. A method for ensuring coverage nodes has been defined by Sen et al. [9] where they have modeled and calculated the minimum number of sensors required to cover all the target points to satisfy all the applications. The working principle can be defined in the following way.

Let P be the set of all target points that need to be covered where $P = \{p_1, p_2, p_3, \ldots p_m\}$, $|P| = m$ indicates that there are m target points. An array k_j is defined where $1 \leq j \leq m$ where y_j can be:

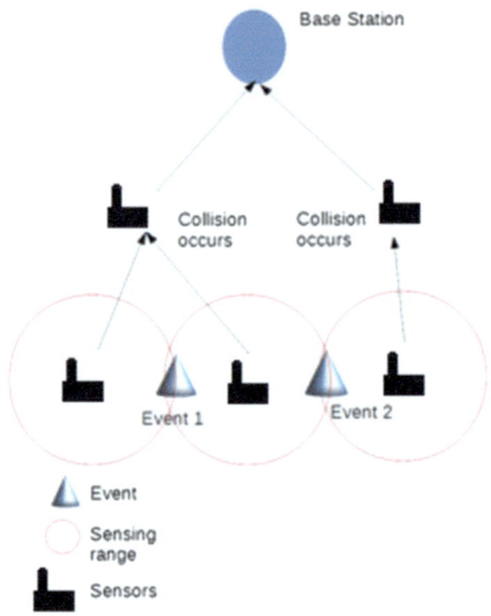

Fig. 2 Improper node deployment

$$y_j = \begin{Bmatrix} 1, \text{ if } pj \in P \\ 0, \text{ otherwise} \end{Bmatrix} \tag{1}$$

Let $\{b_1, b_2, b_3 \ldots b_n\}$ are the set of base stations to which the sensors will relay their data. A subset of the sensors forms a cover when each of the target point's p_j is covered by at least one sensor. The IPP is formulated in a way such that the target points are covered using a minimum number of sensors. A parameter $M_{i,j}$, $1 \le i \le n, 1 \le j \le m$

$$M_{ij} = \begin{Bmatrix} 1, \text{ if } b_i \text{ covers } P_i \\ 0, \quad \text{otherwise} \end{Bmatrix} \tag{2}$$

The variable modeling for IPP can be done in the following manner.

$$x_i = \begin{Bmatrix} 1, \text{ if } b_i \text{ is included} \\ 0, \text{ otherwise} \end{Bmatrix} \tag{3}$$

Now the objective lies in minimizing $Z = \sum_{i=k+1}^{n} x^k$ subject to constraint

$$Z = \sum_{i=k+1}^{n} S_{i,j} x^k \ge y_{i,\text{for all } j=1 \text{ to } m} \tag{4}$$

The required constraint specified ensures that all of the target P points are covered. The IPP formulation gives the minimum no of sensors required to cover all the target points. The authors have modeled the problem into two parts. The first part ensured the minimum number of sensors required to cover all the target points through IPP formulation the second part involves identifying the relay nodes to connect the sensors so that efficient converge casting can be performed discussed thoroughly with the algorithm procedure in [9].

2.2 PSO Based Node Deployment

PSO or particle swarm Optimization falls under the category of Evolutionary Computation or to be precise a Nature-Inspired Optimization usually found in flocking of a group of birds while flying from a source node to a destination node. PSO is a stochastic optimization technique, i.e. a random distribution usually used to solve computationally hard problems. The PSO technique is quite similar to the Genetic Algorithm but the only difference is there is no cross over or mutation. The model composed of candidates known as particles in an n-dimensional hyperspace. The objective lies in searching for the global optimal solution. The working can be explained considering the example of a flock of birds, where each bird can be considered as a particle which is traveling from a source to the destination. Each particle computes the next state position. The global best position is taken into account and next state position is considered with respect to a fitness score. Each particle tries to reach the global best score for the entire group and local best score for its own (Fig. 3).

The PSO Algorithm working can be explained as follows:

(i) X-vector: records the current position of the particle in space
(ii) P-vector: records the location of the best solution so far
(iii) V-Vector: contains a gradient for which particle will travel if undisturbed.

The basic concept of PSO lies in accelerating each particle towards the best position found so far by any particle with a random weighted acceleration at each time step, Fig. 4 depicts the particle movement. This is achieved by simply adding the v-vector to the x-vector to get a new x-vector.

$$x_{i+1} = x_i + v_i \tag{5}$$

As the new x_i is computed the new location is evaluated. Two fitness values are computed, which are named as x-fitness and p-fitness. If the x-fitness better than the p-fitness value $p_{\text{best}} = x_i$ and p-fitness = x-fitness.

The particle velocity can be calculated as follows:

$$\begin{aligned} V_{t+1} = W * v_t + c_1 * r(0,1) * (p_{\text{best}} - x_t) \\ + c_2 * r(0,1) * (g_{\text{best}} - x_t) \end{aligned} \tag{6}$$

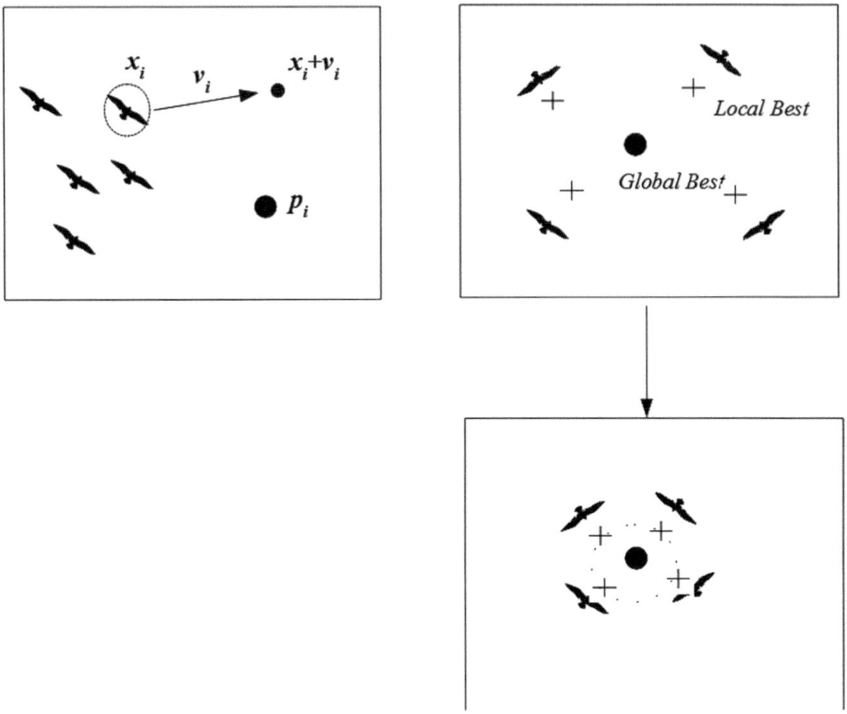

Fig. 3 PSO overview

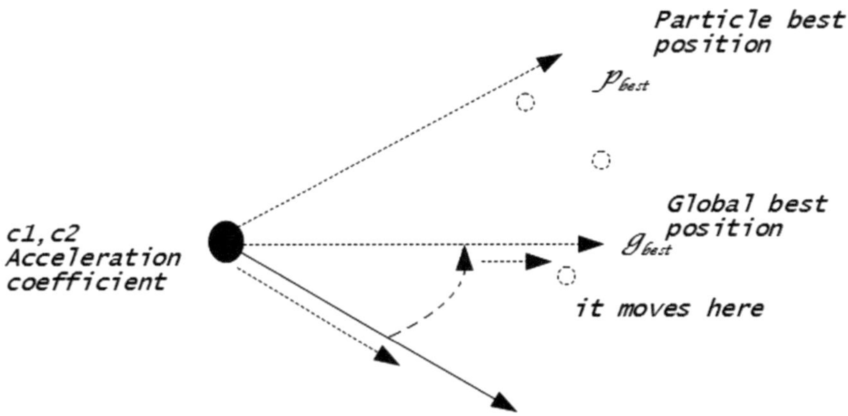

Fig. 4 Particle movement

The algorithm is depicted in a flow diagram in Fig. 5. Here c_1, c_2 are acceleration coefficients and r is a random number which varies in between 0 and 1. V_t is the previous value of the velocity vector, x_t is the current position and W is the weight parameter. On varying the parameters W, c_1 and c_2 the searching methodology is affected as follows:

- *If W is large it helps in the global search*
- *If W is small it helps in local search*
- If $c_1 > c_2$ *then it helps in the global search*
- *If* $c_1 < c_2$ *then it helps in local search.*

The problem of node deployment can be tackled using the PSO algorithm. A model proposed by Aziz et al. [10, 24] where the author distributed each of the

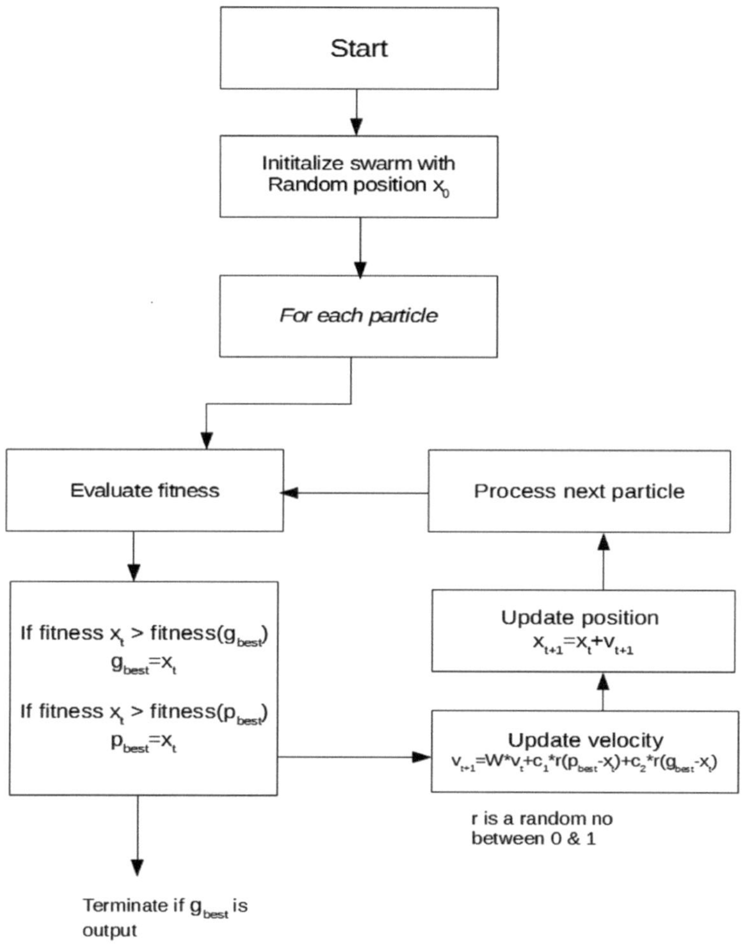

Fig. 5 PSO flowchart

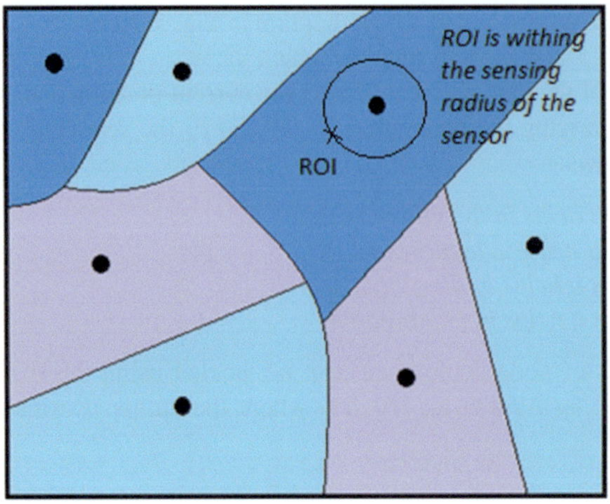

Fig. 6 Voronoi diagram

locations using Voronoi diagram [11]. The PSO algorithm is used in searching the optimal node deployment in the given search space. A random deployment of sensor nodes is depicted in [12], where the Voronoi diagram is formed automatically by the sensors where they broadcast their location and based on the diagram it is decided whether the sensors should be moved or repositioned. The main objective here lies in minimizing the number of sensor nodes to cover the required target area location. The partition achieved by Voronoi diagram is depicted in Fig. 6.

The positioning of sensors to cover a particular target area is very crucial so that the coverage is maximized. Each Region of Interest (ROI) is clustered into polygons called the Voronoi polygons. A point inside the polygon is covered by a sensor s_i if the Region of Interest inside the polygon is within the sensors sensing radius. The authors have considered a homogenous network where all the sensors have the same sensing radius. Initially, the sensors are placed at particular regions and sensor localization [26, 30] is pre-done, i.e. that is each sensor knows its current position. The PSO algorithm is utilized to optimize the position of the sensors in their respective regions of interest. The fitness is calculated using the Voronoi diagram, which is originally created on an unbounded region; the positioning of sensors is done on a specific ROI—bounded region, so the authors took into account the points along the boundary.

Therefore a set of points are selected randomly along each of the boundary, the points, and the Voronoi vertices are called interest points. The distance of the interest points with their nearest sensors is used to evaluate the fitness. An example of Voronoi diagram for 7 ROI is given in Fig. 6 and the proposed sensor placement is given in Fig. 7.

Algorithm 1 Fitness Calculation

Function: fitness_calculation

Interest_point ← *{polygon_vertices,n randomly selected points*
Along the boundary}

For each interest_point
begin
Find the distance of the interest_point to its nearest sensor
If distance > sensing radius
Fitness += distance – sensing radius

For End

Algorithm 2 PSO Voronoi Procedure

Function: PSO Voronoi

Initialize particles population with random deployment $\mathbf{x}_i$
while (Ideal fitness is not attained) ***do***
begin
Calculate fitness values of each particles using fitness function

Update $\mathbf{p}_{best}$ *if the current fitness value is better than* $\mathbf{p}_{best}$*;*
Determine $\mathbf{g}_{best}$ *: choose the particle position with the best fitness value of all the neighbors as the* $\mathbf{g}_{best}$*;*
For each particle do
begin
Calculate particle velocity according to ***(Eq. 2)***
Update particle position according to ***(Eq. 1)***
end
Endwhile

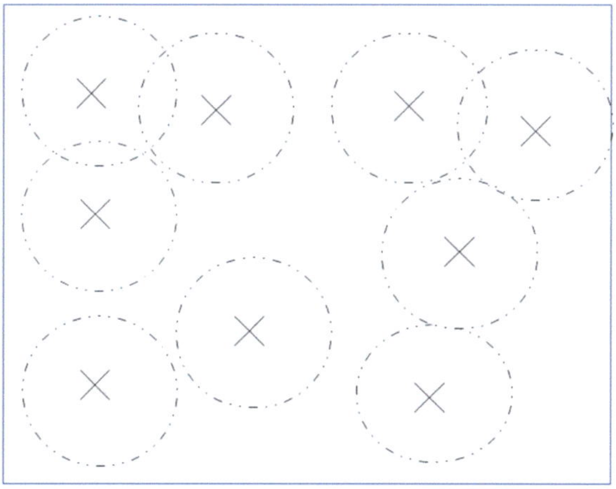

Fig. 7 Proposed sensor placement

The PSO algorithm should be processed carefully so that the sensing range doesn't overlap and the whole ROI is covered. The algorithm to calculate the fitness function and carry out the PSO Voronoi procedure is given in Algorithm 1 and 2.

2.3 ACO in Node Deployment and Load Balancing

Ant Colony Optimization [28, 29] is a meta heuristic-based approach that is inspired by the way the ants search and collect food from an area; the process is called foraging behavior of ant species. The ants communicate with each other by pheromones. The ants use a probabilistic approach of whether to follow the path or not [13]. The path which has more pheromones has a higher chance to be considered in processing. Deneuboroug et. al. have investigated the pheromone behavior of ants by proposing and experiment of the double bridge as depicted in Fig. 8.

There are two bridges from an ant nest to a Food source by two bridges of equal length as well as unequal length as depicted in Fig. 9. As the ant traverses, they release pheromones along the paths. Initially, we consider the two bridges are of equal length. We will be able to observe that the ant choose the path randomly along the way but as the more and more ant traverses and releases pheromones along the path. The path with most pheromones count is usually opted by the later ants. The optimal path is selected as the system progresses. The above Fig. 8 can be modeled into a graph as shown in Fig. 9. Let the probability of transition of ants from a particular node j to node i is p_{ij} which is calculated using a heuristic information H_{ij} with respect to the trail of pheromones K_{ij} of the move, where $i, j = 1, 2, \ldots n$.

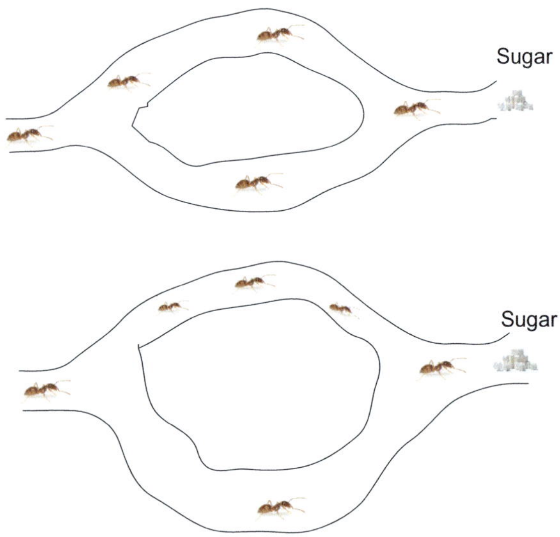

Fig. 8 Double bridge experiment

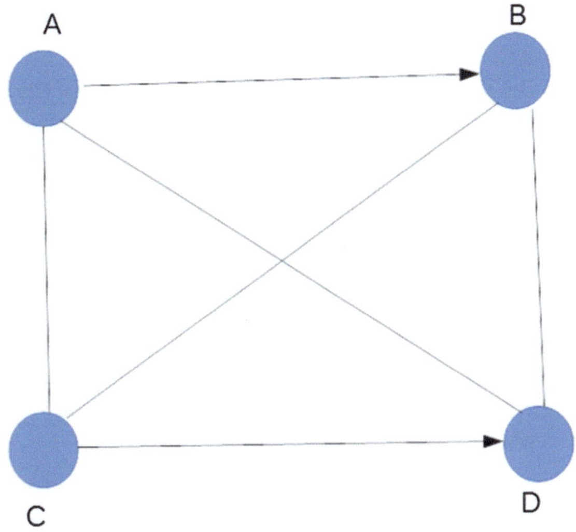

Fig. 9 Graphic representation

$$p_{ij} = \frac{K_{ij}^{\propto} H_{ij}^{\beta}}{\sum_{k \in \text{allowed}} K_{ik}^{\alpha} H_{ik}^{\beta}} \tag{7}$$

The path with more pheromone value is one best to be selected. ACO is widely used in various assignments and routing problems but it can also be modeled to optimize the sensor node deployment and perform load balancing [31–34] in a sensor network.

There are various types of ACO procedures one of such procedure is the MAX-MIN System where the upper bound and the lower bound for the pheromone trail amount is restricted ($K_{\min}$ *to* $K_{\max}$). Authors Findanova et al. in [14] have depicted the use of such procedure in sensor node deployment. A strong exploration of the search space for sensor node deployment is carried which is done by allowing a single ant to add pheromone at each iteration level. After the very first iteration, the pheromone trail is initialized to $K_{\max},$ *for* the next iteration the transition is performed for the best solution receives a pheromone. The pheromone update rule is given as follows:

$$K_{ij} = \rho K_{ij} + \partial K_{ij} \tag{8}$$

$$\partial K_{ij} = \begin{cases} \frac{1}{C(V_{\text{best}})} & \text{if } i, j \in \text{best solution} \\ 0 & \text{otherwise} \end{cases} \tag{9}$$

Here ∂K_{ij} is the small change in the pheromone level and V_{best} is the best solution result and $i, j = 1, 2 \ldots, n$, $\rho \in [0, 1]$ models evaporated in nature. They have considered the WSN layout as a graph $\{g_{ij}\}_{N \times M}$ which is pre-initialized. The location site is represented as $P = \{P_{ij}\}_{N \times M}$. The initial value of the pheromone is usually taken a very small value. The point where the base station is located where the gathered data are transferred from each of the coverage sensors is included in the solution of the optimization process as the first point.

The ants create the rest of the solution from a random node that communicates with the base station. The target points that need to cover can be thought of as the ACO target point. The ant traverses with each of the nodes randomly at every iteration. A probabilistic next step is a calculation is performed by each of the ants with respect to the pheromone value. They have addressed the WSN deployment problem by ensuring full coverage and connectivity as constraints, while the objective function is the number of the sensors. Accordingly, ACO is used for optimization.

2.4 Honey Bee Optimization in Sensor Deployment

Honey Bee optimization is similar to Ant Colony optimization where the technique's processing is drawn from food foraging of bees. There are several behaviors amongst bees such as mating, marriage, food gathering, these behaviors are mimicked into various optimization procedures. The Bee Algorithm is quite similar to ACO approach the only difference is instead of using a probabilistic approach for search procedure fitness is calculated for path selection.

The artificial Bee Colony Optimization consists of three types of Bees, namely Experienced Bees, Onlooker Bees, Scout Bees. The Experienced Bees explores randomly the area and gathers information and returns back to the hive. The experienced bees are also called scout bees. The scout bees provide the information by means of the waggle dance. The waggle dance of bees is depicted in Fig. 10. The speed, distance, duration, and frequency indicate the direction of the food. The onlooker bees process the information gathered by the experienced bees and create a probabilistic approach to search the neighborhood, while the scout bees search the area in a random manner and carry out the process. The Bee Algorithm has both local

Fig. 10 Waggle dance

as well as global search procedures. An Algorithm for Bee Colony Optimization is given in [15] demonstrated in Algorithm 3.

Algorithm 3 Bee Algorithm
Function: Bee Algorithm

Input:{ *n* ← *no of scout bees, m* ← *no of best patches , e* ← *no of elite patches,* n_{ep} ← *Bees in elite patches,* n_{sp} ← *Bees in non elite patches,* n_{gh} ← *size of neighborhood ,M_iter* ← *no of iterations, diff* ← *difference between the first and last iteration* }

Initialize size of population as n, and Initial population
Initialize i=0
Evaluate fitness value of initial population
Sort the initial population

while (i<= M_iter new(fitness)-old(fitness)<=Error) ***do***
begin
i=i+1
Select the elite patches and non-elite best patches for neighborhood Search.
Recruit the forager bees to the elite patches and non-elite best patches.
Evaluate the fitness value of each patch.
Sort the results based on their fitness.
Allocate the rest of the bees for global search to the non-best locations.
Evaluate the fitness value of non-best patches.
Sort the overall results based on their fitness.
Endwhile

A novel Bee swarm optimization algorithm is proposed in [16] for sensor node deployment. The sensor's position in an area can be depicted in a 3D coordinate system comprising of (*x, y, z*) parameters. Each of the bees' experienced onlooker and scout moves in a different pattern in search of an optimal solution. The first phase of the algorithm is Initialization is done where bees are randomly generated, that is the sensors node position is randomly generated. The non-dominated bees are specified, the target points are the flowers, where the sensors are to be deployed optimally for maximizing the coverage space or minimizing the energy consumption. An update is performed in the second phase with a loop in size of *M_iter*. In each iteration type of bee will be specified and then bees, according to their types, move in the state space with particular patterns of movement. The flow of the algorithm and the comparison of the optimized deployment have been thoroughly done in [16].

3 Load Balancing in Sensor Network

Optimal node deployment is very essential for both energy-efficient routing as well as increasing the lifetime of the sensor nodes. As the coverage nodes relay the information towards the base station it takes help of the connecting nodes (i.e. the relay nodes) to channel the information through proper route. Proper scheduling of data packets is done according to the choice of MAC. We know that MAC layer along with the data link control is responsible for physical addressing. The communication in WSN occurs from various source nodes coming from multiple different service providers. Technology is moving towards a single goal where all the devices are connected and can share information with them at ease. The sensor cloud environment is gaining the limelight where the users work on a more abstracted level possible through a mobile app or a browser and are least concerned about the sensor devices and their respective service providers. They are more concerned about the application and its usage [25]. Figure 11 depicts multiple integrated heterogeneous sensors attached to a cloud platform for information sharing.

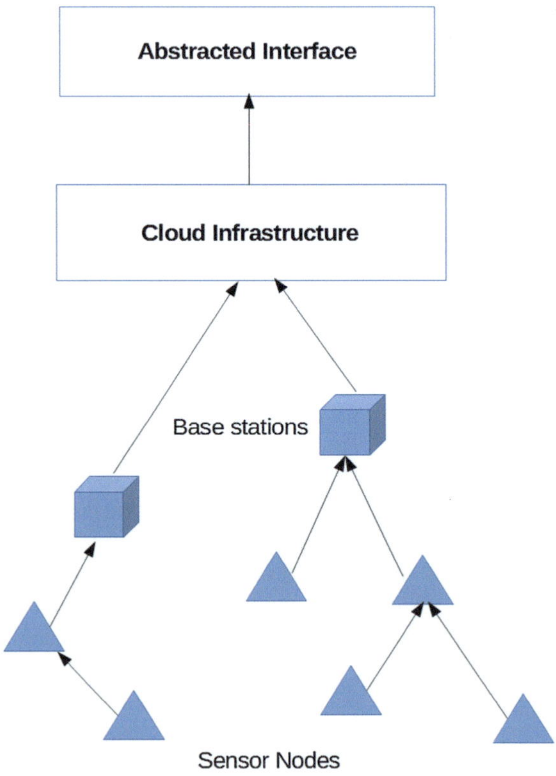

Fig. 11 WSN with cloud infrastructure

Integration among numerous sensor service providers adds more complexity in maintaining power consumption as the sensors have different energy utilities. Furthermore as the sensing tree size from the base station to the coverage nodes increases the number of slots required for data transmission also increases. An example of an unbalanced coverage tree from a base station is depicted in Fig. 12. Convergecasting is the most fundamental operation that any WSN performs. Data Aggregation, on the other hand, summarizes the data collection in an efficient manner. Usually converge casting happens over raw data or periodic data collection and accordingly, aggregation is also classified as no aggregation, partial aggregation, and full aggregation. Figure 12 is considered over data with partial aggregated converge casting where the TDMA protocol is used for data transmission. Data Packets are sent in each time slot, the only thing that is ensured is sending and receiving of packets takes place in different time slots hence no adjacent edge is given the same time slot and accordingly it is varied. It has been observed that as the tree size increases the number of time slots

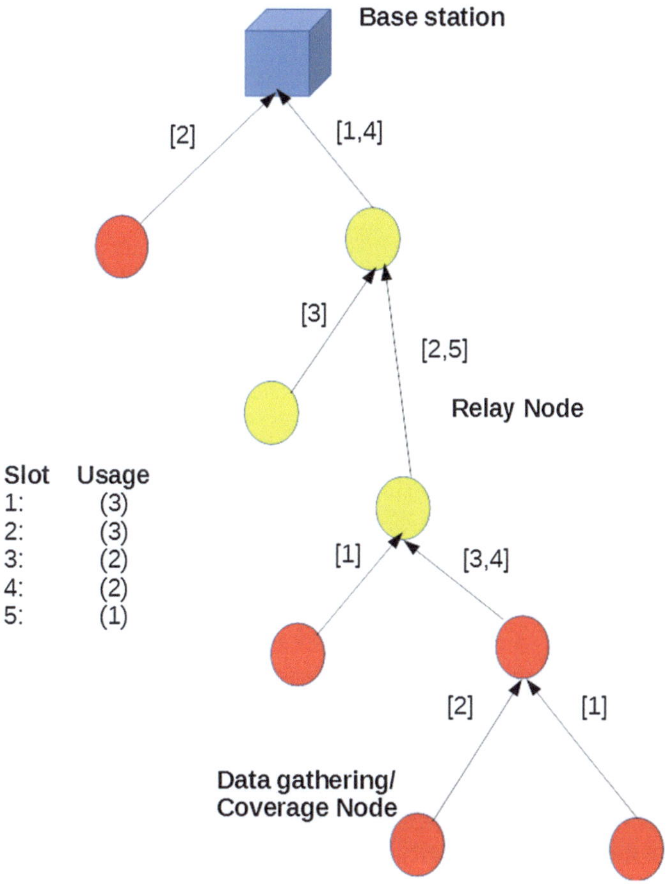

Fig. 12 Unbalanced coverage tree

also increases. In this example a payload value of 2 is used. That is a single time slot and carries 2 data packets. The nodes marked red are the coverage or data gathering nodes hence data they produce data while the nodes marked yellow are relay nodes. The duty of the relay node is to transfer the data collected by its underlying node towards the next hop to the base station (relay nodes do not generate data).

It is being observed that if the coverage sensors are well distributed among other base station then the number of slot usage would have been reduced. A simple way of distributing the coverage nodes equally is to calculate a proximity matrix of every coverage sensors to the base stations. Let the no of base station count for a particular Sensor Network be n_b, let n_c be the number of coverage sensor count. The maximum coverage allocation on any particular base station should be $\lceil n_c/n_b \rceil$.

Load balancing problem cannot be simply addressed by distributing the coverage sensors among the base station. It is also very important to distribute the task and select the best possible sensors for a particular application. The problem of distributing the load over the base station is a multiclass problem. One of the solutions to the problem is given by Zuhairy et al. In [17], where they have generalized the logistic regression model by predicting the sensor Ids. They have proposed a predicting model that chooses the best set of coverage sensors for a particular job, considering optimal network coverage for that particular application. The main idea behind the project is to increase the lifetime of the sensor network and perform dynamic load balancing according to application-specific need.

The task of the proposed algorithm is to select the sensors. Mathematically it can model as follows:

An observation from a past selection of sensor nodes for the respective task is D in which a given feature is said to have values in between $\{0, 1 \ldots K\}$.

Let $S = \left\{s_1, s_2, s_3 \ldots s_j\right\}$ be the sensors, the task of prediction is to choose the set

$E = \{s_i, s_{i+1}, \ldots, s_{i+n}\}$ in which $n >= 2$ and $E \subseteq S$. The probability that the ith sensor is picked for jth category is given by:

$$\Pr\{S_i = j\} = \beta_{ij} \tag{10}$$

AND

$$\sum_{j=1}^{J} \beta_{ij}, \quad \text{for all } i \in \{1, 2 \ldots n\} \tag{11}$$

The value β_{ij} represents the value of probability function Pr for the jth category. Equation 11 depicts that β_{ij} is distributed normally that is the sum of all probability values is equal to 1. The probability function for sensor selection is given in [17]. The optimized selection of sensor nodes is achieved through k clustering algorithm. As a user request the service for a particular task, first checking is performed whether the submitted task is a subtask within the other task in the request. A particular threshold is calculated that determines whether the remaining power of the cluster's sensor will be able to handle the task or not. If the remaining power is less than a certain threshold then choose an alternative sensor and proceed.

Another Load balancing solution is proposed Toumpis et al. Where they have proposed a way of achieving load balancing with the help of Kirchoff's voltage law in [18]. The analysis is based on the Wireless Minimum Cost Problem [19], where optimizing is performed with the help of an arc cost function to optimize the traffic flowing through nearby link. They have presented an algorithm that optimizes the flow by changing the traffic along cycles constructed during an initialization phase. The operation can be viewed as Kirchoff's Voltage law where energy is conserved that contains interferistors. It is robust, distributed, with minimal communication overhead, and converges quickly. As the traffic changes the algorithm they have proposed can also be used for congestion control.

In [20], The WSN nodes are divided into clusters of sensors the cluster member is selected by considering the maximum transmission power of the nodes, its membership depends on the communication cost. In paper [21], their previous work has been upgraded so that the choice of the cluster member is done by considering a comprehensive weight value composed of the distance between the cluster head member and the residual energy. For developing the cluster balance the algorithm considers load equalization.

In this paper [22], for intra- and inter-cluster communication layered approach is used. This algorithm considers a similar network. In this paper [23], network lifetime is increased by clustering the network heads. The cluster heads used the transmission range reconfiguration to balance the clusters that are based on the number of general nodes in the cluster and the number of cluster heads. They have modeled the algorithm to provide effective data aggregation.

4 Conclusion

The wireless sensor network is becoming an intricate part of our day to day life in many applications such as e-healthcare, farming, banking, etc. Although it has its limitations with respect to its processing power and energy utilization. Energy-efficient data gathering is one of the crucial tasks in any WSN design. Furthermore relaying and routing the information to its respective base station is again one of the problems that have to be addressed properly. During the design and development of protocols for data transmission various criteria such as congestion, interference, etc. has to be considered for energy-efficient routing. In order to increase the lifetime of the WSN, there are four major domains where researchers can contribute—Node deployment, Load balancing, Relaying and Routing. Researchers have worked and developed various models and designs for the very purpose of proper energy utilization in WSN, but still there is room for improvement.

References

1. Das SK, Samanta S, Dey N, Kumar R (2019) Design frameworks for wireless networks. In: Lecture notes in networks and systems. Springer, Berlin. pp 1–439. ISBN: 978–981-13-9573-4
2. Roy S et al (2016) Symmetric key encryption technique: a cellular automata based approach in wireless sensor networks. Procedia Comput Sci 78:408–414
3. Das SK, Tripathi S (2019) Energy efficient routing formation algorithm for hybrid ad-hoc network: a geometric programming approach. Peer-to-Peer Networking Appl 12(1):102–128
4. Sujoy S et al (2011) Post disaster management using delay tolerant network. In: Recent trends in wireless and mobile networks. Springer, Berlin, Heidelberg, pp 170–184
5. Mukherjee A et al (2019) Delay tolerant network assisted flying ad-Hoc network scenario: modeling and analytical perspective. Wirel Netw 25(5):2675–2695
6. Mukherjee A et al (2019) A disaster management specific mobility model for flying ad-hoc network. In: Emergency and disaster management: concepts, methodologies, tools, and applications. IGI Global, pp 279–311
7. Yadav AK, Das SK, Tripathi S (2017) EFMMRP: design of efficient fuzzy based multi-constraint multicast routing protocol for wireless ad-hoc network. Comput Netw 118:15–23
8. Das SK, Yadav AK, Tripathi S (2017) IE2M: design of intellectual energy efficient multicast routing protocol for ad-hoc net work. Peer-to-Peer Networking Appl 10(3):670–687
9. Sen BK, Khatua S, Das RK (2015) Target coverage using a collaborative platform for sensor cloud. In: 2015 IEEE international conference on advanced networks and telecommuncations systems (ANTS). IEEE
10. Ab Aziz NAB, Mohemmed AW, Daya Sagar BS (2007) Particle swarm optimization and Voronoi diagram for wireless sensor networks coverage optimization. In: 2007 international conference on intelligent and advanced systems. IEEE
11. Chew LP (1990) Building Voronoi diagrams for convex polygons in linear expected time
12. Zhang Q, Huang J, Wang J, Jin C, Ye J, Zhang W, Hu J (2008) A two-phase localization algorithm for wireless sensor network. In: 2008 international conference on information and automation. IEEE, pp 59–64
13. Dorigo M, Birattari M (2010) Ant colony optimization. Springer, US
14. Fidanova S, Marinov P, Alba E (2012) Ant algorithm for optimal sensor deployment. In: Computational intelligence. Springer, Berlin, Heidelberg, pp 21–29
15. Yuce B, Packianather M, Mastrocinque E, Pham D, Lambiase A (2013) Honey bees inspired optimization method: the bees algorithm. Insects 4(4):646–662
16. Hajizadeh N, Jahanbazi P, Javidan R (2018) Controlled deployment in wireless sensor networks based on a novel Multi Objective Bee Swarm Optimization algorithm. In: 2018 3rd conference on swarm intelligence and evolutionary computation (CSIEC). IEEE
17. Zuhairy RM, Al Zamil MG (2018) Energy-efficient load balancing in wireless sensor network: an application of multinomial regression analysis. Int J Distrib Sens Netw 14(3):1550147718764641
18. Toumpis S, Gitzenis S (2009) Load balancing in wireless sensor networks using kirchhoff's voltage law. In: IEEE INFOCOM 2009. IEEE, pp 1656–1664
19. Duan Q et al (2013) Minimum cost blocking problem in multi-path wireless routing protocols. IEEE Trans Comput 63(7):1765–1777
20. Gupta G, Younis M (2003) Load-balanced clustering of wireless sensornetworks. In: IEEE international conference on communications, 2003. ICC'03, vol 3. IEEE, pp 1848–1852
21. Zhang H, Li L, Yan X-F, Li X (2011) A load-balancing clustering algorithm of WSN for data gathering. In: 2011 2nd International conference on artificial intelligence, management science and electronic commerce (AIMSEC). IEEE, pp 915–918
22. Israr N, Awan I (2006) Multi-hop clustering algo. For load balancing in WSN. Int J Simul 8(1)
23. Kim N, Heo J, Kim HS, Kwon WH (2008) Reconfiguration of cluster heads for load balancing in wireless sensor networks. Comput Commun 31(1):153–159
24. Sarobin MVR, Ganesan R (2015) Swarm intelligence in wireless sensor networks: a survey. Int J Pure Appl Math 101(5):773–807

25. Choi M, Kim J, Yang S, Ha N, Han K (2008) Load balancing for efficient routing in wireless sensor networks. In: 2008 international multi-symposiums on computer and computational sciences. IEEE, pp 62–68
26. Nan G-F, Li M-Q, Li J (2007) Estimation of node localization with a real-coded genetic algorithm in WSNs. In: 2007 international conference on machine learning and cybernetics. vol 2. IEEE
27. Wang G, Cao G, Porta TL (2003) Movement-assisted sensor deployment. IEEE INFOCOM 2004 4:2469–2479
28. Dorigo M, Di Caro G (1999) Ant colony optimization: a new meta-heuristic. In: Proceedings of the 1999 congress on evolutionary computation-CEC99 (Cat. No. 99TH8406), vol 2. IEEE
29. Sim KM, Sun WH (2003) Ant colony optimization for routing and load-balancing: survey and new directions. IEEE Trans Syst Man Cybern-Part A: Syst Hum 33(5):560–572
30. Mnasri S, Thaljaoui A, Nasri N, Val T (2015) A genetic algorithm based approach to optimize the coverage and the localization in thewireless audio-sensors networks. In: 2015 international symposium on networks, computers and communications (ISNCC), pp 1–6
31. Kacimi R, Dhaou R, Beylot AL (2013) Load balancing techniques for lifetime maximizing in wireless sensor networks. Ad Hoc Netw 11(8):2172–2186
32. Mahdavi M, Ismail M, Jumari K (2009) Load balancing in energy efficient connected coverage wireless sensor network. In: 2009 international conference on electrical engineering and informatics, vol 2. IEEE, pp 448–452
33. Zeynali M, Khanli LM, Mollanejad A (2010) Fuzzy based approach for load balanced distributing database on sensor networks. Int J Future Gener Commun Networking 3(2)
34. Low CP, Fang C, Ng JM, Ang YH (2007) Load-balanced clustering algorithms for wireless sensor networks. In: 2007 IEEE international conference on communications. IEEE, pp 3485–3490
35. Dey N et al (eds) (2018) Internet of things and big data analytics toward next-generation intelligence. Springer, Berlin
36. Binh HTT, Nguyen TH, Dey N (2018) Improved cuckoo search and chaotic flower pollination optimization algorithm for maximizing area coverage in wireless sensor networks. Neural Comput Appl 30(7):2305–2317
37. Das SK, Tripathi S (2018) Adaptive and intelligent energy efficient routing for transparent heterogeneous ad-hoc network by fusion of game theory and linear programming. Appl Intell 48(7):1825–1845

Chapter 12
Bio-inspired Algorithm for Multi-objective Optimization in Wireless Sensor Network

Anindita Raychaudhuri and Debashis De

1 Introduction

In recent days, the domain of bio-inspired computing is gradually getting recognition in several application areas including Wireless Sensor Networks (WSN). Researchers are working for several years to resolve issues of WSN for better quality of service in a variety of applications [1–5]. Bio-inspired algorithms inspired by the striking behavior of animals, birds, insects, plants, etc. can be used in various application areas including WSN. In this paper, some of the recent bio-inspired algorithms namely ACO, Artificial bee colony, Bat algorithm, Biogeography-based Optimization (BBO), Cat Swarm Optimization (CSO), Cuckoo search algorithm, Chicken Swarm Optimization Algorithm (CSOA), Elephant herding Optimization (EHO), Moth flame Optimization Algorithm, Fish Swarm Optimization, Grey Wolf Optimization (GWO) algorithm, PSO, GSO, Whale Optimization Algorithm (WOA), etc. have been presented and analyzed. The application areas of each algorithm have been discussed. The objective of this paper is to provide the insight story of the different bio-inspired algorithms to the readers so that they can take on the challenges to solve key problems with the help of a bio-inspired algorithm. Table 1 summarizes the recent publications using above mentioned bio-inspired algorithm in different popular journals.

A. Raychaudhuri (✉)
Department of Computer Science, Sarojini Naidu College for Women,
30 Jessore Road, Kolkata 700028, West Bengal, India
e-mail: aninrc@gmail.com

D. De
Department of Computer Science and Engineering, Maulana Abul Kalam Azad
University of Technology, Haringhata, Nadia 721249, West Bengal, India
e-mail: dr.debashis.de@gmail.com

D. De et al. (eds.), *Nature Inspired Computing for Wireless Sensor Networks*,
Springer Tracts in Nature-Inspired Computing,
https://doi.org/10.1007/978-981-15-2125-6_12

Table 1 Recent publications based on the major bio-inspired algorithms

S. No.	Algorithm	Principal authors	Subject areas	Popular journals	Year of publication
1	Ant colony optimization (ACO)	Hashem-Parvin et al.	Trust	Expert systems with applications	2019
		Li, Y. et al.	Green vehicle routing	Journal of cleaner production	2019
		Sun, Z et al.	Secure routing for WSN	Applied soft computing	2019
		Wang, J. et al.	Energy-efficient WSN	The Journal of supercomputing	2018
		Guleria, K. et al.	Clustering for WSN	Wireless personal communications	2019
		Ghosh, N.et al.	Mobile data collection in WSN	Wireless networks	2019
2	Artificial bee colony (ABC)	Yue, Y. et al.	Coverage and connectivity of WSN	Wireless personal communications	2019
		Lu, Y. et al.	Energy efficiency in WSN	IEEE access	2019
		Mann, P.S. et al.	Energy-efficient clustering in WSN.	Artificial intelligence review	2019
		Saad, E. et al.	Efficiency and robustness	Applied soft computing	2019
		Zhang, X. et al.	Energy-efficient internet of things network	IEEE access	2019
3	Bat algorithm (BA)	Hong, W.C.et al.	Platform motion forecasting	Applied mathematical modeling	2019
		Osaba, E. et al.	Vehicle routing Problem	Swarm and evolutionary computation	2019
		Ng, C.K.et al.	Sensor deployment	IEEE communications letters	2018
		Lyu, S.et al.	Optimization and convergence	Journal of computational science	2019
		Sharma, S. et al.	Routing for WSN	In Soft computing and signal processing	2019
		Cui, Z. et al.	Big data sensing systems in internet of things	Journal of parallel and distributed computing	2018
4	Biogeography-based optimization (BBO)	Gupta, G.P. et al.	Coverage and connectivity in WSN	Wireless networks	2018

(continued)

Table 1 (continued)

S. No.	Algorithm	Principal authors	Subject areas	Popular journals	Year of publication
		Lalwani, P., et al.	Energy saving routing for WSN	Soft computing	2018
		Elhoseny, M.et al.	Coverage optimization	Expert systems with applications	2018
		Senniappan, V. et al.	Energy-efficient clustering in WSN	Journal of ambient intelligence and smart environments	2018
5	Cat swarm optimization (CSO)	Tsai, P.W.	Numerical optimization problem	Expert systems with applications	2012
		Temel, S. et al.	Maximize coverage in WSN	IEEE transactions on systems	2013
		Kong, L. et al.	Energy-aware routing for WSN	In advanced technologies, embedded and multimedia for human-centric computing	2014
		Kong, L. et al.	Minimize power consumption in WSN	International journal of distributed sensor networks	2015
		Soto, R. et al.	Solve manufacturing cell design problem	Computational intelligence and neuroscience	2019
6	Cuckoo search algorithm	Ghosh,A. et al.	Signal attenuation minimization for WSN	Engineering optimization	2019
		Yu, X. et al.	Node localization in WSN	Wireless personal communications	2019
		Meng, X., et al.	Multi-objective hydropower station operation	Energy	2019
		Chi, R.et al.	Hybrid optimization problem	Neural computing and applications	2019
		Wu, Z. et al.	Electrical load forecasting	Applied energy	2019
7	Chicken Swarm optimization algorithm (CSOA)	Yu, X.et al.	Localization in WSN	Computer networks	2019
		Aziz, A. et al.	Optimize energy consumption in IoT based sensor network	Journal of network and computer applications	2019

(continued)

Table 1 (continued)

S. No.	Algorithm	Principal authors	Subject areas	Popular journals	Year of publication
		Movva, P. et al.	Energy-efficient routing in WSN.	IEEE access	2019
		Al Shayokh, M. et al.	WSN node localization problem	Wireless personal communications	2017
8	Elephant herding optimization (EHO)	Correia, S. et al.	Energy-based localization in WSN	Sensors	2018
		Strumberger, I., et al.	Node localization in WSN	Sensors	2019
		Tuba, E.et al.	Data clustering	Information and communication technology for intelligent systems	2019
		Li, J.et al.	Global optimization	Mathematics	2019
9	Fish swarm optimization algorithm (FSOA)	Zheng, Z.X.et al.	Minimize power consumption of chillers and cooling towers	Mathematics and computers in simulation	2019
		Qin, N. et al.	Mobile coverage in WSNs	Wireless communications and mobile computing	2018.
		Li, X., et al.	Energy-efficient hybrid routing protocol for WSN	Sensors	2018
		Yin, H., et al.	WSN nodes placement optimization	Algorithms	2018
10	Grey wolf optimization (GWO) algorithm	Kaushik, A. et al.	network stability and energy efficiency	Wireless personal communications	2019
		Kaushik, A., et al.	coverage maximization	Wireless personal communications	2019
		Zapotecas-Martínez, S. et al.	Multi-objective optimization	Expert systems with applications	2019
		Tu, Q. et al.	Feature selection	Applied soft computing	2019.
11	Glowworm swarm optimization (GSO)	Liao, W.H. et al.	sensor deployment	Expert systems with applications	2011
		Ray, A. et al.	Energy efficiency in WSN	Simulation modeling practice and theory	2016

(continued)

Table 1 (continued)

S. No.	Algorithm	Principal authors	Subject areas	Popular journals	Year of publication
		Wang, Y., et al.	Multi-objective optimization	Algorithms	2019
		Salkuti, S.R. et al.	Congestion management	Journal of electrical engineering and technology	2019
		Song, L. et al.	Node location in WSN	Journal of sensors	2019
12	Moth flame optimization(MFO) algorithm	Miloud, M. et al.	Node localization in decentralized WSN	International journal of distributed systems and technologies	2019
		Mittal, N.et al.	Energy efficiency in WSN	Wireless personal communications	2019
		Khan, M.F., et al.	Clustering algorithm for the internet of vehicle	IEEE access	2019
13	Particle Swarm Optimization (PSO)	Jordehi, A.R.	Optimal scheduling of home appliances	Applied soft computing	2019
		Nouiri, M. et al.	Job-shop scheduling	Journal of intelligent manufacturing	2018
		Lynn, N. et al.	Multiobjective optimization	Swarm and evolutionary computation	2018
		Tam, N.T. et al.	Improve lifetime and network in WSN	wireless networks	2018
		Vijayalakshmi, K. et al.	Cluster head selection in WSN	Cluster computing	2018
		Kaur, T. et al.	Fault tolerant WSN	IEEE sensors journal	2018
14	Whale optimization algorithm (WOA)	Ahmed, M.M. et al.	Maximizing lifetime of large-scale WSN	Telecommunication Systems.	2019.
		ValayapalayamKittusamy, S.R., et al.	vehicular communication networks	International journal of communication systems	2019.
		Hassan, M.K. et al.	Real-time remote monitoring frameworkfor patients	Future generation computer systems	2019
		Verma, G.K. et al.	Repair partitioned heterogeneous WSN	Journal of grid and distributed computing	2018

2 Review of Bio-Inspired Algorithms

Day by day bio-inspired algorithms are becoming increasingly popular for its high quality of solutions, flexibility, accuracy, ease of implementation and high convergence speed [6–8]. Some of the popular bio-inspired algorithms which are applied in several application areas especially wireless sensor network are presented in brief.

2.1 *Ant Colony Optimization (ACO)*

Ant Colony Optimization is a bio-inspired algorithm based on the behavior of real ants. It has been observed that a group of ants can cooperatively figure out the shortest route between their nest and their food. The communication between ants happens through volatile chemical substances known as pheromones. Algorithm 1 describes the ACO algorithm [9].

Algorithm 1: The Ant Colony Optimization Algorithm

Represent the solution space by a construction graph
Set ACO parameters and initialize pheromone trails
While termination condition are not met

    Construct Ant solutions
    Apply local search (optional)
    Update pheromones

end while

2.2 *Artificial Bee Colony (ABC)*

It is an important bio-inspired algorithm inspired by the collective foraging behavior of honey bees [11]. Like ACO it also searches for an optimal numerical solution among large data sets. Through the social cooperation of three types of bees namely employed bees, onlooker bees, and scout bees they can complete their task. The employed bees take the responsibility of searching food around the food source and share the information with the onlooker bees and finally these types of bees select good food sources from those discovered by the employed bees. ABC can address the multi-dimensional numeric problem [12–16].

2.3 Bat Algorithm (BA)

Bat algorithm is a bio-inspired algorithm developed by Xin-She Yang in 2010, inspired by the echolocation features of microbats [17]. Bat is the only mammal that has the capability of sensing echolocation that helps them to find the prey. Bat algorithm has the following three idealized rules:

i. All bats apply echolocation property. By sensing the reflection of sound they can easily differentiate between different objects like food, prey and background barriers.
ii. Bats can naturally adjust the frequency of their radiated pulses and regulate the rate of pulse emissions based on the proximity of their target.
iii. The loudness can vary from a large positive value to a minimum constant value.

According to the basic Bat algorithm each bat is associated with a velocity v_i^t and a location x_i^t at iteration t in a d-dimensional solution space. Among all the bats there exists a current best solution x_*. Initially each bat is randomly assigned a frequency which is uniformly from $[f_{\min}, f_{\max}]$. The following three equations update the frequency, velocity and location as follows:

$$f_i = f_{\min} + (f_{\max} - f_{\min})\beta \text{ where } \beta \in [0, 1] \tag{1}$$

$$v_i^t = v_i^{t-1} + \left(x_i^{t-1} - x_*\right) f_i \tag{2}$$

$$x_i^t = x_i^{t-1} + v_i^t \tag{3}$$

Bat algorithm is one of the important bio-inspired algorithms which are gaining popularity day by day for its simple working principle and efficiency. Since the development of Bat algorithm it has been applied in more or less every area of optimization including clustering, image processing, feature selection, medical data analysis [18], coverage optimization and connectivity preservation of wireless sensor network [19–26], etc. but it is very useful for complex high dimensional problem where convergence is a challenge. As a metaheuristic optimization algorithm, it has great importance because of its usefulness in a variety of real-world applications.

2.4 Biogeography-Based Optimization (BBO)

BBO is based on the concept of biogeography which is conceived of the geographical distribution of biological species. It is an algorithm inspired by biological evolution which optimizes a function through the iterative refinement of candidate solutions concerning a given measure of fitness function. BBO was motivated by the natural process and was originally proposed by Dan Simon in 2008 [21]. It is well known to have a fairly good performance to solve continuous problems in a variety of applications [28–32].

2.5 *Cat Swarm Optimization (CSO)*

It is a swarm optimization algorithm inspired by the two common behavior of cats known as Seeking and Tracing [33]. Seeking determines where to move next based on the higher fitness value of choosing points. In Tracing Mode cat traces some target. Both Seeking and Tracing mode will be continuously used by the cat until a terminating condition is met. In the last few years, CSO algorithm has been used to find an optimal solution for some applications [34–38].

2.6 *Cuckoo Search Algorithm*

Cuckoo birds usually lay their eggs in the nests of other host birds. Based on this natural behavior Xin-she Yang and Suash Deb developed a bio-inspired optimization algorithm known as the Cuckoo search algorithm in 2009 [39]. According to this algorithm, each egg in a nest represents a solution whereas a new solution is represented by a cuckoo egg. The purpose of this algorithm is to apply the novel and potentially better solutions to supersede a not-so-good solution in the nests [40–46].

2.7 *Chicken Swarm Optimization Algorithm (CSOA)*

It is a bio-inspired algorithm that mimics the behaviors of the chicken swarm, including roosters, hens, and chicks [47]. CSOA can efficiently implement the chickens' swarm intelligence property to optimize problems [48–52].

2.8 *Elephant Herding Optimization (EHO)*

It is one of the most recent bio-inspired algorithms proposed by Wang et al. [53]. EHO algorithm is based on the herding behavior of elephants. The populations of elephants are composed of some clans. In each clan, the elephants live together under the leadership of one matriarch. In general, the matriarch is considered to be the eldest one or the fittest one in the clan. In each generation, a fixed number of male elephants leave their group and live far away. These two main behaviors can be mapped in two basic steps named "clan updating operator" and "separating operator" which can be used for global optimization [54–58].

2.9 Fish Swarm Optimization Algorithm (FSOA)

In recent days optimization techniques based on the collective activities of social species have been massively used for solving a variety of real-world optimization problems. The social behavior of fish colonies like food searching, dealing with dangers has been explored to develop a novel algorithm named Fish Swarm Optimization Algorithm [59]. According to the researchers, this algorithm has many advantages including high accuracy, flexibility, fault tolerance and high convergence speed [60–64].

2.10 Grey Wolf Optimization (GWO)

A new metaheuristic called Grey Wolf Optimizer (GWO) inspired by grey wolves have been proposed by Mirjalili et al. in [65]. The GWO algorithm mimics the leadership hierarchy and hunting mechanism of grey wolves in nature [66].

2.11 Glowworm Swarm Optimization (GSO)

This algorithm has been introduced by Krishnanand and Ghose in 2005 inspired by the behavior of glowworm [67]. Though the algorithm shares a few features with a well-known swarm intelligence based algorithm known as ant colony optimization (ACO), significant differences exist. Researchers have proved that the GSO algorithm can provide efficient solutions for the optimization of multiple optima continuous functions [68, 69].

In the GSO algorithm [68–70] every glowworm determines a probabilistic value for selecting its adjacent glowworm having a higher intensity of luciferin than its own. The glowworm decides to move towards the chosen glowworm based on this probability value. For each glowworm i, the probability of movement towards a neighbor k is as follows,

$$p_{ik}(t) = \frac{f_k(t) - f_i(t)}{\sum_{j \in G_i(t)} f_j(t) - f_i(t)} \quad (4)$$

Here $f_i(t)$ = luciferin level of glowworm i at time t.
$G_i(t)$ = Set of neighbors of glowworm i at time t.

2.12 Moth Flame Optimization (MFO) Algorithm

It is a novel nature-inspired optimization paradigm based on the navigation method of moths which was proposed by Mirjalili in [71]. This type of navigation method of moths in nature is known as transverse orientation. MFO algorithm has been widely used in both science and industry for its usefulness [72, 73].

2.13 Particle Swarm Optimization (PSO)

Particle Swarm Optimization is one of the most popular metaheuristic algorithms used for optimization. PSO was introduced by Kennedy, J. and Eberhart, R in 1995 [74]. Like other metaheuristic algorithms, PSO optimizes a problem by iteratively trying to improve a candidate solution for the given fitness function. Algorithm 1 describes PSO algorithm.

Algorithm 2: PSO Algorithm

Step1: Initialize particles population
Step2: Calculate fitness values of each particles using fitness function
Step3: Update P_{id} if the current fitness value is better than P_{id}
Step4: Determine P_{gd}, the particle position with its best fitness value of all the neighbors
Step5: Update the velocities of all the particles using Eq. (5)
Step6: Update position of the particles using Eq. (6)
Step7: Repeat steps 2–6 until maximum iteration or ideal fitness is not attained.

PSO algorithm is essentially described by the velocity and position update equation as follows:

$$V_{id}(t+1) = V_{id}(t) + c_1 R_1 (P_{id}(t) - X_{id}(t)) + c_2 R_2 \big(P_{gd}(t) - X_{id}(t)\big) \quad (5)$$

$$X_{id}(t+1) = X_{id}(t) + V_{id}(t+1) \quad (6)$$

where

V_{id} = Velocity of the ith particle in the dth dimension and t denotes the iteration counter.
X_{id} = Position of the ith particle.
P_{id} = Best position of the ith particle in the dth dimension.
P_{gd} = Particle position with its best fitness value of all the neighbors.
R_1 and R_2 are two n-dimensional vectors with random numbers uniformly selected in the range of [0.0, 1.0].
C_1 and C_2 are positive constant weighting parameters also called the cognitive and social parameters, respectively.

Initially, PSO was intended for simulating the social behavior of birds and fish. But due to its high quality of solutions, computational efficiency and speed of convergence, PSO has been enormously applied in several application areas including wireless sensor networks [74–86]. The efficacy of the PSO algorithm is due to the interaction of particles with its neighbors.

2.14 Whale Optimization Algorithm (WOA)

In recent days more and more modern metaheuristic algorithms are emerging. The Whale optimization algorithm is one of the most recently proposed metaheuristic optimization algorithms introduced by Mirjalili, S. and Lewis, A. in 2016 [87]. WOA mimics the hunting mechanism of humpback whales in nature. In this short period due to its easy implementation, this algorithm has been used in a wide range of problems in several domains including wireless sensor networks [88–93] (Fig. 1).

3 Domains of Applications

Bio-inspired Computing has emerged as one of the most studied research topics in recent days. During the last decades, hundreds of novel approaches have been recorded [94]. The growing attention in this topic is due to the adaptive nature of different bio-inspired algorithms which produces near-optimal solution over a wide range of real-world applications covering all most all areas including, production engineering, computer networks, biomedical engineering, data mining, security, power systems and many more. Some of the interesting applications in different domains using some important bio-inspired algorithms are summarized in this section.

Ant colony optimization is one of the most studied bio-inspired algorithms which attracted a great number of researchers. As a result, a significant number of research papers are available in a variety of application areas like scheduling, digital image processing and power engineering [10]. It has a significant contribution in the area of clustering, security, data collection and routing algorithm of WSN [95–101]. During the last decades, artificial bee colony (ABC) proves to be a promising algorithm due to its potential advantage of being hybridized to different metaheuristic algorithms which makes it more robust. As a result, researchers have published a considerable amount of research papers in diverse application areas [102].

Researchers have already proved that bio-inspired algorithms like Bat algorithm, artificial bee colony, FSO, PSO, and cuckoo search algorithm are playing a major role in medical data analysis [18, 103, 104]. A novel hybrid approach based on discrete particle swarm optimization and statistical method shows a fairly good result for excavating breast cancer patterns [78].

Fig. 1 Bio-inspired algorithms useful for wireless sensor network

Amongst the other bio-inspired algorithms, BBO, CSO, EHO [105], Grey Wolf Optimization (GWO) algorithm don't have a major contributions in medical data analysis but they are contributing significantly in a variety of domains [106–110]. Though Chicken swarm optimization algorithm, Glowworm swarm optimization (GSO) algorithms are yet not applied in the area of medical science they are playing important role in other areas of optimization including clustering, coverage, and connectivity of wireless sensor network [111–118]. Also, Moth flame optimization (MFO) Algorithm has a significant number of applications in different areas [119–121].

We have analyzed the usage of bio-inspired algorithms based on the number of published publications across various subject domains and summarized in Table 2. In the time of real-world application development, one of the big matters of question is which algorithm to choose in which type of problem domain. Figure 2 shows

Table 2 Usage of bio-inspired algorithm across various subject domains

S. No.	Bio-inspired algorithm	Usage in various subject areas			
		applied science	Engineering	Medical science	Multidisciplinary
1	Ant colony optimization (ACO)	Medium	High	Low	Low
2	Artificial bee colony	High	High	Medium	Medium
3	Bat algorithm	Medium	Medium	Medium	Medium
4	Biogeography-based optimization (BBO)	Medium	Low	Very low	Low
5	Cat swarm optimization (CSO)	Low	Medium	Low	Low
6	Cuckoo search algorithm	Low	Medium	Medium	Low
7	Chicken Swarm optimization algorithm (CSOA)	Low	Low	NO usage till date	Very low
8	Elephant herding optimization (EHO)	Low	Medium	Low	Low
9	Fish swarm optimization	Medium	Very high	Medium	Low
10	Grey wolf optimization (GWO) algorithm	Low	Medium	Low	Very low
11	Glowworm swarm optimization (GSO)	Low	Medium	NO usage till date	Very low
12	Moth flame optimization algorithm	Low	Low	Very low	Very low
13	Particle swarm optimization (PSO)	High	High	Medium	Medium
14	Whale optimization algorithm (WOA)	Low	Low	Very low	Very low

the advantages and limitations of different bio-inspired algorithms which help us to choose the right algorithm.

4 Application of Bio-Inspired Algorithms in Different Areas of Wireless Sensor Network

During the last decades, Wireless Sensor Networks are playing a very important role in all application areas. Some crucial issues of WSN are clustering, optimal routing, dynamic allocation of motes, energy and lifetime optimization [31, 84, 90,

Bio-inspired Algorithm	Advantages	Limitations
Ant Colony Optimization (ACO)	• Continuous optimization problem solving • Can be used in dynamic application	• Fault tolerance • slow convergence
Artificial bee colony (ABC)	• Single and multi-objective optimization • Coverage, connectivity, routing in WSN	• Scalability • Slow convergence
Bat Algorithm (BA)	• Complex high-dimensional problem solving capability where convergence is a challenge. • Network path analysis • Adaptive learning problems in WSN.	• Scalability • Fault tolerance
Biogeography-based optimization (BBO)	• optimization of multidimensional real-valued functions	• Scalability • Fault tolerance
Cat Swarm Optimization (CSO)	• Numerical optimization problem solving • Energy aware routing • Coverage optimization in WSN	• Scalability • Fault tolerance
Cuckoo search algorithm	• Multi-objective allocation and scheduling. • Signal attenuation minimization • Node localization for WSN	• Scalability • Fault tolerance
Chicken Swarm optimization Algorithm (CSOA)	• Feature selection • Energy efficient routing • Node localization in WSN.	• Scalability • Fault tolerance
Elephant herding optimization (EHO)	• Global optimization, • Data clustering • Energy efficient node localization	• Fault tolerance
Fish Swarm Optimization (FSO)	• Job scheduling, • Clustering, energy efficient routing & coverage in WSN.	• Fault tolerance
Grey Wolf Optimization (GWO) algorithm	• Feature selection • Multi-objective optimization • Network stability • Energy efficiency in WSN	• Fault tolerance
Glowworm swarm optimization (GSO)	• Multi-objective optimization, • energy and coverage optimization of WSN	• Scalability • Fault tolerance
Moth flame optimization Algorithm	• Parameter estimation and extraction • Node Localization • Energy optimization in decentralized WSN.	• Scalability • Benchmark functions
Particle Swarm Optimization (PSO)	• Optimal scheduling • Multi-objective problem solving capability • maximize Life time and Network of WSN • Fault tolerant WSN.	• Requires large amount of memory • Not effective for high speed real time applications
Whale optimization algorithm (WOA)	• Suitable for vehicular communication networks • maximize lifetime of large-scale WSN	• Fault tolerance

Fig. 2 Advantages and limitations of the different bio-inspired algorithm

122]. Bio-inspired algorithms played an important role in a variety of fields including wireless sensor networks during the last decade [12–121]. The contributions of some important bio-inspired algorithms in different areas of WSN are summarized in Table 3. These algorithms can be used in various parameter optimizations in WSN which are as follows:

Table 3 Contribution of Bio-inspired algorithm in different areas of WSN

Serial No.	Algorithm	Different area of WSN				
		Energy optimization	Coverage	Communication issues/Connectivity	Fault-tolerant/Security	Network lifetime
1	ACO	✓	✓	✓	✓	✓
2	Artificial bee colony	✓	✓	✓	✓	✓
3	Bat algorithm	×	✓	✓	×	×
4	Biogeography-based optimization (BBO)	✓	✓	✓	×	✓
5	Cat swarm optimization (CSO)	✓	✓	✓	×	✓
6	Cuckoo search algorithm	×	✓	✓	×	×
7	Chicken swarm optimization algorithm (CSOA)	✓	✓	✓	×	✓
8	Elephant herding optimization	✓	✓	×	×	✓
9	Fish swarm optimization algorithm (FSOA)	✓	✓	✓	×	✓
10	Grey wolf optimization (GWO) algorithm	✓	✓	×	×	✓
11	GSO	✓	✓	✓	×	✓
12	PSO	✓	✓	✓	✓	✓
13	Moth flame optimization algorithm	✓	×	✓	✓	✓
14	Whale optimization algorithm (WOA)	✓	✓	✓	×	✓

(i) For finding the optimal path in routing.
(ii) For minimizing the complexity of problems.
(iii) Enhances security in Wireless networks by detecting the faulty nodes in a wireless network.
(iv) It optimizes energy consumption.
(v) It helps in increasing the network's lifetime.

5 Challenges and Key Issues of Bio-Inspired Computing

Different numerical simulations and observations have revealed that bio-inspired algorithms can work remarkably well, but in most cases, we don't have a clear idea about why they perform under certain circumstances for a particular kind of problem. Although there are some improvements in the theoretical analysis the mathematical insight of their working principles is still not fully discovered. Hence it is highly required to use more systematic approaches to analyze a bio-inspired algorithm for the estimation of their rates of convergence and can help to select the correct algorithm for a certain type of problem to resolve it competently.

Some key issues of Bio-inspired computing are summarized in Fig. 3. This gives direction to the researchers about the possible future research in this topic.

6 Bio-Inspired Computation and Its Future

In bio-inspired computation, it is very hard to decide whether we need to refine methods revealed so far by the community, or instead search for a new bio-inspired algorithm. We have made a classification of bio-inspired algorithms based on our literature review. This classification will give some guidelines to the researchers to choose algorithms in the future. The algorithms are mainly classified into four classes having different scope for applications as shown in Table 4.

In class 1 type of algorithms, we observed that a lot of opportunities are present in terms of theoretical development of the algorithm. Also, there are plenty of opportunities for doing exhaustive comparative analysis amongst the algorithms for deep understanding. In class 2 type of algorithms, what we observed that these algorithms are fairly matured in terms of theory development. Therefore, it would be really exciting to apply them in different areas like industrial engineering, financial management and supply chain in novel engineering and commercial perspectives. Class 3 represents the algorithms, which were introduced, but in some way failed to draw the attention of the research community. It may be interesting to try and mix these

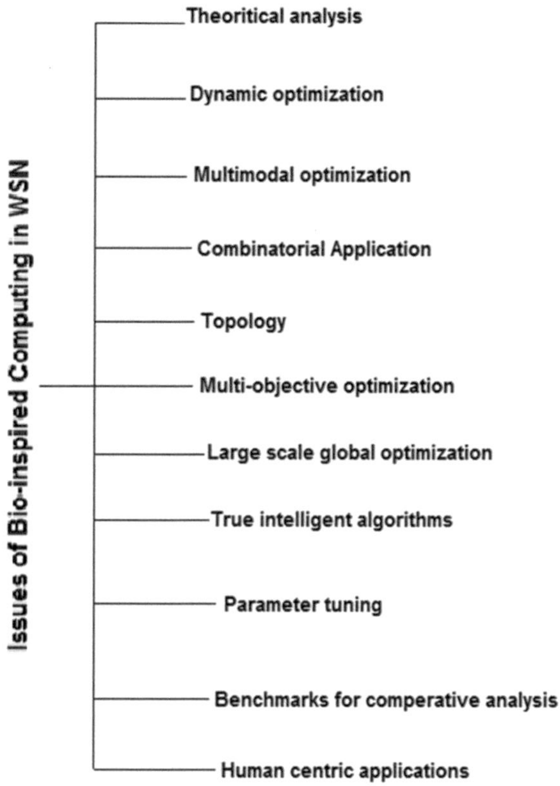

Fig. 3 Issues of bio-inspired computing

Table 4 Classification of Bio-inspired algorithms based on scope of research

Class 1 Scope of theoretical development and comparison	Class 2 Theoretically matured, need to apply in multidisciplinary areas
• *Grey wolf optimization (GWO) algorithm* • *Glowworm swarm optimization (GSO)* • *Moth flame optimization (MFO) algorithm* • *Whale optimization algorithm (WOA)*	• *Artificial bee colony* • *Bat Algorithm* • *Biogeography-based optimization (BBO)* • *Cuckoo search algorithm*
Class 3 Need rediscovery	Class 4 Adopted by scientific community
• *Cat swarm optimization (CSO)* • *Chicken swarm optimization algorithm (CSOA)* • *Elephant herding optimization (EHO)*	• *Ant colony optimization (ACO)* • *Fish swarm optimization (FSO)* • *Particle swarm optimization (PSO)*

algorithms with other concepts and explore the consequences. Class 4 represents algorithms that have been undoubtedly accepted and implemented by a large group of scientific community. A lot of studies have already been implemented using these algorithms in a variety of application areas.

7 Conclusion

In this chapter, some bio-inspired algorithms and its applications in a variety of field's especially wireless sensor networks are reviewed. It has been observed that these algorithms are playing a major role not only in WSN but also in some other important fields like medical data analysis. The role of Bio-inspired algorithms has been illustrated in achieving the quality of service in wireless sensor networks for several important parameters like energy, security, network life, and coverage. We have briefly reviewed these algorithms with respect to its applications starting from its inception to the recent one. In addition, we have classified these algorithms based on their scope of research which will guide the researchers to choose the right algorithm and attain the objective in the future. Even though bio-inspired computation has gained great popularity, still some of the algorithmic properties such as the rate of convergence, computational complexity need to be proved through rigorous mathematical formulation.

References

1. Das SK, Samanta S, Dey N, Kumar R (2020) Design frameworks for wireless networks. In: Lecture notes in networks and systems. Springer, pp 1–439. ISBN: 978-981-13-9573-4
2. Mukherjee A, Dey N, Kausar N, Ashour AS, Taiar R, Hassanien AE (2019) A disaster management specific mobility model for flying ad-hoc network. In: Emergency and disaster management: concepts, methodologies, tools, and applications. IGI Global, pp 279–311
3. Das SK, Tripathi S (2018) Intelligent energy-aware efficient routing for MANET. Wirel Netw 24(4):1139–1159
4. Das SK, Tripathi S (2017) Energy efficient routing formation technique for hybrid ad hoc network using fusion of artificial intelligence techniques. Int J Commun Syst 30(16):e3340
5. Dey N, Ashour AS, Shi F, Fong SJ, Sherratt RS (2017) Developing residential wireless sensor networks for ECG healthcare monitoring. IEEE Trans Consum Electron 63(4):442–449
6. Darwish A (2018) Bio-inspired computing: algorithms review, deep analysis, and the scope of applications. Future Comput Inf J 3(2):231–246
7. Kar AK (2016) Bio inspired computing–a review of algorithms and scope of applications. Expert Syst Appl 59:20–32
8. Del Ser J, Osaba E, Molina D, Yang XS, Salcedo-Sanz S, Camacho D, Das S, Suganthan PN, Coello CAC, Herrera F (2019) Bio-inspired computation: where we stand and what's next. Swarm Evol Comput 48:220–250
9. Dorigo M, Birattari M (2010) Ant colony optimization. Springer, US, pp 36–39
10. Mohan BC, Baskaran R (2012) A survey: ant colony optimization based recent research and implementation on several engineering domain. Expert Syst Appl 39(4):4618–4627

11. Karaboga D (2005) An idea based on honey bee swarm for numerical optimization. Technical report-tr06, Erciyesuniversity, engineering faculty, computer engineering department, vol 200, pp 1–10
12. Yue Y, Cao L, Luo Z (2019) Hybrid artificial bee colony algorithm for improving the coverage and connectivity of wireless sensor networks. Wirel Pers Commun 1–14
13. Lu Y, Sun N, Pan X (2019) Mobile sink-based path optimization strategy in wireless sensor networks using artificial bee colony algorithm. IEEE Access 7:11668–11678
14. Mann PS, Singh S (2019) Improved artificial bee colony metaheuristic for energy-efficient clustering in wireless sensor networks. Artif Intell Rev 51(3):329–354
15. Saad E, Elhosseini M, Haikal AY (2019) Culture-based Artificial Bee Colony with heritage mechanism for optimization of wireless sensors network. Appl Soft Comput
16. Zhang X, Zhang X, Han L (2019) An energy efficient internet of things network using restart artificial bee colony and wireless power transfer. IEEE Access 7:12686–12695
17. Yang XS (2010) A new metaheuristic bat-inspired algorithm. In: Nature inspired cooperative strategies for optimization (NICSO 2010). Springer, Berlin, Heidelberg, pp 65–74
18. Menad H, Amine A (2018) Bio-inspired algorithms for medical data analysis. In: Handbook of research on biomimicry in information retrieval and knowledge management. IGI Global, pp 251–275
19. Hong WC, Li MW, Geng J, Zhang Y (2019) Novel chaotic bat algorithm for forecasting complex motion of floating platforms. Appl Math Model
20. Osaba E, Yang XS, FisterJr I, Del Ser J, Lopez-Garcia P, Vazquez-Pardavila AJ (2019) A discrete and improved bat algorithm for solving a medical goods distribution problem with pharmacological waste collection. Swarm Evol Comput 44:273–286
21. Ng CK, Wu CH, Ip WH, Yung KL (2018) A smart bat algorithm for wireless sensor network deployment in 3-D environment. IEEE Commun Lett 22(10):2120–2123
22. Lyu S, Li Z, Huang Y, Wang J, Hu J (2019) Improved self-adaptive bat algorithm with step-control and mutation mechanisms. J Comput Sci 30:65–78
23. Sharma S, Verma S, Jyoti K (2019) A new bat algorithm with distance computation capability and its applicability in routing for WSN. In: Soft computing and signal processing. Springer, Singapore, pp 163–171
24. Cui Z, Cao Y, Cai X, Cai J, Chen J (2018) Optimal LEACH protocol with modified bat algorithm for big data sensing systems in Internet of Things. J Parallel Distrib Comput
25. Gandomi AH, Yang XS, Alavi AH, Talatahari S (2013) Bat algorithm for constrained optimization tasks. Neural Comput Appl 22(6):1239–1255
26. Yang XS (2013) Bat algorithm: literature review and applications. arXiv preprint arXiv:1308.3900
27. Simon D (2008) Biogeography-based optimization. IEEE Trans Evol Comput 12(6):702–713
28. Gupta GP, Jha S (2018) Biogeography-based optimization scheme for solving the coverage and connected node placement problem for wireless sensor networks. Wirel Netw 1–11
29. Lalwani P, Banka H, Kumar C (2018) BERA: a biogeography-based energy saving routing architecture for wireless sensor networks. Soft Comput 22(5):1651–1667
30. Elhoseny M, Tharwat A, Yuan X, Hassanien AE (2018) Optimizing K-coverage of mobile WSNs. Expert Syst Appl 92:142–153
31. Kaushik A, Indu S, Gupta D (2018) Optimizing and enhancing the lifetime of a wireless sensor network using biogeography based optimization. International conference on application of computing and communication technologies. Springer, Singapore, pp 260–272
32. Senniappan V, Subramanian J (2018) Biogeography-Based Krill Herd algorithm for energy efficient clustering in wireless sensor networks for structural health monitoring application. J Ambient Intell Smart Environ 10(1):83–93
33. Chu SC, Tsai PW, Pan JS (2006) Cat swarm optimization. In: Pacific Rim international conference on artificial intelligence. Springer, Berlin, Heidelberg, pp 854–858
34. Tsai PW, Pan JS, Chen SM, Liao BY (2012) Enhanced parallel cat swarm optimization based on the Taguchi method. Expert Syst Appl 39(7):6309–6319

35. Temel S, Unaldi N, Kaynak O (2013) On deployment of wireless sensors on 3-D terrains to maximize sensing coverage by utilizing cat swarm optimization with wavelet transform. IEEE Trans Syst Man Cybern: Syst 44(1):111–120
36. Kong L, Chen CM, Shih HC, Lin, CW, He BZ, Pan JS (2014) An energy-aware routing protocol using cat swarm optimization for wireless sensor networks. In: Advanced technologies, embedded and multimedia for human-centric computing. Springer, Dordrecht, pp 311–318
37. Kong L, Pan JS, Tsai PW, Vaclav S, Ho JH (2015) A balanced power consumption algorithm based on enhanced parallel cat swarm optimization for wireless sensor network. Int J Distrib Sens Netw 11(3):729680
38. Soto R, Crawford B, Aste Toledo A, Castro C, Paredes F, Olivares R (2019) Solving the manufacturing cell design problem through binary cat swarm optimization with dynamic mixture ratios. Comput Intell Neurosci
39. Yang XS, Deb S (2009) Cuckoo search via Lévy flights. In: 2009 world congress on nature & biologically inspired computing (NaBIC). IEEE, pp 210–214
40. Ghosh A, Chakraborty N (2019) Cascaded cuckoo search optimization of router placement in signal attenuation minimization for a wireless sensor network in an indoor environment. Eng Optim 1–20
41. Yu X, Hu M (2019) Hop-count quantization ranging and hybrid cuckoo search optimized for DV-HOP in WSNs. Wirel Pers Commun 1–16
42. Meng X, Chang J, Wang X, Wang Y (2019) Multi-objective hydropower station operation using an improved cuckoo search algorithm. Energy 168:425–439
43. Chi R, Su YX, Zhang DH, Chi XX, Zhang HJ (2019) A hybridization of cuckoo search and particle swarm optimization for solving optimization problems. Neural Comput Appl 31(1):653–670
44. Wu Z, Zhao X, Ma Y, Zhao X (2019) A hybrid model based on modified multi-objective cuckoo search algorithm for short-term load forecasting. Appl Energy 237:896–909
45. Shehab M, Khader AT, Al-Betar MA (2017) A survey on applications and variants of the cuckoo search algorithm. Appl Soft Comput 61:1041–1059
46. Binh HTT, Hanh NT, Dey N (2018) Improved cuckoo search and chaotic flower pollination optimization algorithm for maximizing area coverage in wireless sensor networks. Neural Comput Appl 30(7):2305–2317
47. Meng X, Liu Y, Gao X, Zhang H (2014) A new bio-inspired algorithm: chicken swarm optimization. In: International conference in swarm intelligence. Springer, Cham, pp 86–94
48. Yu X, Zhou L, Li X (2019) A novel hybrid localization scheme for deep mine based on wheel graph and chicken swarm optimization. Comput Netw 154:73–78
49. Deb S, Gao XZ, Tammi K, Kalita K, Mahanta P (2019) Recent studies on chicken swarm optimization algorithm: a review (2014–2018). Artif Intell Rev 1–29
50. Al Shayokh M, Shin SY (2017) Bio inspired distributed WSN localization based on chicken swarm optimization. Wirel Pers Commun 97(4):5691–5706
51. Aziz A, Singh K, Osamy W, Khedr AM (2019) Effective algorithm for optimizing compressive sensing in IoT and periodic monitoring applications. J Netw Comput Appl 126:12–28
52. Movva P, Rao PT (2019) Novel two-fold data aggregation and MAC scheduling to support energy efficient routing in wireless sensor network. IEEE Access 7:1260–1274
53. Wang GG, Deb S, Coelho LDS (2015) Elephant herding optimization. In: 2015 3rd international symposium on computational and business intelligence (ISCBI). IEEE, 1–5
54. Strumberger I, Beko M, Tuba M, Minovic M, Bacanin N (2018) Elephant herding optimization algorithm for wireless sensor network localization problem. In: technological innovation for resilient systems: 9th IFIP WG 5.5/SOCOLNET advanced doctoral conference on computing, electrical and industrial systems, DoCEIS 2018, Costa de Caparica, Portugal, May 2–4, 2018, Proceedings 9. Springer International Publishing, pp 175–184
55. Correia S, Beko M, da Silva Cruz L, Tomic S (2018) Elephant herding optimization for energy-based localization. Sensors 18(9):2849
56. Strumberger I, Minovic M, Tuba M, Bacanin N (2019) Performance of elephant herding optimization and tree growth algorithm adapted for node localization in wireless sensor networks. Sensors 19(11):2515

57. Tuba E, Dolicanin-Djekic D, Jovanovic R, Simian D, Tuba M (2019) Combined elephant herding optimization algorithm with K-means for data clustering. In: Information and communication technology for intelligent systems. Springer, Singapore, pp 665–673
58. Li J, Guo L, Li Y, Liu C (2019) Enhancing elephant herding optimization with novel individual updating strategies for large-scale optimization problems. Mathematics 7(5):395
59. Li XL (2002) An optimizing method based on autonomous animats: fish-swarm algorithm. Syst Eng-Theory Pract 22(11):32–38
60. Neshat M, Sepidnam G, Sargolzaei M, Toosi AN (2014) Artificial fish swarm algorithm: a survey of the state-of-the-art, hybridization, combinatorial and indicative applications. Artif Intell Rev 42(4):965–997
61. Zheng ZX, Li JQ, Duan PY (2019) Optimal chiller loading by improved artificial fish swarm algorithm for energy saving. Math Comput Simul 155:227–243
62. Qin N, Xu J (2018) An adaptive fish swarm-based mobile coverage in WSNs. Wirel Commun Mob Comput
63. Li X, Keegan B, Mtenzi F (2018) energy efficient hybrid routing protocol based on the artificial fish swarm algorithm and ant colony optimisation for WSNs. Sensors 18(10):3351
64. Yin H, Zhang Y, He X (2018) WSN nodes placement optimization based on a weighted centroid artificial fish swarm algorithm. Algorithms 11(10):147
65. Mirjalili S, Mirjalili SM, Lewis A (2014) Grey wolf optimizer. Adv Eng Softw 69:46–61
66. Kohli M, Arora S (2018) Chaotic grey wolf optimization algorithm for constrained optimization problems. J Comput Des Eng 5(4):458–472
67. Krishnanand KN, Ghose D (2005) Detection of multiple source locations using a glowworm metaphor with applications to collective robotics. In: Proceedings 2005 IEEE swarm intelligence symposium, 2005. SIS 2005, IEEE. pp 84–91
68. Krishnanand KN, Ghose D (2009) Glowworm swarm optimization for simultaneous capture of multiple local optima of multimodal functions. Swarm Intell 3(2):87–124
69. Krishnanand KN, Ghose D (2008) Theoretical foundations for rendezvous of glowworm-inspired agent swarms at multiple locations. Robot Auton Syst 56(7):549–569
70. Liao WH, Kao Y, Li YS (2011) A sensor deployment approach using glowworm swarm optimization algorithm in wireless sensor networks. Expert Syst Appl 38(10):12180–12188
71. Mirjalili S (2015) Moth-flame optimization algorithm: a novel nature-inspired heuristic paradigm. Knowl-Based Syst 89:228–249
72. Zhao H, Zhao H, Guo S (2016) Using GM (1, 1) optimized by MFO with rolling mechanism to forecast the electricity consumption of inner mongolia. Appl Sci 6(1):20
73. Khalilpourazari S, Pasandideh SHR (2017) Multi-item EOQ model with nonlinear unit holding cost and partial backordering: moth-flame optimization algorithm. J Ind Prod Eng 34(1):42–51
74. Kennedy J, Eberhart R (1995) Particle swarm optimization (PSO). In: Proceeding IEEE international conference on neural networks, Perth, Australia, pp 1942–1948
75. Poli R (2008) Analysis of the publications on the applications of particle swarm optimisation. J Artif Evol Appl
76. Kulkarni RV, Venayagamoorthy GK (2010) Particle swarm optimization in wireless-sensor networks: a brief survey. IEEE Trans Syst Man Cybern Part C (Appl Rev) 41(2):262–267
77. Wachowiak MP, Smolíková R, Zheng Y, Zurada JM, Elmaghraby AS (2004) An approach to multimodal biomedical image registration utilizing particle swarm optimization. IEEE Trans Evol Comput 8(3):289–301
78. Yeh WC, Chang WW, Chung YY (2009) A new hybrid approach for mining breast cancer pattern using discrete particle swarm optimization and statistical method. Expert Syst Appl 36(4):8204–8211
79. Muthukaruppan S, Er MJ (2012) A hybrid particle swarm optimization based fuzzy expert system for the diagnosis of coronary artery disease. Expert Syst Appl 39(14):11657–11665
80. Jordehi AR (2019) Binary particle swarm optimisation with quadratic transfer function: a new binary optimisation algorithm for optimal scheduling of appliances in smart homes. Appl Soft Comput

81. Nouiri M, Bekrar A, Jemai A, Niar S, Ammari AC (2018) An effective and distributed particle swarm optimization algorithm for flexible job-shop scheduling problem. J Intell Manuf 29(3):603–615
82. Lynn N, Ali MZ, Suganthan PN (2018) Population topologies for particle swarm optimization and differential evolution. Swarm Evol Computation 39:24–35
83. Aydoğan EK, Delice Y, Özcan U, Gencer C, Bali Ö (2019) Balancing stochastic U-lines using particle swarm optimization. J Intell Manuf 30(1):97–111
84. Tam NT, Hai DT (2018) Improving lifetime and network connections of 3D wireless sensor networks based on fuzzy clustering and particle swarm optimization. Wirel Netw 24(5):1477–1490
85. Vijayalakshmi K, Anandan P (2018) A multi objective Tabu particle swarm optimization for effective cluster head selection in WSN. Cluster Comput 1–8
86. Kaur T, Kumar D (2018) Particle swarm optimization-based unequal and fault tolerant clustering protocol for wireless sensor networks. IEEE Sens J 18(11):4614–4622
87. Mirjalili S, Lewis A (2016) The whale optimization algorithm. Adv Eng Softw 95:51–67
88. El Aziz MA, Ewees AA, Hassanien AE (2017) Whale optimization algorithm and moth-flame optimization for multilevel thresholding image segmentation. Expert Syst Appl 83:242–256
89. Nasiri J, Khiyabani FM (2018) A whale optimization algorithm (WOA) approach for clustering. Cogent Math Stat 5(1):1483565
90. Ahmed MM, Houssein EH, Hassanien, AE, Taha A, Hassanien E (2019) Maximizing lifetime of large-scale wireless sensor networks using multi-objective whale optimization algorithm. Telecommun Syst 1–17
91. Valayapalayam Kittusamy SR, Elhoseny M, Kathiresan S (2019) An enhanced whale optimization algorithm for vehicular communication networks. Int J Commun Syst p.e3953
92. Hassan MK, El Desouky AI, Elghamrawy SM, Sarhan AM (2019) A Hybrid Real-time remote monitoring framework with NB-WOA algorithm for patients with chronic diseases. Future Gener Comput Syst 93:77–95
93. Verma GK, Ranga V (2018) Whale optimizer to repair partitioned heterogeneous wireless sensor networks. Int J Grid Distrib Comput 11(5):11–28
94. Yang XS (2014) Nature-inspired optimization algorithms. Elsevier
95. Parvin H, Moradi P, Esmaeili S (2019) TCFACO: trust-aware collaborative filtering method based on ant colony optimization. Expert Syst Appl 118:152–168
96. Li Y, Soleimani H, Zohal M (2019) An improved ant colony optimization algorithm for the multi-depot green vehicle routing problem with multiple objectives. J Cleaner Prod
97. Sun Z, Wei M, Zhang Z, Qu G (2019) Secure routing protocol based on multi-objective ant-colony-optimization for wireless sensor networks. Appl Soft Comput 77:366–375
98. Wang J, Cao J, Sherratt RS, Park JH (2018) An improved ant colony optimization-based approach with mobile sink for wireless sensor networks. J Supercomputing 74(12):6633–6645
99. Guleria K, Verma AK (2019) Meta-heuristic Ant Colony optimization based unequal clustering for wireless sensor network. Wirel Pers Commun 105(3):891–911
100. Ghosh N, Banerjee I, Sherratt RS (2019) On-demand fuzzy clustering and ant-colony optimisation based mobile data collection in wireless sensor network. Wirel Netw 25(4):1829–1845
101. Dahan F, El Hindi K, Mathkour H, AlSalman H (2019) Dynamic flying ant colony optimization (DFACO) for solving the traveling salesman problem. Sensors 19(8):1837
102. Karaboga D, Gorkemli B, Ozturk C, Karaboga N (2014) A comprehensive survey: artificial bee colony (ABC) algorithm and applications. Artif Intell Rev 42(1):21–57
103. Giveki D, Salimi H, Bahmanyar G, Khademian Y (2012) Automatic detection of diabetes diagnosis using feature weighted support vector machines based on mutual information and modified cuckoo search. *arXiv preprint* arXiv:1201.2173
104. Ashour AS, Samanta S, Dey N, Kausar N, Abdessalemkaraa WB, Hassanien AE (2015) Computed tomography image enhancement using cuckoo search: a log transform based approach. J Signal Inform Process 6(03):244
105. Wang GG, Deb S, Gao XZ, Coelho LDS (2016) A new metaheuristic optimisation algorithm motivated by elephant herding behaviour. Int J Bio-Inspired Comput 8(6):394–409

106. Kaushik A, Indu S, Gupta D (2019) A grey wolf optimization approach for improving the performance of wireless sensor networks. Wirel Pers Commun 1–21
107. Kaushik A, Indu S, Gupta D (2019) A grey wolf optimization based algorithm for optimum camera placement. Wirel Pers Commun 1–25
108. Zapotecas-Martínez S, García-Nájera A, López-Jaimes A (2019) Multi-objective grey wolf optimizer based on decomposition. Expert Syst Appl 120:357–371
109. Tu Q, Chen X, Liu X (2019) Multi-strategy ensemble grey wolf optimizer and its application to feature selection. Appl Soft Comput 76:16–30
110. Mirjalili S, Saremi S, Mirjalili SM, Coelho LDS (2016) Multi-objective grey wolf optimizer: a novel algorithm for multi-criterion optimization. Expert Syst Appl 47:106–119
111. Ray A, De D (2016) An energy efficient sensor movement approach using multi-parameter reverse glowworm swarm optimization algorithm in mobile wireless sensor network. Simul Model Pract Theory 62:117–136
112. Wang Y, Cui Z, Li W (2019) A novel coupling algorithm based on glowworm swarm optimization and bacterial foraging algorithm for solving multi-objective optimization problems. Algorithms 12(3):61
113. Salkuti SR, Kim SC (2019) Congestion management using multi-objective glowworm swarm optimization algorithm. J Electr Eng Technol 1–11
114. Song L, Zhao L, Ye J (2019) DV-hop node location algorithm based on GSO in wireless sensor networks. J Sens
115. Antoniou P, Pitsillides A, Blackwell T, Engelbrecht A, Michael L (2013) Congestion control in wireless sensor networks based on bird flocking behavior. Comput Netw 57(5):1167–1191
116. Bharathi MA, Mallikarjuna M, VijayaKumar BP (2012) Bio-inspired approach for energy utilization in wireless sensor networks. Procedia Eng 38:3864–3868
117. Saleem M, Ullah I, Farooq M (2012) BeeSensor: an energy-efficient and scalable routing protocol for wireless sensor networks. Inf Sci 200:38–56
118. Miloud M, Abdellatif R, Lorenz P (2019) Moth flame optimization algorithm range-based for node localization challenge in decentralized wireless sensor network. Int J Distrib Syst Technol (IJDST) 10(1):82–109
119. Mittal N (2019) Moth flame optimization based energy efficient stable clustered routing approach for wireless sensor networks. Wirel Pers Commun 104(2):677–694
120. Khan MF, Aadil F, Maqsood M, Bukhari SHR, Hussain M, Nam Y (2019) Moth flame clustering algorithm for internet of vehicle (MFCA-IoV). IEEE Access 7:11613–11629
121. Sapre S, Mini S (2018) Moth flame based optimized placement of relay nodes for fault tolerant wireless sensor networks. In: 2018 9th international conference on computing, communication and networking technologies (ICCCNT), IEEE. pp 1–6
122. Ray A, De D (2016) Energy efficient clustering protocol based on K-means (EECPK-means)-midpoint algorithm for enhanced network lifetime in wireless sensor network. IET Wirel Sensor Syst 6(6):181–191

Chapter 13
TLBO Based Cluster-Head Selection for Multi-objective Optimization in Wireless Sensor Networks

Madhuri Malakar and Shweta

1 Introduction

Wireless Sensor Network (WSN) is one type of wireless network that consists of multiple sensor nodes that are scattered in an environment [1]. The purpose of this scattering or deploying in particular field of interest to sense different conditions of the environment in terms of humidity and temperature to achieve the goal. It is basically used to monitor and detect different events and forward it to the Base Station (BS) with the help of sink node. In terms of real-life applications some of the examples are as follows:

(i) **Biodiversity mapping**: The sensor nodes efficiently sense the environmental information and helps to analyze it. In wildlife applications, this is rapidly used to detect and monitor the behavior of the animals.
(ii) **Disaster relief operations**: It is efficiently used in disaster management to rescue different situations. It helps to save the human as well as animal from wildfire or different types disaster situations.
(iii) **Intelligent buildings**: It is also used in intelligent building design or bridge design. It helps to proper measure of air conditioning, ventilation, humidity and reduction of energy. Moreover, it helps to determine the temperature consumption and flow of the rooms.

M. Malakar (✉)
Department of Computer Science and Engineering, National Institute of Technology, Durgapur, West Bengal 713209, India
e-mail: malakar.madhuri2018@gmail.com

Shweta
IBM Pvt Ltd., 137, Galway Drive, Apt 302, Mooresville, NC 28117, USA
e-mail: shwetapr@in.ibm.com

D. De et al. (eds.), *Nature Inspired Computing for Wireless Sensor Networks*, Springer Tracts in Nature-Inspired Computing,
https://doi.org/10.1007/978-981-15-2125-6_13

(iv) **Machine surveillance**: In machine prevention and surveillance, the WSN also plays a vital role in measuring sensing and controlling information for cable and parts of the machine.
(v) **Medicine and health care**: The WSN also plays a vital role in medical science for medicine and health care. It helps to intensive care as well as post-operation system. Apart from these, this is also used in surveillance of patients and doctors.
(vi) **Precision agriculture**: In agriculture, WSN also used rapidly to precise agriculture which needed fertilizer, irrigation, pesticides, etc.

Apart from the above-mentioned applications, the WSN also plays a vital role in different other applications such as entertainment, education, business, marketing, etc. Although WSN has several advantages in terms of real-life applications, it has also some limitations which are as follows.

(i) **Battery issue**: Each node of the WSN is consists of limited capacity of battery which is insufficient during any mission or operation. Battery charge or replacement is not possible during any operation. Hence, it is insecure for any transmission or operation.
(ii) **Distraction**: The WSN is consists of several type parameters where nature of the parameters is conflicting one to another. So, here raise distraction by other devices of the network.
(iii) **Low speed**: Speed of the network is very low compared to wired network. So, its processing speed and capabilities are soft in terms of operation. But it efficiently monitors, detects and transmits the information to the proper destination.
(iv) **Security**: At the time of operation, sensor nodes are scattered in any region without any verification. So, it is less secure in comparison to wired network. Hence, easily trap the required information of the network.

Therefore, there is a need for some evolutionary algorithms to optimize network parameters efficiently. In last few decades, several evolutionary algorithms are derives based on nature applications some of them Artificial Immune Algorithm (AIA) [2], Bacteria Foraging Optimization (BFO) [3], Differential Evolution (DE) [4, 5], Evolution Programming (EP) [6], Evolution Strategies (ES) [7], and Genetic Algorithm (GA) [8].

Some of the recognized evolutionary algorithms are: Genetic Algorithms (GA) [2] which works on the principle of the Darwinian theory of the survival of the fittest and the theory of evolution of the living beings, Differential Evolution (DE) [3, 4] which is similar to GA with specialized crossover and selection method, Evolution Strategies(ES) [5], Evolution Programming (GP) [6], Artificial Immune Algorithm (AIA) [7] which works on the immune system of the human being, and Bacteria Foraging Optimization (BFO) etc. [8].

Apart from evolutionary algorithms, there are various swarm-based techniques also proposed in last few decades to reduce the issues and problems in our real life. Some of them are described as follows. Ant Colony Optimization (ACO) [9] that inspired by the behavior of the birds. Artificial Bee Colony (ABC) [10–12] that inspired the activities of the bees. Biogeography-Based Optimization (BBO) [13, 14] that derive the principle of emigration and immigration of the species. Gravitational Search Algorithm (GSA) [15] derives from the gravitational force technique that works between the bodies. Some others algorithms also derives to enhance the real-life operation or system such as Grenade Explosion Method (GEM) [16], Harmony Search (HS) [17], Particle Swarm Optimization (PSO) [18], Shuffled Frog Leaping (SFM) [19] and combination of PSO and multimodal function [20], etc.

In the above discussions, several pros and cons are mentioned that contain several types of issues and uncertainties. This causes some security and privacy issues [21]. Several works exist for outlier detection or anomaly detection, fault diagnosis, intrusion detection, mobility prediction [22, 23]. Most of the outlier detection and prediction algorithms are outperform with the help of machine learning algorithms [24, 25]. The reason is the robustness of the algorithm in order to optimization [26–28]. The current proposal will help the researchers to guide new direction of the optimization in terms of meta-heuristic method Teaching–Learning-Based Optimization (TLBO). In the proposed method, an intelligent Cluster-head (CH) selection technique is proposed for WSN. The purpose of this technique to optimize multiple conflicting objectives in terms of enhancing the network's lifetime. The application of CH here to select data and information from all other sensor nodes and transmit it to the BS. It also helps to manage and control the sensor nodes.

The remaining of the paper is illustrated as Sect. 2, i.e., literature review which highlights some of the existing proposals which are a combination of both heuristic and meta-heuristic methods. Section 3, i.e., preliminaries. It discussed the basic key elements of the proposal, i.e., TLBO and its inherent components. Section 4 is the main proposed work that highlights the basic proposal in chronological ways. The last section, i.e., Section 5 concludes the proposal by discussing the main outcomes of the proposal.

2 Literature Review

In the last few decades, several works have been proposed. In this section, some of the works are discussed as follows. Mostafaei and Menth [1] proposed a survey on Wireless Sensor Network (WSN). The author's work explains the basics of WSN and Software-Defined Network (SDN). It also describes the fundamentals and the operations of WSN and challenges the need to investigate in more detail in the field of WSN. Rao et al. [29] proposed Teaching–learning-based optimization: an optimization method for continuous non-linear large-scale problems that explains the use of TLBO for global optimization. The work also explains the entire concept of the population-based algorithm and how it provides better solutions when compared with other evolutionary algorithms. The traditional techniques often failed to solve large scale problems or optimization problems with many local optima. This technique helps overcome the limitations and the work is applicable in overcoming many limitations in the field of WSN. Rao and Patel [30] proposed TLBO bases unconstrained optimization technique for solving a classroom problem. It is based on multidimensional and combination of non-linear and linear issues. It consists of teachers with different factors having combination nature of adaptive and self-learning. Finally, it validated with some existing proposals in terms of some metrics. In last decade, several works have been proposed based on ACO technique [31–34]. Sun et al. [35] proposed a technique for optimization of the nodes deployment for WSN using ACO. In this technique, the authors apply two techniques like culture algorithm with ant colony algorithm. The proposed technique is used to solve coverage and connectivity both issues and enhance the network lifetime of the WSN. Sharma and Grover [36] proposed a model in WSN for energy-efficient routing protocol. The key technique of this model is modified ACO which is an extension of earlier ACO. The proposed technique handles the utilization of the node batteries of the WSN. This efficient utilization helps to control non-uniform signal transformation during operation. Kaur and Mahajan [37] proposed a meta-heuristic based efficient routing protocol for WSN. The proposed method is the fusion of two meta-heuristic techniques as ACO and PSO. This efficient routing protocol is based on energy capacity of the nodes. The proposed method helps in data aggregation of the network by reducing residual energy capacity of the nodes. In last decade, several works have been proposed based on ANN technique [38–40]. Some of them as follows. Alarifi and Tolba [41] proposed an adaptive neural network-based optimization technique for WSN. This is based on reinforcement based learning technique. So, it uses adaptive Q-learning method. It helps to design clustering in the network that helps to reduce energy consumption of the nodes and data collection from sensor nodes to the BS. Eldhose and Jisha [42] proposed a node aggregation method for WSN. This is based on clustering method where each active cluster is decided by ANN. This neural network method helps to process the data packet periodically and send to the BS efficiently. Chang et al. [43] proposed a frailty detection system for WSN using ANN. It is totally home-based system. The main purpose of this proposal to automatically manage the personal information of a home. It consists of several information such as ePad, eChair, eScale, etc. Finally, it helps to predict frailty and test overall system performance. Serpen and Gao [44] proposed a perceptron method for WSN using ANN. This method is based on embedded techniques. This method is based on parallel as well as distributed technique which is works as neurocomputing method. This prediction method helps to enhance the network metrics as well as network lifetime. In last decade, several works have been proposed based on PSO technique [45, 46]. Phoemphon et al. [47] proposed a localization method for WSN. The proposed method is a hybrid method using fusion of fuzzy logic and PSO. The proposed method is used to detect global position of the sensor nodes. It helps to reduce cost as well as battery energy of the nodes. The fuzzy logic is used to reduce uncertainty of the network where PSO is used to inspire the neighbor nodes. Sun et al. [48] proposed a multi-objective technique for WSN. This technique is used binary PSO techniques for multi-objective optimization. It helps to optimize several objectives such as load balancing, energy consumption, task execution time. Several constraints are designed for each objective function. All objective functions help to enhance network performance. Cao et al. [49] proposed a deployment optimization technique for WSN using PSO. This method is used parallelism in the context of distributed system. The proposed method is based on NP-hard problems. So, it uses PSO as a cooperative co-evolutionary system. It helps to reduce computational time and enhance message passing as well as data transmission. Yan et al. [50] proposed a positioning system for optical WSN using PSO. The main parameter of this technique is residual energy. Residual energy indicates difference between initial energy and energy used during sending/receiving data packets. In this proposal, decision maker first considers the location of the nodes then optimizes it. Finally, it helps to reduce the energy consumption of the nodes. Some of the swarm intelligence based algorithms described as follows. Zahedi et al. [51] proposed a clustering-based hybrid method for WSN. This method is a fusion of fuzzy logic and swarm intelligence. The role of swarm intelligence is to inspire the different clustering of the network. And the role of fuzzy logic is to reduce uncertainty of the network by controlling the conflicting parameters of the network. Bruneo et al. [52] proposed a modeling technique for WSN using swarm intelligence method. The main aim of this proposal to optimize route of the network between source node to sink node by using different conflicting parameters. The key element of this network markovian agent that helps to trap the issues and enhance the message passing system. Ari et al. [53] proposed an efficient clustering technique for WSN using swarm intelligence. In this technique, honeybees nature-inspired method is used as a swarm intelligence for managing clustering of the network. The proposed method is based on multi-objective optimization technique where each objective employs linear programming and manage trade-off between energy-efficiency and cost function of the network. Sreelaja and Pai [54] proposed a sinkhole attack detection technique for WSN using swarm intelligence. In this method, ACO is used as a swarm intelligence technique. The proposed method is based on voting method which is used to generate a notification for identifying attacks within the network. Finally, it outperforms the network metrics in terms of validation and simulation. Fuzzy logic is also used to reduce uncertainty of the network with the help of multi-value logic. This logic is inherent with the above-mentioned nature-inspired algorithms to make the process more efficient and robustness. Here, some of the algorithms are described bases on fuzzy logic in the context of linear and non-linear optimizations. In [55–62], the authors described several works based on ad-hoc networks. These works are based on energy-efficient routing. Some of them are used fuzzy logic or some of them are used its extension, or some of them are used fuzzy logic with optimization. Apart from ad-hoc network, fuzzy logic also used in WSN. Some of the proposals are described as follows. In [63, 64], the authors are used soft computing techniques for enhancing the network performance. Fuzzy logic is used in these proposals as a soft computing technique. In these proposals, fuzzy logic is used to design input and output parameters of the route using triangular membership functions. Finally, it helps to evaluate all feasible routes as well as optimal route of the network. Amri et al. [65] proposed a fuzzy-based method for node localization system in WSN. The main aim of this method to precisely finds the location of the sensor nodes in WSN. It also helps in precision enhancement of the network. This enhancement of the network helps in energy-saving and network lifetime enhancement of the WSN. Mazinani et al. [66] proposed a fuzzy-based clustering method in WSN. It is a multi-cluster based routing technique which is used a threshold value for determining the cluster head. The proposed method helps to avoid control message and possibility of collision in WSN. Finally, it helps to find cluster heads and enhance network lifetime as well as message transmission. Das and Das [63] proposed a soft computing based routing protocol for WSN. In this proposal, the authors basically used fuzzy logic technique to reduce the uncertainty of the network using fuzzy inference system. There are two basic parameters are considered here first is energy and second is distance to evaluate the output parameter reward. It helps to reduce uncertainty related to route selection of the WSN. In last decade, several works have been proposed based on GA technique [40, 67–69]. Hanh et al. [70] proposed an intelligent and efficient routing protocol for area coverage in WSN. The basic key element of this routing protocol is GA which is a meta-heuristic algorithm. This algorithm helps to design a fitness function or objective function by approximation method for maximizing coverage of the network. Finally, this algorithm helps to produce best performance in terms of solution quality. Somauroo and Bassoo [71] proposed a GA based efficient variant algorithm for 3D WSN. The proposed method uses chain-based technique as PEGASIS technique. This technique is used to enhance the network lifetime of the WSN. Moreover, this approach also used to design clustering of the network that helps to reduce number of sensor nodes in a chain. This process helps to reduce delays in data transmission. Wang et al. [72] proposed an energy-efficient routing and clustering method for WSN. The proposed method is based on GA technique. In this proposed routing protocol, optimal solution is obtained by comparing previous round with current round. In each iteration, round is improved in terms of fitness function and their constraints. Finally, it helps to improve overall network performance. Al-Shalabi et al. [73] proposed an efficient multi-hop routing method in WSN using GA. This is basically optimal multi-path finding system. It is based on clustering method which basic aim to find optimal path between source which is a cluster-head and destination node which is a BS. Kumar et al. [74] proposed a GA based distributed zone approach in WSN. The proposed method is based on green communication system that includes two major things first is communication optimization and second is energy optimization. The main aim of this proposal for finding optimal path between source nodes to sink node. It also helps to enhance convergence speed of the solution. Finally, it shows outperform results compared to existing methods. Das et al. [60] provide a details book for wireless network. It consists of several frameworks of wireless networks along with WSN and ad-hoc networks. Some of works are [61] describe intelligent routing technique using fuzzy petri net and strategy management using non-linear formulation technique. Both works help to estimate uncertainty parameters of the network. Apart from the above-mentioned literature, there are several works [75–78] have been proposed for WSN that uses some other technique to achieve the purpose of the WSN efficiently.

3 Preliminary: Teaching-Learning-Based Optimization (TLBO)

Teaching-Learning-Based optimization (TLBO), plays a vital role in today's world of optimization. TLBO is a recently proposed population-based evolutionary optimization algorithm. The main idea behind TLBO is the simulation of a classical school learning process consisting of Learner's learning or gaining information from teachers or teacher's broadcasting information to the learners in the classroom. TLBO considers group of students (i.e., learners) as the population and subjects are treated as the design variables of the optimization problem. Teaching–learning is a process of improving one-self by learning from other individuals may be from a teacher or by interacting with the other individuals in the classroom. It is entirely based on the effect of influence of the teacher on the output of learners in the class. The output is obtained in terms of mean, best, standard deviation, error, and number of function evaluations (FEs) while conducting experiments on algorithms which mainly depends on the quality of the teacher. The best solution is represented as the teacher in the entire population. The fitness value of the optimization problem is analogous to the result of the learners.

3.1 Teacher Phase

In the Teacher Phase, the teacher plays a vital role in the entire phase. The resultant mean of the learner and up-gradation of their result is completely dependent on teacher. Teachers impart knowledge to all the learners in the class so as to improve their average grades. The best learner from among the population is chosen as teacher: (XTeacher) according to the fitness value and the mean position is calculated for updating the position of the learners.

3.2 Learner Phase

In the learner's phase, learners acquire knowledge from the teacher as well as by interacting among themselves. The major difference between teacher and the learner phase is that in teacher phase teacher imparts knowledge to the learners and in learners, phase learners enhance their knowledge by learning. It may be either by teacher or by mutual interaction among themselves. It helps learners in enhancing their knowledge and brooding themselves up by knowledge which can be judged in terms of resultant mean after algorithmic experiments as fitness values.

4 Proposed Method

The proposed method is divided into three phases such as (i) Network model, (ii) Parameters formulation, and (iii) TLBO formulation. Network model is used to illustrate the network information that highlights how the basic paradigms of the WSN are arranged in order to achieve the goal. The parameter formulation phase is used for discussing basic input parameters that are used in this proposal. Last phase, i.e., TLBO formulation is used to enhance the network lifetime by optimizing several objectives of the WSN.

4.1 Network Model

In the proposed method the network model shown in Fig. 1. In this figure, different types of sensor nodes are available to achieve the purpose of the network of taking participate in the operation. Each node has its own radio range. The nodes communicate or move within this radio range.

The figure contains five types of node-level where each level depends on the basic parameters of the network that help to achieve the main goal of the network. Here, total radio range is six and total node level category is five. Both are different due to variation of sensor nodes that may change based on the requirement. Each category of node is colored for identifying the similarity and dissimilarity of the nodes. Blue colored node indicates Cluster-head (CH) of the sensor nodes that depend on some properties which is discussed later phases.

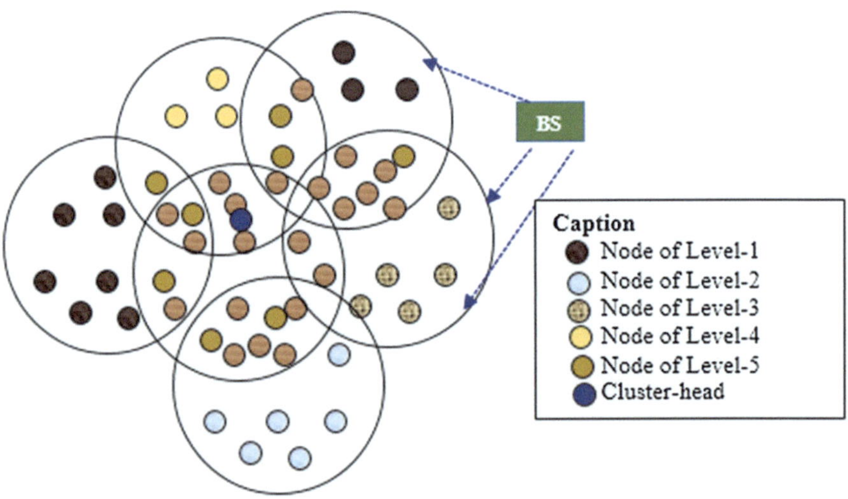

Fig. 1 Network model

4.2 Parameter Formulation

The network contains several types of parameters that may be similar type nature or conflicting nature one to another. Each node resides within a specific radio range based on Euclidian distance shown in Eq. 1.

$$\mathrm{Dist}(n_i, n_j) = \sqrt{(x_j - x_i)^2 + (y_j - y_i)^2}. \tag{1}$$

where (x_i, y_i) and (x_j, y_j) are the address of nodes n_i and n_j.

In this proposal, there are two basic parameters are considered first is residual energy and second is distance. Residual energy of the node is calculated with the help of initial energy and consumed energy which shown in Eq. 2 and consumed energy is calculated with the help of energy required during transmit and received packets which are shown in Eq. 3.

$$\mathrm{Res_{eng}} = \mathrm{Ini_{eng}} - \mathrm{Con_{eng}} \tag{2}$$

$$\mathrm{Con_{eng}} = \mathrm{Eng_t} + \mathrm{Eng_r} \tag{3}$$

4.3 TLBO Formulation

The proposed method basically enhance the network lifetime of the WSN by optimizing two type objectives (i) Category-A, (ii) Category-B. Category-A objective indicates maximization nature whose enhancement helps to enhancement of the network lifetime. Category-B objective indicates minimization nature whose detraction helps to enhancement of the network lifetime. Its details categorization shown in Fig. 2.

Network lifetime indicates time duration between two events (i) when network is started for particular operation, and (ii) most of the nodes are exhaust. Both category objectives defined are as follows.

(a) **Throughput**: It indicates successfully received data packet at the receiver end in particular unit of time. It is shown in Eq. 4.

$$\text{Throughput} = \frac{(\text{Delivered packet} * \text{Packet size})}{\text{Total duration of the simulation}} \tag{4}$$

(b) **Packet Delivery Ratio**: It indicates the actual ratio between delivered and sent packets at receiver end. It is shown in Eq. 5.

$$\text{Packet Delivery Ratio} = \frac{\text{Sum of all received data by the destination node}}{\text{sum of all sent data by the source node}} \times 100 \tag{5}$$

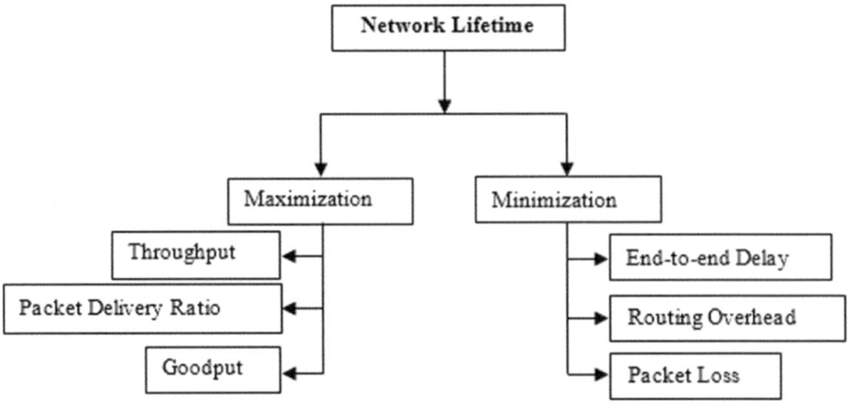

Fig. 2 Category of objective for optimization

(c) **Goodput**: It indicates successfully received data packet at the destination node in particular unit of time. It does not include several overhead such as retransmission and undesirable information. It shows in Eq. (6).

$$\text{Goodput} = \frac{\text{Sum of all received data by the destination node-overhead}}{\text{Sum of all sent data by the source node}} \times 100 \tag{6}$$

(d) **End-to-end Delay**: It indicates total time taken by a packet to reach at the destination node. It also include latency and buffering delays. It shown in Eq. 7.

$$\text{End - to - end Delay} = \frac{\text{Sum of all difference between arrive and send time for packet}}{\text{Total number of connections}} \tag{7}$$

(e) **Routing Overhead**: It indicates total number of control packets is required to send one successfully received data packet shown in Eq. 8.

$$\text{Routing Overhead} = \frac{\text{Sum of all control packets by source node}}{\text{Sum of all received data packet by destination node}} \tag{8}$$

(f) **Packet Loss**: It indicates total number of packets dropped during a successful data transmission shown in Eq. 9.

$$\text{Packet Loss} = \frac{\text{Sum of all packet dropped}}{\text{Sum of all data packet sent}} \times 100 \tag{9}$$

The proposed method is based on TLBO that helps to enhance the network lifetime by proper selection of CH. This selection helps to optimize both category network metrics. In this proposal, Res_{eng} is considered as 10 J and Dist. considered as 50 M.

Table 1 Calculated node-level of sensor nodes for round 1

Nodes	Res_{eng}	Dist.	Node_Level
n1	10	45	**27.50**
n2	8	40	24.00
n3	7	32	19.50
n4	6	35	20.50
n5	5	36	20.50
n6	6	42	24.00
n7	3	25	14.00
n8	2	30	16.00
n9	4	33	18.50
n10	3	49	26.00
Mean	**5.4**	**36.7**	
Max			**27.50**

Here, two basic input parameters residual energy and distance act as two decision variables and output parameter, i.e., node level act as $f(x)$ which is the main objective of the formulation. Sensor nodes considered as n_i where $i = 1$ to 10. Node_Level is calculated by using Eq. 10. Details calculation shown in Table 1 for round 1.

$$\text{Node\_Level} = \text{Mean}(\text{Res}_{\text{eng}},\ \text{Dist.}) \tag{10}$$

Based on Table 1, selected CH is node-1 having highest node level (i.e., 27.50). The updated table of Table 1 for round 1 shown in Table 3 which contains some updation using TLBO method. In this table, Difference_Mean is calculated for both parameters, i.e., residual energy and distance. Node level is also updated by using these two parameters. Details of "Difference_Mean" logic shown in Algorithm 1 and its parameters description shown in Table 2.

Table 2 Parameters descriptions

Parameter	Description
r1	Random number for first difference mean, i.e., residual energy
10	Residual energy of node n1
5.4	Mean residual energy of node n1 to n10
Res_{eng}	Residual energy of node
r2	Random number for second difference mean, i.e., distance
45	Distance of node n1 to cluster-head
36.7	Mean distance of node n1 to n10
Dist.	Distance of node

Algorithm 1 Difference mean calculation.

```
/* For parameter 1 i.e. Res_eng */
Difference_mean(Res_eng)
if(r1*(10-5.4)>=0)
then
Res_eng=r1*(10-5.4)          /*r1=5 any random number*/
else
Res_eng=0                    /*which indicate node is exhaust*/


/* For parameter 2 i.e. Dist.*/
Difference_mean(Dist.)
if(r2*(45-36.7)>=0)          /*r2=5 any random number*/
then
Dist.= r2*(45-36.7)
else
Dist.=0                      /*which also indicate node is exhaust*/
```

In updatedTable 3 for round 1, Node_Level of nodes n2 to n9 are rejected due to less efficiency and Node_Level of n10 is also rejected because here new energy becomes zero that means the node is dead. So, by comparing maximum Node_Level from Tables 1 and 3 updated table shown as final Node_Level in Table 4.

By observing data of Table 4, again node n1 is the CH due to highest node level. The same procedure is repeated in the WSN for selecting the CH of the nodes. In this proposal, just two round iterations are shown. Hence, Tables 5 and 6 show same analysis for round 2.

Table 3 Updated calculated node-level of sensor nodes for round 1

Nodes	Energy	Distance	Node_Level
n1	23.00	41.50	32.25
n2	13.00	16.50	***14.75***
n3	8.00	0.00	***4.00***
n4	3.00	0.00	***1.50***
n5	0.00	0.00	***0.00***
n6	3.00	26.50	***14.75***
n7	0.00	0.00	***0.00***
n8	0.00	0.00	***0.00***
n9	0.00	0.00	***0.00***
n10	0.00	61.50	**30.75**

Table 4 Final node-level for round 1

Nodes	Energy	Distance	Node_Level
n1	23.00	41.50	**32.25**
n2	8.00	40.00	24.00
n3	7.00	32.00	19.50
n4	6.00	35.00	20.50
n5	5.00	36.00	20.50
n6	6.00	42.00	24.00
n7	3	25	14.00
n8	2	30	16
n9	4	33	18.5
n10	3	49	26
Mean	**6.7**	**36.35**	
Max			**32.25**

Table 5 Updated calculated node-level of sensor nodes for round 2

Nodes	Energy	Distance	Node_Level
n1	81.50	25.75	53.63
n2	6.50	18.25	***12.38***
n3	1.50	0.00	***0.75***
n4	0.00	0.00	***0.00***
n5	0.00	0.00	***0.00***
n6	0.00	28.25	***14.13***
n7	0.00	0.00	***0.00***
n8	0.00	0.00	***0.00***
n9	0.00	0.00	***0.00***
n10	0.00	63.25	31.63

Table 6 Final node-level for round 2

Nodes	Energy	Distance	Node_Level
n1	81.50	25.75	53.63
n2	8.00	40.00	24.00
n3	7.00	32.00	19.50
n4	6.00	35.00	20.50
n5	5.00	36.00	20.50
n6	6.00	42.00	24.00
n7	3.00	25.00	14.00
n8	2.00	30.00	16.00
n9	4.00	33.00	18.50
n10	0.00	63.25	31.63

In round 2, node n1 is also selected as CH. This cluster-head selection procedure helps to enhance the network lifetime efficiently by optimizing two types of conflicting metrics shown in Fig. 2.

5 Conclusion

In this paper, TLBO is used as meta-heuristic technique inspired by the concepts of students and teachers. It helps to optimize several conflicting metrics such as throughput, packet delivery ratio, goodput in terms of maximization and end-to-end delay, routing overhead, packet loss in terms of minimization. The process is repeated only two times, i.e., round-1 and round-2. In each round cluster-head is efficiently selected by output parameter which is $f(x)$, i.e., node level. This output parameter consists of two crucial parameters of the nodes first is energy and second is distance. Finally, this procedure is repeated continuously until the finish main operation. The future work is to enhance this work in terms of real-life scenarios and compare this with some existing meta-heuristic works for outperforming the network metrics.

References

1. Mostafaei H, Menth M (2018) Software-defined wireless sensor networks: a survey. J Netw Comput Appl 119:42–56
2. Farmer JD, Packard NH, Perelson AS (1986) The immune system, adaptation, and machine learning. Phys D: Nonlinear Phenom 22(1–3):187–204
3. Passino KM (2002) Biomimicry of bacterial foraging for distributed optimization and control. IEEE Control Syst Mag 22(3):52–67
4. Storn R, Price K (1997) Differential evolution–a simple and efficient heuristic for global optimization over continuous spaces. J Global Optim 11(4):341–359
5. Price K, Storn RM, Lampinen JA (2006) Differential evolution: a practical approach to global optimization. Springer Science & Business Media
6. Fogel LJ, Owens AJ, Walsh MJ (1966) Artificial intelligence through simulated evolution. Willey, New York
7. Runarsson TP, Yao X (2000) Stochastic ranking for constrained evolutionary optimization. IEEE Trans Evol Comput 4(3):284–294
8. Holland JH (1975) Adaptation in natural and artificial systems. University of Michigan Press, Ann Arbor, USA
9. Dorigo M (1992) Optimization, learning and natural algorithms. Ph.D. Thesis, Politecnico di Milano
10. Akay B, Karaboga D (2012) Artificial bee colony algorithm for large-scale problems and engineering design optimization. J Intell Manuf 23(4):1001–1014
11. Karaboga D, Akay B (2011) A modified artificial bee colony (ABC) algorithm for constrained optimization problems. Appl Soft Comput 11(3):3021–3031
12. Akay B, Karaboga D (2012) A modified artificial bee colony algorithm for real-parameter optimization. Inf Sci 192:120–142
13. Simon D (2008) Biogeography-based optimization. IEEE Trans Evol Comput 12(6):702–713
14. Ma H, Simon D (2011) Blended biogeography-based optimization for constrained optimization. Eng Appl Artif Intell 24(3):517–525

15. Rashedi E, Nezamabadi-Pour H, Saryazdi S (2009) GSA: a gravitational search algorithm. Inf Sci 179(13):2232–2248
16. Ahrari A, Atai AA (2010) Grenade explosion method—a novel tool for optimization of multimodal functions. Appl Soft Comput 10(4):1132–1140
17. Geem ZW, Kim JH, Loganathan GV (2001) A new heuristic optimization algorithm: harmony search. Simulation 76(2):60–68
18. Kennedy J, Eberhart R (1995) Particle swarm optimization (PSO). In: Proceedings of the IEEE international conference on neural networks, Perth, Australia, pp 1942–1948
19. Eusuff MM, Lansey KE (2003) Optimization of water distribution network design using the shuffled frog leaping algorithm. J Water Resour plann Manage 129(3):210–225
20. Li S, Tan M, Tsang IW, Kwok JTY (2011) A hybrid PSO-BFGS strategy for global optimization of multimodal functions. IEEE Trans Syst Man Cybern Part B (Cybern) 41(4):1003–1014
21. Karati A, Islam SH, Biswas GP (2018) A pairing-free and provably secure certificateless signature scheme. Inf Sci 450:378–391
22. Jain PK, Pamula R (2019) Two-step anomaly detection approach using clustering algorithm. International conference on advanced computing networking and informatics. Springer, Singapore, pp 513–520
23. Mishra G, Agarwal S, Jain PK, Pamula R (2019) Outlier detection using subset formation of clustering based method. International conference on advanced computing networking and informatics. Springer, Singapore, pp 521–528
24. Kumari P, Jain PK, Pamula R (2018) An efficient use of ensemble methods to predict students academic performance. In: 2018 4th international conference on recent advances in information technology (RAIT). IEEE, pp 1–6
25. Punam K, Pamula R, Jain PK (2018) A two-level statistical model for big mart sales prediction. In: 2018 international conference on computing, power and communication technologies (GUCON). IEEE, pp 617–620
26. Das SP, Padhy S (2018) A novel hybrid model using teaching–learning-based optimization and a support vector machine for commodity futures index forecasting. Int J Mach Learn Cybernet 9(1):97–111
27. Das SP, Padhy S (2017) Unsupervised extreme learning machine and support vector regression hybrid model for predicting energy commodity futures index. Memetic Comput 9(4):333–346
28. Das SP, Padhy S (2017) A new hybrid parametric and machine learning model with homogeneity hint for European-style index option pricing. Neural Comput Appl 28(12):4061–4077
29. Rao RV, Savsani VJ, Vakharia DP (2012) Teaching–learning-based optimization: an optimization method for continuous non-linear large scale problems. Inf Sci 183(1):1–15
30. Rao RV, Patel V (2013) An improved teaching-learning-based optimization algorithm for solving unconstrained optimization problems. Sci Iranica 20(3):710–720
31. Kaliannan J, Baskaran A, Dey N, Ashour AS (2016) Ant colony optimization algorithm based PID controller for LFC of single area power system with non-linearity and boiler dynamics. World J Model Simul 12(1):3–14
32. Kaliannan J, Baskaran A, Dey N (2015) Automatic generation control of Thermal-Thermal-Hydro power systems with PID controller using ant colony optimization. Int J Service Sci Manag Eng Technol (IJSSMET) 6(2):18–34
33. Jagatheesan K, Anand B, Dey N, Ashour AS (2018) Effect of SMES unit in AGC of an interconnected multi-area thermal power system with ACO-tuned PID controller. In: Advancements in applied metaheuristic computing. IGI Global, pp 164–184
34. Jagatheesan K, Anand B, Dey KN, Ashour AS, Satapathy SC (2018) Performance evaluation of objective functions in automatic generation control of thermal power system using ant colony optimization technique-designed proportional–integral–derivative controller. Electr Eng 100(2):895–911
35. Sun X, Zhang Y, Ren X, Chen K (2015) Optimization deployment of wireless sensor networks based on culture–ant colony algorithm. Appl Math Comput 250:58–70
36. Sharma V, Grover A (2016) A modified ant colony optimization algorithm (mACO) for energy efficient wireless sensor networks. Opt-Int J Light Electron Optics 127(4):2169–2172

37. Kaur S, Mahajan R (2018) Hybrid meta-heuristic optimization based energy efficient protocol for wireless sensor networks. Egypt Inform J 19(3):145–150
38. Chatterjee S, Sarkar S, Dey N, Ashour AS, Sen S, Hassanien AE (2017) Application of cuckoo search in water quality prediction using artificial neural network. Int J Comput Intell Stud 6(2–3):229–244
39. Hore S, Chatterjee S, Sarkar S, Dey N, Ashour AS, Balas-Timar D, Balas VE (2016) Neural-based prediction of structural failure of multistoried RC buildings. Struct Eng Mech 58(3):459–473
40. Chatterjee S, Sarkar S, Hore S, Dey N, Ashour AS, Shi F, Le DN (2017) Structural failure classification for reinforced concrete buildings using trained neural network based multi-objective genetic algorithm. Struct Eng Mech 63(4):429–438
41. Alarifi A, Tolba A (2019) Optimizing the network energy of cloud assisted internet of things by using the adaptive neural learning approach in wireless sensor networks. Comput Ind 106:133–141
42. Eldhose EK, Jisha G (2016) Active cluster node aggregation scheme in wireless sensor network using neural network. Procedia Technol 24:1603–1608
43. Chang YC, Lin CC, Lin PH, Chen CC, Lee RG, Huang JS, Tsai TH (2013) eFurniture for home-based frailty detection using artificial neural networks and wireless sensors. Med Eng Phys 35(2):263–268
44. Serpen G, Gao Z (2014) Complexity analysis of multilayer perceptron neural network embedded into a wireless sensor network. Procedia Comput Sci 36:192–197
45. Chatterjee S, Hore S, Dey N, Chakraborty S, Ashour AS (2017) Dengue fever classification using gene expression data: a PSO based artificial neural network approach. In: Proceedings of the 5th international conference on frontiers in intelligent computing: theory and applications. Springer, Singapore, pp 331–341
46. Jagatheesan K, Anand B, Samanta S, Dey N, Ashour AS, Balas VE (2017) Particle swarm optimisation-based parameters optimisation of PID controller for load frequency control of multi-area reheat thermal power systems. Int J Adv Intell Paradigms 9(5–6):464–489
47. Phoemphon S, So-In C, Niyato DT (2018) A hybrid model using fuzzy logic and an extreme learning machine with vector particle swarm optimization for wireless sensor network localization. Appl Soft Comput 65:101–120
48. Sun Z, Liu Y, Tao L (2018) Attack localization task allocation in wireless sensor networks based on multi-objective binary particle swarm optimization. J Netw Comput Appl 112:29–40
49. Cao B, Zhao J, Lv Z, Liu X, Kang X, Yang S (2018) Deployment optimization for 3D industrial wireless sensor networks based on particle swarm optimizers with distributed parallelism. J Netw Comput Appl 103:225–238
50. Yan Z, Goswami P, Mukherjee A, Yang L, Routray S, Palai G (2019) Low-energy PSO-based node positioning in optical wireless sensor networks. Optik 181:378–382
51. Zahedi ZM, Akbari R, Shokouhifar M, Safaei F, Jalali A (2016) Swarm intelligence based fuzzy routing protocol for clustered wireless sensor networks. Expert Syst Appl 55:313–328
52. Bruneo D, Scarpa M, Bobbio A, Cerotti D, Gribaudo M (2012) Markovian agent modeling swarm intelligence algorithms in wireless sensor networks. Perform Eval 69(3–4):135–149
53. Ari AAA, Yenke BO, Labraoui N, Damakoa I, Gueroui A (2016) A power efficient cluster-based routing algorithm for wireless sensor networks: honeybees swarm intelligence based approach. J Netw Comput Appl 69:77–97
54. Sreelaja NK, Pai GV (2014) Swarm intelligence based approach for sinkhole attack detection in wireless sensor networks. Appl Soft Comput 19:68–79
55. Das SK, Tripathi S (2018) Intelligent energy-aware efficient routing for MANET. Wirel Netw 24(4):1139–1159
56. Yadav AK, Das SK, Tripathi S (2017) EFMMRP: design of efficient fuzzy based multi-constraint multicast routing protocol for wireless Ad-hoc network. Comput Netw 118:15–23
57. Das SK, Tripathi S (2018) Adaptive and intelligent energy efficient routing for transparent heterogeneous Ad-hoc network by fusion of game theory and linear programming. Appl Intell 48(7):1825–1845

58. Das SK, Tripathi S (2017) Energy efficient routing formation technique for hybrid Ad hoc network using fusion of artificial intelligence techniques. Int J Commun Syst 30(16):e3340, 1–16
59. Das SK, Yadav AK, Tripathi S (2017) IE2M: design of intellectual energy efficient multicast routing protocol for Ad-hoc network. Peer-to-Peer Networking Appl 10(3):670–687
60. Das SK, Samanta S, Dey, N, Kumar R (2019) Design frameworks for wireless networks. In: Lecture notes in networks and systems. Springer, pp 1–439. ISBN: 978-981-13-9573-4
61. Das SK, Tripathi S (2020) A nonlinear strategy management approach in software-defined Ad hoc network. In: Design frameworks for wireless networks. Springer, Singapore, pp 321–346
62. Samantra A, Panda A, Das SK, Debnath S (2020) Fuzzy petri nets-based intelligent routing protocol for Ad hoc network. In: Design frameworks for wireless networks. Springer, Singapore, pp 417–433
63. Das SK, Kumar A, Das B, Burnwal AP (2013) Ethics of reducing power consumption in wireless sensor networks using soft computing techniques. Int J Adv Comput Res 3(1):301
64. Das SK, Das B, Burnawal AP (2014) Intelligent energy competency routing scheme for wireless sensor network. Int J Res Comput Appl Rob 2(3):79–84
65. Amri S, Khelifi F, Bradai A, Rachedi A, Kaddachi ML, Atri M (2017) A new fuzzy logic based node localization mechanism for wireless sensor networks. Future Gener Comput Syst
66. Mazinani A, Mazinani SM, Mirzaie M (2019) FMCR-CT: An energy-efficient fuzzy multi cluster-based routing with a constant threshold in wireless sensor network. Alexandria Eng J 58(1):127–141
67. Karaa WBA, Ashour AS, Sassi DB, Roy P, Kausar N, Dey N (2016) Medline text mining: an enhancement genetic algorithm based approach for document clustering. In: Applications of intelligent optimization in biology and medicine. Springer, Cham, pp 267–287
68. Dey N, Ashour A, Beagum S, Pistola D, Gospodinov M, Gospodinova E, Tavares J (2015) Parameter optimization for local polynomial approximation based intersection confidence interval filter using genetic algorithm: an application for brain MRI image de-noising. J Imaging 1(1):60–84
69. Chatterjee S, Sarkar S, Dey N, Ashour AS, Sen S (2018) Hybrid non-dominated sorting genetic algorithm: II-neural network approach. In: Advancements in applied metaheuristic computing. IGI Global, pp 264–286
70. Hanh NT, Binh HTT, Hoai NX, Palaniswami MS (2019) An efficient genetic algorithm for maximizing area coverage in wireless sensor networks. Inf Sci 488:58–75
71. Somauroo A, Bassoo V (2019) Energy-efficient genetic algorithm variants of PEGASIS for 3D wireless sensor networks. Appl Comput Inf
72. Wang T, Zhang G, Yang X, Vajdi A (2018) Genetic algorithm for energy-efficient clustering and routing in wireless sensor networks. J Syst Softw 146:196–214
73. Al-Shalabi M, Anbar M, Wan TC, Alqattan Z (2019) Energy efficient multi-hop path in wireless sensor networks using an enhanced genetic algorithm. Inf Sci
74. Kumar S, Kumar V, Kaiwartya O, Dohare U, Kumar N, Lloret J (2019) Towards green communication in wireless sensor network: GA enabled distributed zone approach. Ad Hoc Networks 101903
75. Mukherjee A, Dey N, Kausar N, Ashour AS, Taiar R, Hassanien AE (2019) A disaster management specific mobility model for flying Ad-hoc network. In: Emergency and disaster management: concepts, methodologies, tools, and applications. IGI Global, pp 279–311
76. Fong S, Li J, Song W, Tian Y, Wong RK, Dey N (2018) Predicting unusual energy consumption events from smart home sensor network by data stream mining with misclassified recall. J Ambient Intell Humaniz Comput 9(4):1197–1221
77. Roy S, Karjee J, Rawat US, Dey N (2016) Symmetric key encryption technique: a cellular automata based approach in wireless sensor networks. Procedia Comput Sci 78:408–414
78. Elhayatmy G, Dey N, Ashour AS (2018) Internet of Things based wireless body area network in healthcare. In: Internet of things and big data analytics toward next-generation intelligence. Springer, Cham, pp 3–20

Chapter 14
Nature-Inspired Algorithms for Reliable, Low-Latency Communication in Wireless Sensor Networks for Pervasive Healthcare Applications

Prateeti Mukherjee and Ankur Das

1 Introduction

The rapid growth of wireless technology in the recent years has resulted in an unexpected acceleration in the market for Internet of things, with the total number of IoT devices that are in use reaching the colossal figure of 7 billion, as suggested in the "State of the IoT 2018–Analyst Insights from Q1/Q2 2018" reports. With such speedy progress of wireless communication technologies, the popularity of wireless sensor networks (WSNs) has attracted the attention of both industry and academia in researching and developing promising real-world solutions to aid industries such as healthcare, transportation, agriculture, telecommunication, and education, to name a few. A wireless sensor network consists of a number of interconnected nodes that perform distributed sensing tasks. Reference [1] describes a WSN as a "powerful infrastructure-less network" that comprises of thousands of autonomous low-power sensors organized in an ad hoc manner and displaying collaborative behavior to serve a varied set of purposes. The nodes gather information from the environment, relay the data through a network, and depending on available resources, may even be capable of processing the collective information from their peers. These devices form the sensing layer in IoT-based architectures. WSNs are considered the most appropriate choice in disciplines that involve monitoring, sensing, and collaborative decision-making. The amalgamation of detection systems, signal processing, and data communication enables such networks to form a powerful platform that manipulates, and processes data collected from the environment [2, 3].

P. Mukherjee (✉)
Computer Science and Engineering, Institute of Engineering and Management,
Salt Lake City 700091, India
e-mail: mukherjeeprateeti01@gmail.com

A. Das
Department of Nanoscience and Nanotechnology, University of Siegen, Siegen, Germany
e-mail: ankur.das@student.uni-siegen.de

D. De et al. (eds.), *Nature Inspired Computing for Wireless Sensor Networks*,
Springer Tracts in Nature-Inspired Computing,
https://doi.org/10.1007/978-981-15-2125-6_14

With such great potential, it is important that the algorithms and protocols designed for these networks take into account all possible performance issues that might occur during initialization, normal processing, as well as emergency situations. In these networks, every sensor is equipped with a sensing unit, a transmission module for transfer of messages, a power supply unit, and a processing and storage facility either embedded in the device or existing externally. These components are designed with multiple resource constraints that include dimensional limitations—since, for certain applications, the devices cannot exceed a specific size limit; heating and energy restraints in healthcare scenarios, and limited computing capabilities. The power supply issue is arguably the most restricting aspect of WSN design. Several factors contribute to this impediment, such as reduced space in portable technology and hostile nature of sensing environment that make it difficult to replace the batteries of the humungous number of deployed sensing equipment. Further, in pervasive healthcare applications, a real-time monitoring system is required to collect immediate data about the patient's activities. Often, the devices are placed directly on the patient's body, enhancing the limitations related to power usage, heating, and size that directly impact energy restrictions on the device. Further, the monitoring systems are expected to identify emergency situations such as cardiac arrest and sudden falls, which makes it crucial for the network to be reliable, and the devices to be in an active state at all times. Thus, a low latency, energy saving scheme must be employed for routing data in such networks to facilitate an increased lifespan.

The issue of power consumption in nodes could, in theory, be minimized at the time of transmission, reception, idle listening, or idle overhearing. Designing a suitable route for the transmission of data packets is a feasible solution to overcome energy consumption issues [4]. In [5], the dilemma of route formation for message passing is explained. The work mentions that while short routes lead to depletion of resources in the intermediate nodes, resulting in decreased network lifetime, longer routes may comprise of a huge number of nodes increasing the chances of network delay problems. It is, therefore, crucial to find the shortest route among nodes to get good results in terms of low-energy consumption and network lifetime.

Nature-inspired computing, a discipline that strives to develop novel computing techniques by observing the behavior of naturally occurring phenomena, has succeeded in solving a number of complex problems and generating new avenues of research such as neural networks, swarm optimization techniques, genetic algorithms, evolutionary computation, to name a few. Numerous routing algorithms have been developed in the past few years, inspired by biological optimization patterns, such as ant colony optimization (ACO), particle swarm optimization (PSO), among the others discussed in detail in Sect. 4. A select few are then explained in detail, considering various circumstances, for the purposes of comparison and identification of the most appropriate nature-inspired routing mechanism applicable to specific use-cases in ubiquitous healthcare architectures.

The paper is organized as follows: Sect. 2 provides a survey of research work conducted in this field of study and the fascinating protocols existing in literature suitable for WSN design. The wireless sensor network architecture is then briefly explained in Sect. 3. The nature-inspired routing protocols that are applicable to our problem statement concerning U-healthcare are discussed in Sect. 4, and the comparative study is listed in tabular form. Finally, in Sect. 5, conclusions are drawn, followed by the references.

2 Literature Survey

A wireless sensor network (WSN) is composed of a collection of devices constrained by factors such as energy and processing power that gather data about a set of phenomena. When it comes to ubiquitous healthcare, several factors come into play. The existing challenges in this field have driven researchers to develop algorithms that extend the lifetime of the wireless sensor network, while ensuring the system remains unobtrusive, light, and portable. An efficient way to do so, as suggested in [6], is clustering organization to hierarchically structure the sensors in groups and assign a single node as the cluster head. This cluster head is responsible for executing specific tasks such as gathering data from other cluster sensors and relaying it to the base station through the network. Delegating such responsibilities to the cluster head improves the data acquisition process, while increasing lifetime of network. The cluster heads, however, have limited battery, and their energy is spent faster due to the added responsibility. The challenge here is to elect a new cluster head from the remaining sensing motes to enable normal functioning of the network, even when the head is compromised. Several cluster head selection problems exist in literature. Multi-objective evolutionary algorithms, a subset of nature-inspired computing that is commonly known by the abbreviation MOEA, have been extensively developed upon to prepare advanced algorithms that tackle the problem of cluster head selection.

Reference [7] criticizes the class of MOEAs that use nondominated sorting and sharing on account of several factors, including computational complexity, non-elitist approach, and the necessity of specifying a sharing parameter. In their work, the authors suggest a nondominated sorting-based multi-objective EA (MOEA), called nondominated sorting genetic algorithm II (NSGA-II), which alleviates all the aforementioned difficulties. The authors claim that simulation results on difficult test problems prove that their model is capable of finding a better spread of solutions and better convergence near the true Pareto-optimal front compared to Pareto-archived evolution strategy and strength-Pareto EA—two other elitist MOEAs that pay special attention to create a diverse Pareto-optimal front. The algorithm provides a promising solution to clustering issues, and on account of its reduced complexity and elitism, it could very well be implemented in WSN design for pervasive healthcare applications.

Another approach to the design of such networks is the creation of a decentralized architecture, wherein control is fully distributed among the individual cluster units, in a localized fashion. This approach is inspired by the behaviors portrayed by ants and is termed the ant colony optimization (ACO) algorithm [8, 9]. The units are allocated computing resources, and the units form clusters that then interact in simple, localized ways. The beauty of the ACO optimization lies in its simplicity. The algorithm is discussed in detail in Sect. 4.2.

While discussing clustering techniques, the most popular algorithm is the traditional low-energy adaptive clustering hierarchy, otherwise called the LEACH protocol [10, 11]. This method makes use of randomized rotation with uniform clustering of local cluster heads in an attempt to increase the scalability and performance of the network. It is self-adaptive and self-organized. The protocol uses each round as a unit, made up of a cluster set-up stage followed by a steady-state stage. Unnecessary energy costs are avoided by a steady-state stage constituting multiple frames, with the duration being much longer than the set-up stage. There is, however, a major drawback residing in the protocol concept. The randomness of cluster formation causes the cluster heads to have very different energies. Further, the distance between cluster heads and base station is also unpredictable. The cluster head, due to its increased responsibilities, tends to drain its energy much faster than the other nodes in the network. If the current energy of a cluster head is less or its distance from the base station is sizeable, then the cluster head will perish quickly due to a heavy energy burden.

Such issues are addressed in studies that modify the traditional LEACH to improve its performance. LEACH-TLCH or the LEACH Protocol with Two Levels Cluster Head is introduced in [12]. While the cluster formation and head selection procedures are identical, the modified algorithm uses information such as average energy and distance of node from the base station to select secondary cluster heads. Conditions against these values are specified in the algorithm, ensuring the problem of early drainage of selected cluster head is avoided.

Other notable works that are synonymous with clustering techniques include the hybrid energy distributed (HEED) approach [13, 14], wherein the clustering and cluster head selection are based on hybrid energy distributed clustering approach in an attempt to extend the lifetime of network. An energy-efficient hierarchical clustering (EEHC) was proposed in [15], which was successful in increasing lifespan, but caused overload in cluster heads due to the complexity of hierarchical clustering. Paper [16] proposed a distribution scheme of cluster heads to reduce energy dissipation by avoiding unnecessary redundancy. When compared to the traditional LEACH, the authors claim their model performed significantly better in prolonging network lifetime.

A very popular routing protocol that forms a major part of swarm intelligence techniques is the particle swarm optimization (PSO) [17]. PSO is a metaheuristic, inspired by the synchronous flocking, spontaneous direction change, scattering, and regrouping of birds; used for finding maximum and minimum values of a function. The conventional PSO algorithm is a population-based technique exhibiting self-organization through direct communication and exchange of information between a node and its neighbors to find a good solution. The protocol is studied at length in Sect. 4.1. Several modifications to the conventional PSO have also been formulated to suit the needs of WSNs.

Although several research studies have been conducted in the field of WSN design through nature-inspired computing techniques, studies catering to the specific healthcare domain are a definite rarity. Numerous papers and projects exist in literature that focus on building products and smart platforms to cater to the ill, the elderly, or the infants [18, 19]. Further, a plethora of surveys on algorithms suited to WSN design is scattered across the world of academia [20, 21]. However, a comprehensive study that focuses on the requirements of efficient routing strategy in pervasive healthcare services, while studying bio-inspired mechanisms to achieve the same represents the novelty of this research work.

3 Wireless Sensor Network Architecture

Recent advancements in the field of electronics and wireless communication systems have paved the way for the proliferation of wireless sensor networks (WSNs) [22]. Such networks have now emerged as an essential component in many domains such as industrial automation, medicine, agriculture, environment and climate change, and emergency management, to name a few. Clusters of sensors, pervasive computing techniques, and artificial intelligence research have collectively created an interdisciplinary concept termed ambient intelligence. Studies in this field aim to solve challenges we face in our daily lives through the use of technology that will not hinder the everyday tasks of a member of the civilized society.

The healthcare industry is in dire need for technological intervention, on account of the increasing costs, growing records of medical error, inadequate staffing in major clinics, and the continuous elderly population increase in developed nations. In [23], it is stated that as per the forecasts of the Population Reference Bureau, in the next 20 years, the 65-and-over population in the developed countries will be nearly 20% of the overall population. This puts an enormous amount of pressure on healthcare practitioners to provide better services, despite the challenges stacked against them. Delivering quality care to every individual is no easy feat, however, integration of sensing equipment and consumer electronics creates a promising avenue for the application of ambient intelligence in healthcare. Such networks are capable of providing continuous medical monitoring, memory enhancement, hands-free control of home appliances, access to digitized health records, and automated emergency communication. Early detection of possible health issues is the key purpose that drives these constant monitoring systems, which would then lead to reduced costs and improve the individual's quality of life, sparing them the discomfort of regular hospital visits and the extreme possibility of battling death. To enable the existence of these in-home monitoring equipments, the concept of wireless sensor networks comes into play.

The networks primarily comprise of a large number of sensing modules deployed over a geographical region, with the nodes communicating with each other through wireless technology. These nodes must be compact, lightweight, robust, and battery-powered so they may be used in any environment, and do not lose functionality in adverse conditions. The task of the sensors is to monitor physical and environmental conditions and to acquire what is known as contextual information. The structure of a sensor node is depicted in Fig. 1.

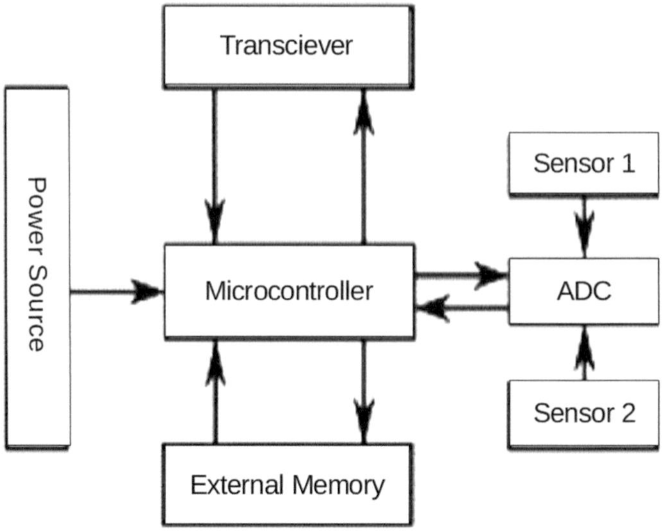

Fig. 1 Basic structure of a sensor node

Each node comprises of four primary components [1]: An external power source, a transceiver unit, a sensing unit comprising of one or more sensors and a processing unit, or a controller. Certain secondary components could be included within the unit that are application-dependent, such as power generator system in addition to or instead of batteries, a location finding system that may be useful for emergency contacts, or a mobilizer. The operating modes of a sensor are as follows: (1) transmission, (2) reception, (3) idle listening, and (4) sleep.

The sensors in ubiquitous healthcare frameworks predominantly act as data providers, sensing biological information from either outside or inside the human body. They communicate the acquired information to a control device worn on the body or placed in the vicinity. Then, the data assembled from the control devices are conveyed to remote destinations with the computing resources for the diagnostic and decision-making process to be executed.

Sensors used in pervasive healthcare applications have certain requirements, as mentioned in [24]. The fundamental concept of ubiquitous sensing lies in the unobtrusive nature of the technology involved. Therefore, the most essential requirement in the design of wireless medical sensors relates to their lightweight and miniature size to allow continuous, non-invasive, and unobtrusive monitoring of the patient. It has been experimentally determined that the size and weight of any electronic sensor rely heavily on the size and weight of its batteries. However, simply decreasing battery size does not solve the issue, since a battery's capacity is directly related to its dimensions. This is a relevant area of research in the development of microelectronics and optimization algorithms. Security and privacy concerns have also proved to be a major challenge in the provision of e-healthcare services. Data infringement and breach of privacy have legal implications, and the development of a secure architecture for transfer and storage of data is crucial for the success of pervasive computing in healthcare. Availability of data as and when required, integrity and authenticity of the information relayed, coupled with measures that ensure confidentiality is a major requirement in the design of WSN for healthcare services.

With the increasing amount of data being transferred between healthcare providers, physicians, and clinics, the process of information exchange often gets chaotic and dysfunctional. To create a method to the madness, data mobility and sharing of resources are a key element in ubiquitous healthcare systems. Hence, the various elements of the network must possess interoperability features, to ensure that the systems and devices can accumulate and interpret data, and record the information in a user-friendly, organized manner. To enable this property, the network must ensure sufficient bandwidth for proper performance of every component, and a sufficient level of fault tolerance must be included for reliable routing, on account of the failure of one or more nodes. Although inexpensive sensing devices are a considerable threat to the reliability of the system, faults in the communication and transmission steps could lead to erroneous diagnosis and failure of the network. The communication constraint varies between nodes since the sampling rates required by the sensors are different. High demands on the communication channel, however, increase the requirements of the system and reduce battery life, thereby increasing the overall expenses. A calculated trade-off between communication and computation is of paramount importance in the design of medical applications relying on WSNs [25].

The scientific community strives to solve these issues and meet the aforementioned requirements to serve the desired applications. Recent studies in this field tend to mimic biological and animal behaviors that can be easily observed in nature to create protocols and algorithms, as discussed in the previous section. Our goal is to investigate major algorithms and create a comparative study to determine the prospective avenues of biological inspiration in WSN design for smart healthcare.

4 Routing Protocols for WSN in Healthcare

While discussing the nature-inspired routing protocols that may be used for WSN design to serve healthcare applications, and it is important to understand the concept of real-time computing. According to [26], there exist three major components that characterize real-time systems. The most precious resource among these elements is time. A real-time application is generally comprised of a set of cooperating tasks that are invoked at regular intervals and are executed within a stipulated deadline. The interactive tasks require that the messages be sent and received in a timely manner. The factor is essential, since the correctness of computation is determined not only on the logical accuracy but also on the time at which the results are produced.

4.1 Deadline Classification

The deadline of a real-time task may be classified as follows:

1. **Hard deadline**—Catastrophic consequences of not meeting the pre-determined deadline
2. **Firm deadline**—Although the consequences of not meeting the deadline do not tend to be disastrous, the results produced by the task fail to be of any use once the deadline has expired
3. **Soft deadline**—A deadline that is neither firm nor hard; the results produced by a task have a utility that reduces overtime after expiration of deadline.

In medical applications, a soft real-time system is generally advised. Identification of emergency situations such as heart failures or sudden falls is useful even if the detection took a few second beyond the deadline. However, if the results are relayed a day later, it is safe to assume that the utility would be null. The soft deadline criterion is an important point to consider when choosing the appropriate network design for the system.

4.2 Architecture Design Objectives

In our quest to create the best possible design, it is important to identify the design considerations on the basis of target application. We can then qualitatively analyze these concerns against the relevant technological trends and estimate the design and cost bottlenecks, while recognizing the ones that may be easily resolved through technological progress.

As suggested in [27], the design considerations for the WSN subsystem in ubiquitous healthcare application are as follows:

1. Power consumption in nodes
2. Power exhausted during transmission
3. Unobtrusive nature of sensing equipment
4. Portability of platform
5. Real-time availability of sensors, processors, and access to emergency healthcare services
6. Reliable communication among the elements
7. Multi-hop routing for relay of acquired data, and
8. Security and confidentiality of sensitive medical information.

5 Nature-Inspired Routing Protocols

Natural systems possess certain extraordinary features that act as a source of inspiration and play a crucial role in determining a path toward the optimal solution for artificial networking systems [28, 29]. Sohail Jabbar and Rabia Iram have discussed the optimization technique of bio-inspired WSN systems in their article "Intelligent Optimization of Wireless Sensor Network through Bio-inspired Computing: Survey and Future Directions" [30]. The work suggests that one can achieve improved performance of computational tasks through flexibility, adaptability, decentralization, and fault tolerance with bio-inspired algorithms, based on the behavior of biological systems in nature. Figure 2 depicts the optimization techniques to solve the prevalent issues based on three categories, namely, search space, problem, and objective function formation. Optimization is one of the main features of biological systems that aids in finding the best solution. In most cases, iteration is needed in the optimization process during the working sessions. Some typical examples of iterative working sessions are demonstrated in foraging behavior in ants and bee clusters, flocking in birds, herding in animals, and schooling in fish.

Figure 3 provides a visual representation of the hierarchical structure of evolutionary algorithms, a broad subset of nature-inspired computing techniques.

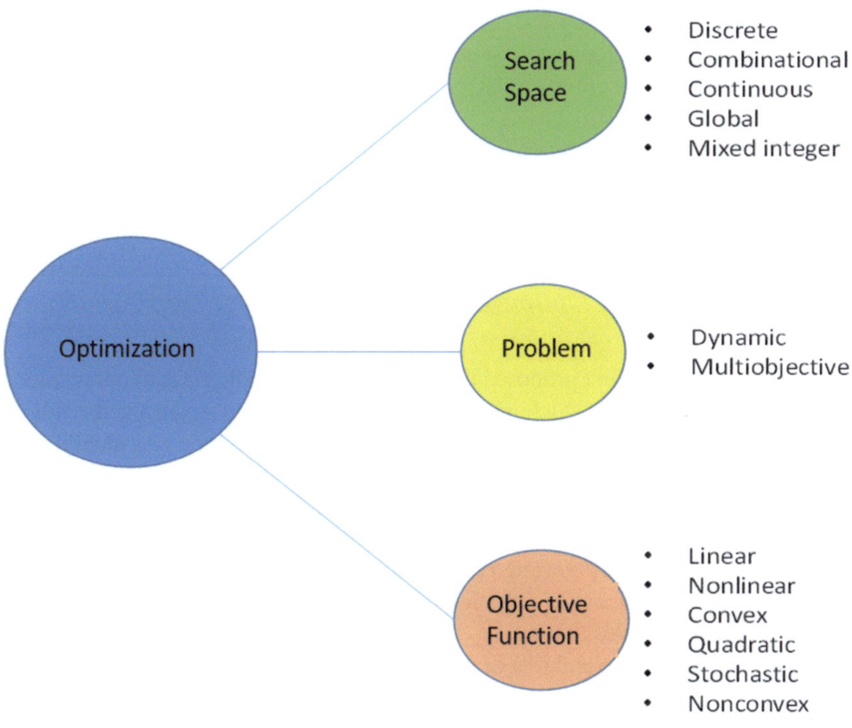

Fig. 2 Optimization categories

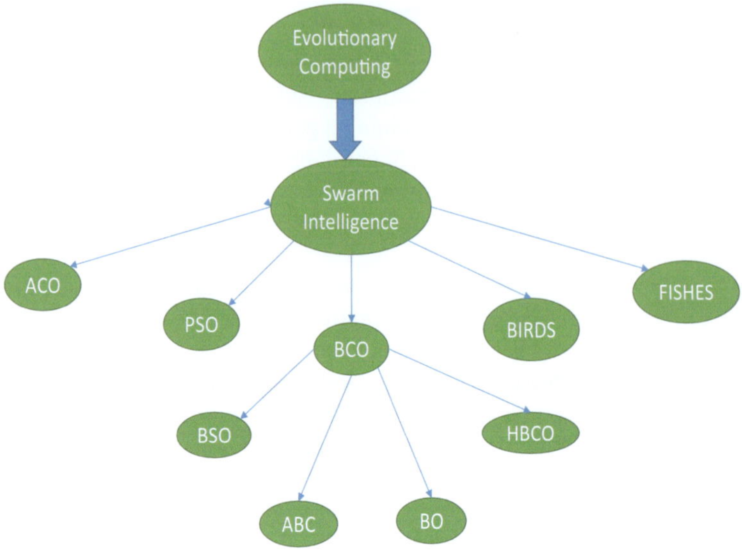

Fig. 3 Hierarchical representation of evolutionary algorithms

The abbreviations displayed in the image represent the following:

ACO	Ant colony optimization
PSO	Particle swarm optimization
BO	Bee optimization
BSO	Bee swarm optimization
ABC	Artificial bee colony
HBCO	Hybrid bee colony optimization.

5.1 Particle Swarm Optimization

Kennedy and Eberhart first proposed the PSO [31], a population-based optimization technique inspired by the flocking behavior depicted by birds during the migration months. A swarm of particles fly around the problem space and exhibit self-organization by responding to positive feedback (neighboring particles' influence directed toward a good or optimum position) and negative feedback (leaving current optimum in pursuit of a better position through social knowledge). The particles, instead of moving directly toward a good position, explore around a good position, thereby adding randomness in particle movement. The dynamics of the swarm are governed by a set of rules that modify the velocity of each particle according to the experience of the particle and that of its neighbors, depending on the social network structure within the swarm.

The key features of PSO techniques are the unique searching mechanism, computational efficiency, and easy implementation. The particle represents the population of members which are ideally mass-less and volume-less. Every particle in the swarm finds a solution in a higher-dimensional space defined by the following four vectors:

1. Velocity of the particle
2. The best position found so far
3. The current position vector of that particle
4. The best position found by neighboring particles.

The adjustment of particle position in space during the search process is based on the best position achieved by itself and that of its neighboring particles. In every step of the iterative process, each particle updates its position and velocity according to the following equations:

$$X_{k+1}^{i} = X_{k}^{i} + V_{k+1}^{i}$$
$$V_{k+1}^{i} = V_{k}^{i} + c_1 r_1 \left(p_{k}^{i} - X_{k}^{i}\right) + c_2 r_2 \left(p_{k}^{g} - x_{k}^{i}\right)$$

X_k^i indicates particle position
V_k^i specifies particle velocity
p_k^i specifies "remembered" position
$c_1 c_2$ indicates cognitive and social parameters
$r_1 r_2$ random numbers from 0 to 1.

A scheme of the conventional PSO algorithm is as follows:

```
ALGORITHM PSO

Initialize location and velocity of particle
repeat
for (each particle)
    evaluate objective function f at particle's
    location
end for loop
for (each particle)
    update the personal best position
    (coordinates of the best value obtained thus
    far)
end for loop
update the global/local best position (best
value obtained so far by any particle in its
topological neighborhood)
for (each particle)
  update the velocity
  compute the new location of the particle
end for loop
until stopping criterion is met
```

Several modifications to this traditional algorithm exist. PSO-clustering methods, also known as PSO-C is used in solving NP-hard optimization problems [32]. The algorithm considers available energy and distance between nodes with respect to cluster heads. Another work [33] proposes a graph-theoretic approach, coupled with PSO for multi-hop sensor networks. Here, for every *i*th round, the cluster head is selected using a weight function denoted by w_i, computed in an iterative manner. Routing of data packets is optimized considering the distance between source and destination nodes, remaining energy, and fitness function. The paper [34] suggests the application of PSO to optimize sensor deployment strategy, in order to maximize network coverage in mobile sensor networks. The work suggests the deployment of nodes in a centralized manner, which results in an increase in the burden of the base station. Such an approach is not advisable in healthcare scenarios, since the presence of a powerful base station that is capable of tackling multiple compute-intensive processes without failures or lags might not be the ideal architecture in portable medical equipment for pervasive sensing.

5.2 Ant Colony Optimization

In nature, ant-like species employ their collaborative behavior to locate food sources for their colony. Pheromone is a chemical substance that helps in acquiring collective knowledge of the environment used by an ant to communicate with other ants with the sole purpose of locating the closest food sources. When an ant locates the food source, a volatile pheromone trail is laid down by it on its way back to the colony. The other ants follow this pheromone trail to locate the food source. Generally, one can observe multiple pheromone trails between the nest and food source, but the population density is highest in the shortest pheromone trail. This social and organized behavior of ant has been used to implement ACO technology. Dorigo and Dicar proposed the ant colony optimization (ACO) which has been experimentally deemed the most successful swarm intelligence technique (SI) in advanced WSN designs [35].

The original algorithm, as proposed by Marco Dorigo in 1992 in his Ph.D. thesis [36, 37], has since been expanded to solve a wider class of numerical problems. The iterative process that directs a virtual ant in the decision graph toward good solutions is explained in the following scheme:

ALGORITHM ACO

```
Initialize pheromone values
repeat
for (ant k ∈ {1, . . ., m})
      construct a solution
end for loop
for (all pheromone values)
      decrease the value by a percentage
      {evaporation of pheromone, negative
      feedback}
end for loop
for (all pheromone values corresponding to good
solutions)
      increase the value {intensification of
      pheromone concentration, positive feedback}
end for loop
until stopping criterion is met
```

The stopping criterion could either be a certain number of iterations (or for a fixed amount of time), or a solution of a given quality. Checking for stagnation or convergence could also be employed as a stopping criterion.

A novel routing algorithm, based on the ACO, for hierarchical WSNs was proposed by Zhenchao Wangin [38]. In his work, heterogenous nodes were first deployed in a cluster to cater to the needs of large-scale WSNs. Then, weighted parameters were defined related to these nodes, and these parameters were mapped against the variants of the ACO. This resulted in the development of a new algorithm with inspiration from the ACO that was capable of finding the best possible route among cluster heads. To enhance the validity of data transmission, the concept of multiple routing was employed and through simulations and it was claimed to produce better results in terms of probability and network balance.

In another work [39], enhancements to the ant-based routing protocols were suggested to increase energy efficiency in routing. The suggested modifications were as follows:

1. Intelligently analyze routing tables and determine the importance of neighboring nodes
2. Update routing tables in case of link failures or node shutdowns
3. Decrease flooding ability of ants to enhance congestion control.

Overall, the proposed algorithm avoided the use of node energy while traversing the optimal path, thereby prolonging the lifetime of network while preserving connectivity among elements. The authors evaluated the performance of their algorithm against the following metrics:

1. Success rate
2. Energy consumption

3. Energy efficiency
4. Standard deviation
5. Network lifetime
6. Network latency.

The results of their simulation suggested that their proposed method increased the efficiency of the original ant colony optimization for wireless sensor network design.

5.3 *Artificial Immune System*

In the year 2002, De Castro and Timis proposed artificial immune systems in WSN design, algorithms that are inspired by the fundamental behavior of immunology in mammals. Falko Dressler has described the working principle of bio-inspired AIS in WSN [40]. The key features of AIS which enable the creation of self-optimizing and self-learning processes are defined by its capabilities, such as autonomous recognition, robust nature, diversity of application, reinforcement learning abilities, and the possibility of memorizing observations. The main target of AIS is to efficiently detect changes in the environment or subtle deviations from the normal system behavior. The function of the immune system in a mammal is to protect the body from infections. There are two kinds of factors responsible for this phenomenon. The first constitutes the generation of a response to attack pathogens, which then leads to an unspecified retort (using leucocytes). The second factor represents the immunologic memory and allows faster and very specific response to the adversities using B-cells and T-cells. A very important function of the human body is immune recognition. The binding region of a receptor and a portion of an antigen called epitode are collectively assigned this task. Since antibodies have a single type of reactor and antigens may show several epitodes, it is possible for different antibodies to recognize a single antigen. The immune systems are able to differentiate self and non-self-cells. As cell death may be induced by antigenic encounters, some kind of positive and negative selections are established by the immune system. A list of components existing under different names in biological and artificial systems is provided below:

1. Body: the entire mobile ad hoc network
2. Self-cells: well-behaving nodes
3. Non-self-cells: misbehaving nodes
4. Antigen: the sequence of observed DSR protocol events recognized in the sequence of packet headers. Examples of events are "data packet sent," "data packet received," "data packet received followed by data packet sent," "route request packet received followed by route reply sent"
5. Antibody: a pattern with the same format as the compact representation of Antigen
6. Negative Selection: antibodies are generated during an offline learning phase. In a deployed system, negative selection would be done in a test bed with nodes deployed by an operator.

AIS-based algorithms are primarily used for topology control in sensor networks. Various theories regarding the functioning and organizational structure of the natural immune system (NIS) exist in literature. Multiple studies have relayed the modeling of the NIS into artificial immune systems, for application in non-biological scenarios [41]. A varied set of AIS algorithm models has also been built, including classical view models, clonal selection theory models, network theory models, and danger theory models [42]. Such systems have been experimentally found to solve issues in multiple domains, including network intrusion, anomaly detection, concept learning, clustering of data, and data mining. Developments in these systems are a promising solution to the needs of secure transfer of sensitive medical records from light, portable sensing platforms to private databases.

5.4 Plant Biology-Inspired Framework for WSN

Vasaki Ponnusamy and Azween Abdullah discussed the implementation of plant biology-inspired framework in WSN for mobile sensing architectures [43]. Figure 4 shows the general architecture of mobile sensor network functionality. The framework, when held in contrast to plant-inspired architectures, essentially comprises of three layers for a WSN system. The first layer holds the plant's root mechanism and its coordination rules with the rhizosphere, while the second layer possesses the plant's communication protocols with the microbe organism's rhizosphere. The third layer is concerned with the communication mechanism existing between the microbe organisms and the base station.

A detailed depiction of plant biology-inspired mechanism for WSN design is provided in Fig. 5. The function of the leaves is to continually search for a particular event from the sensing environment, comparable to the quest for acquiring benefits

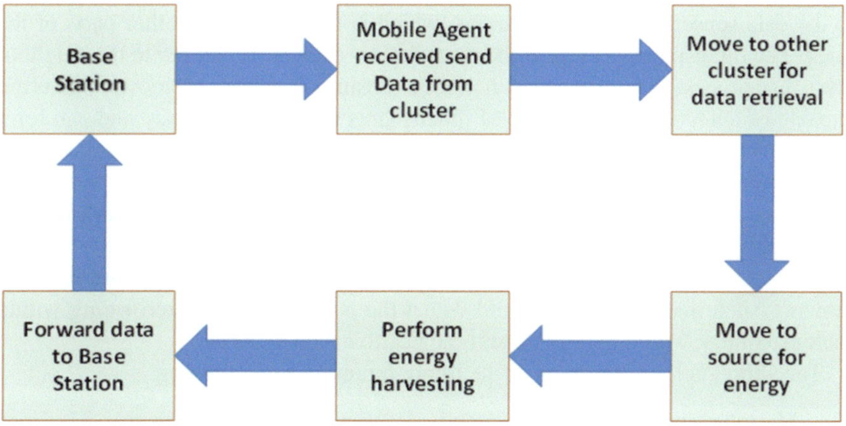

Fig. 4 General architecture of mobile sensor network functionality

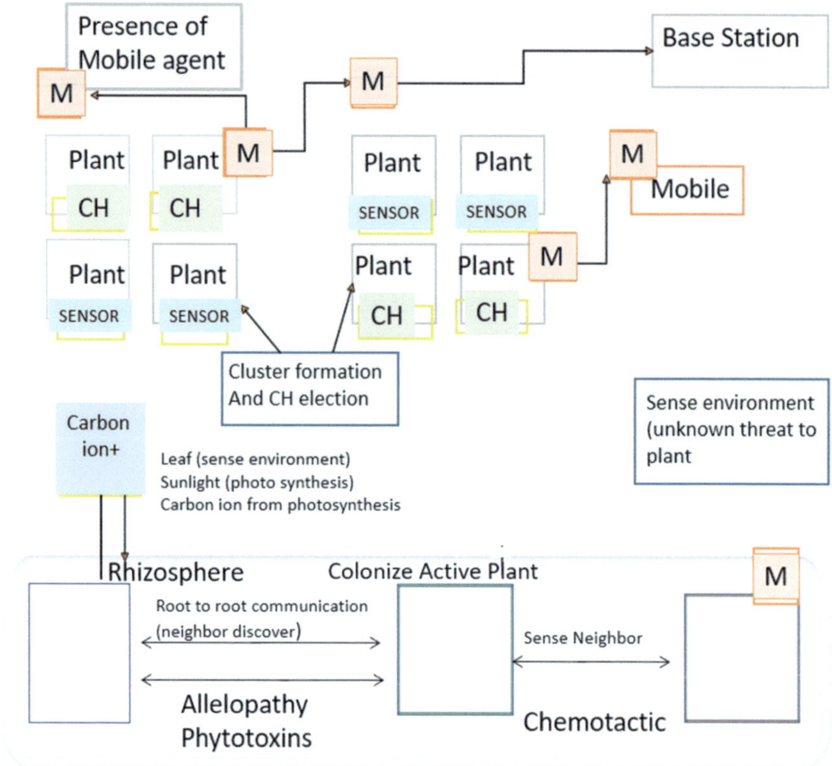

Fig. 5 Plant biology-inspired framework for WSN

of sunlight and natural resources in the real world. Plants absorb carbon-dioxide and water molecules to prepare carbohydrates. Here, the individual atoms are analogous to the data sensed in artificial systems, which is then relayed to other parts of the plant. The rhizosphere present at the roots detects each other, similar to the neighbor discovery process that takes place in mobile sensing platforms. Moreover, plants are capable of detecting the presence of other plants in the same colony and can form clusters, wherein they can send and receive messages and detect atmospheric adversities. This fundamental concept may be applied to the artificial sensing system, to encourage interaction between nodes, discussing the possibility of network failures and threats to effective routing. The roots of a plant are highly sensitive to overcrowding and scarcity of resources. Employing such effective sensitivity in WSN systems, such that every node could detect the possibility of overcrowding within their proximity, would greatly benefit the health of the network.

The abbreviations displayed in the figure represent the following:

M Microbes (mobile agents).
CH Carbohydrates.

The chemotactic relationship between organisms in the soil is either positive or negative in nature. Symbiotic relationship constitutes the positive relation of plants with the rhizosphere. Amino acids are the main reagents of the reaction exhibited by the roots of the plant and function as a communication agent. Parasitic relationships between the host plant and other flora or fauna form the negative relationship. These concepts could be applied to WSN architectures, wherein positive communication would be associated with the transfer of information among sensor nodes and mobile agents, while adversities could be given the label of negative communication.

The incredible behavioral patterns existing in biological systems provide great inspiration for efficient transfer of data packets in artificial networks. While there is an abundance of computer algorithms that are inspired by the marvels of nature, for the purposes of our research, we narrowed down to a select few to explain in detail. However, a comprehensive analysis of the existing and impending challenges in the wireless network architectures for ambient sensing technologies in the medical field, along with the biological patterns that are applicable to solve these issues is provided in Table 1.

6 Conclusions

Wireless sensor networks have received substantial attention of researchers and academics due to their potential utilization for a large spectrum of applications, ranging from intrusion detection and weather monitoring to ambient technologies. While the application scenarios might be appealing, energy-efficient routing of sensed data from node to base station is a major challenge for WSN architectures. Further, the protocols must also distribute the traffic over multiple, intelligently discovered paths, to deplete battery levels at the nodes at an equal rate. The adversities and challenges prevalent in these frameworks, when deployed in a real-world scenario, are surfeit.

In this paper, we discussed a specific field of application of the wireless sensing architectures, namely pervasive computing in medical applications. While ubiquitous healthcare has indeed emerged as a new frontier of smart, e-healthcare technology, the design of such advanced architectures is complex. The intricacies of the network and the intertwined set of design considerations and performance metrics have been discussed in detail in this chapter. Researchers have been fascinated by the symbiotic nature of biological systems and believe that studying such phenomenon shall result in valuable knowledge for computer network design. Thus, inspired by the beautiful symphonies existing in co-dependent biological organisms, several optimization protocols for artificial networks have been developed over time. The paper surveys multiple such protocols that are relevant to networks developed primarily for the purpose of ambient sensing and routing of sensitive data. The biological processes that are pertinent to solving issues in WSN design for U-healthcare are listed in tabular format, in an effort to encourage the reader to target the perfect natural processes for inspiration, when attempting to solve a particular challenge in this field.

Table 1 Analysis of challenges in networking for pervasive healthcare schemes

Challenges in networking for pervasive healthcare	Major issues	Biological principles
Large-scale networking	1. Number of devices, sensor networks: 10^2–10^6 nodes 2. Service space 3. Traffic load incurred 4. Larger search space for optimal route	1. Ant colony optimization technique (ACO) 2. Epidemic spreading: transmission mechanisms of viruses
Dynamic nature	1. Highly dynamic architectures • Node behaviors and demand • Highly dynamic link qualities • Varying load 2. Mobile ad hoc networks 3. Real-time tracking of mobile targets 4. Dynamic spectrum access	1. Artificial immune system 2. Activator-inhibitor systems
Resource constraints	1. Mismatch between the increase in demand and supply • Set of available services • Bandwidth capacity • Network lifetime 2. Scarcity of resources vary with network • Power, processing in WSN • Spectrum in CRN • Capability and size in nanonetworks	1. Foraging processes in ant colonies 2. Cellular signaling network
Autonomous operation	1. With increased network scales centralized control becomes unpractical 2. Nodes/links/paths may die out 3. Some networks are infrastructure-less • MANET, WSN, wireless mesh network 4. Unattended autonomous operation required, such as • Self-organization, self-evaluation, survivability	1. Epidemic spreading mechanisms 2. Insect colonies 3. Synchronization of fireflies 4. AIS

(continued)

Table 1 (continued)

Challenges in networking for pervasive healthcare	Major issues	Biological principles
Heterogeneous architecture	1. Heterogeneity and asymmetry • Node types and capabilities • Link capabilities • Network characteristics 2. RFID device to mobile vehicles	1. Homeostatic system 2. Insect colonies composed of heterogeneous individuals
Micro- and nanolevels	1. NEMS and MEMS 2. Nanonetworks 3. RF and acoustic inapplicable • Antenna size • Channel limitations • Different rules of physics	1. Living cells • Sense the environment • Receive the external signal • Perform at nanoscale 2. Cellular signaling network

References

1. Matin MA, Islam MM (2012) Overview of wireless sensor network, wireless sensor networks—technology and protocols. In: Matin MA (ed). IntechOpen. https://doi.org/10.5772/49376
2. Das SK, Tripathi S (2018) Energy efficient routing formation algorithm for hybrid ad-hoc network: a geometric programming approach. Peer-to-Peer Networking Appl 12. https://doi.org/10.1007/s12083-018-0643-3
3. Das SK, Tripathi S (2017) Adaptive and intelligent energy efficient routing for transparent heterogeneous ad-hoc network by fusion of game theory and linear programming. Appl Intell 48:1–21. https://doi.org/10.1007/s10489-017-1061-6
4. Das SK, Tripathi S (2016) Intelligent energy-aware efficient routing for MANET. Wirel Networks 24. https://doi.org/10.1007/s11276-016-1388-7
5. Abouzar P, Michelson DG, Hamdi M (2016) RSSI-based distributed self-localization for wireless sensor networks used in precision agriculture. IEEE Trans Wirel Commun 15(10):6638–6650
6. Miranda K, Zapotecas-Martínez S, López-Jaimes A, García-Nájera A (2019) A comparison of bio-inspired approaches for the cluster-head selection problem in WSN. https://doi.org/10.1007/978-3-319-96451-5_7
7. Deb K, Pratap A, Agarwal S, Meyarivan TAMT (2002) A fast and elitist multiobjective genetic algorithm: NSGA-II. IEEE Trans Evol Comput 6:182–197. https://doi.org/10.1109/4235.996017
8. Arif M, Rani T (2012) ACO based routing for MANETs. Int J Wirel Mobile Networks 4. https://doi.org/10.5121/ijwmn.2012.4211
9. Liu Y, Zhang H, Ni Q, Zhou Z, Zhu G (2008) An effective ant-colony based routing algorithm for mobile ad-hoc network. In: IEEE international conference on circuits and systems for communications, 2008, ICCSC 2008, vol 4, pp 100–103
10. Li Y, Ding L, Liu F (2011) The improvement of LEACH protocol in WSN. In: Proceedings of 2011 international conference on computer science and network technology, Harbin, 2011, pp 1345–1348
11. Hu G, Xie D, Wu Y (2007) Research and improvement of LEACH for wireless sensor networks. Chin J Sens Actuators 6(20):1391–1396
12. Mohammed A (2018) Improvement of an energy balance method of leach based on genetic algorithm

13. Chand S, Singh S, Kumar B (2014) Heterogeneous HEED protocol for wireless sensor networks. Wirel Pers Commun 77:2117
14. Younis O, Fahmy S (2004) HEED: a hybrid, energy-efficient, distributed clustering approach for ad hoc sensor networks. IEEE Trans Mobile Comput 03(4):366–379
15. Kumar D, Aseri TC, Patel RB (2009) EEHC: energy efficient heterogeneous clustered scheme for wireless sensor networks. Comput Commun 32:662–667. https://doi.org/10.1016/j.comcom.2008.11.025
16. Siddiqi MZ, Ilyas N, Aziz A, Kiran H, Arif S, Tahir J, Qasim U, Khan Z, Javaid N (2017) Optimized energy efficient routing using dynamic clustering in wireless sensor networks
17. Chopard B, Tomassini M (2018) Particle swarm optimization. In: An introduction to metaheuristics for optimization. Natural computing series. Springer, Cham
18. Nandyala S, Kim H-K (2016) From cloud to fog and IoT-based real-time U-healthcare monitoring for smart homes and hospitals. Int J Smart Home 10:187–196. https://doi.org/10.14257/ijsh.2016.10.2.18
19. Albahri OS, Zaidan A, Bahaa B, Hashim M, Albahri AS, Alsalem M, Mohsin A, Mohammed K, Alamoodi A, Enaizan O, Alyousif S, Zughoul O, Momani F, Hammed K, Alqaysi M, Abbas K, Almahdi E, Shafei G, Jafar M (2019) Based multiple heterogeneous wearable sensors: a smart real-time health monitoring structured for hospitals distributor. IEEE Access. pp 1–1. https://doi.org/10.1109/ACCESS.2019.2898214
20. Alemdar H, Ersoy C (2010) Wireless sensor networks for healthcare: a survey. Comput Netw 54(15):2688–2710
21. Cook DJ, Augusto JC, Jakkula VR (2009) Ambient intelligence: technologies, applications, and opportunities. Pervasive Mobile Comput 5:277–298
22. Dey N, Ashour AS (2015) Antenna design and direction of arrival estimation in meta-heuristic paradigm: a review. Int J Serv Sci Manage Eng Technol (IJSSMET) 7. https://doi.org/10.4018/IJSSMET.2016070101
23. Kinsella K, Phillips DR (2005) Global aging: the challenge of success, vol 60. Population Bulletin
24. Andreu J, Leff D, Ip H, Yang G-Z (2015) From wearable sensors to smart implants—toward pervasive and personalized healthcare. IEEE Trans Bio-med Eng 62. https://doi.org/10.1109/tbme.2015.2422751
25. Arastouie N, Sabaei M, Hakami V, Soltanali S (2013) A novel trade-off between communication and computation costs for data aggregation in wireless sensor networks. Int J Ad Hoc Ubiquitous Comput 12(4):245–253
26. Shin KG, Ramanathan P (1994) Real-time computing: a new discipline of computer science and engineering. Proc IEEE 82(1):6–24. https://doi.org/10.1109/5.259423
27. Boitier V, Tajan P, Dilhac JM (2016) WSN nodes: design considerations and energy management. J Phys Conf Ser 773:012043. https://doi.org/10.1088/1742-6596/773/1/012043
28. Dey N (2017) Advancements in applied metaheuristic computing
29. Dey N, Ashour AS, Bhattacharyya S (2019) Applied nature-inspired computing: algorithms and case studies
30. Shehzad S (2013) Intelligent optimization of wireless sensor networks through bio-inspired computing: survey and future directions. Int J Distrib Sens Netw 2013:13
31. Kennedy J, Eberhart R (1995) Particle swarm optimization. IEEE Int Conf Neutral Networks 4
32. Kulkarni RV, Venayagamoorthy GK (2011) Particle swarm optimization in wireless-sensor networks: a brief survey. IEEE Trans Syst Man Cybern Part C 41(2):262–267. https://doi.org/10.1109/tsmcc.2010.2054080
33. Vimalarani C, Subramanian R, Sivanandam SN (2016) An enhanced PSO-based clustering energy optimization algorithm for wireless sensor network. Sci World J 2016:11, Article ID 8658760. https://doi.org/10.1155/2016/865876
34. Gopakumar A, Jacob L (2008) Localization in wireless sensor networks using particle swarm optimization. In: Proceedings of the IET international conference on wireless, mobile and multimedia networks, Jan 2008, Beijing, China. IEEE, pp 227–230

35. Bonabeau E, Dorigo M, Theraulaz G (1999) Swarm intelligence. Oxford University Press, Oxford
36. Dressler F (2007) Self-organization in autonomous sensor actuator networks. Wiley, Inc., USA
37. Jabbar S, Iram R, Minhas A, Shafi I, Khalid S, Ahmad M (2013) Intelligent optimization of wireless sensor networks through bio-inspired computing: survey and future directions. Int J Distrib Sens Netw. https://doi.org/10.1155/2013/421084
38. Shen Z, Zhu Y, Tian X, Tang Y (2008) An ant colony system based energy prediction routing algorithms for wireless sensor networks, pp 1–4. https://doi.org/10.1109/wicom.2008.888
39. Saleem M, Di Caro G, Farooq M (2011) Swarm intelligence based routing protocol for wireless sensor networks: survey and future directions. Inf Sci 181:4597–4624. https://doi.org/10.1016/j.ins.20.10.07.005
40. Dressler F (2008) Bio-inspired networking, pp 1–20
41. Bidgoli AM, Nikdel A. AISTC: a new artificial immune system-based topology control protocol for wireless sensor networks
42. Engelbrecht AP (2007) Computational intelligence an introduction, 2nd edn. Wiley, London
43. Ponnusamy V, Abdullah A (2010) A bio-inspired framework for wireless sensor network using mobile sensors. pp 1–6